Two Thousand
Formulas, Recipes
& Trade Secrets

Two Thousand
Formulas, Recipes
& Trade Secrets

The Classic "Do-It-Yourself" Book of Practical Everyday Chemistry

By Harry Bennett, F.A.I.C.

FERAL HOUSE

Feral House
PO Box 39910
Los Angeles, CA 90039

www.FeralHouse.com

10 9 8 7 6 5 4 3 2 1

ABBREVIATIONS

THE FOLLOWING ARE THE ABBREVIATIONS USED IN "THE PRACTICAL
BOOK OF FORMULAS."

A. S. .... Alcohol Soluble	GM. or G. .... Gram
AM. .... Ammonium	INF. .... Infusion
AMP. .... Ampere	KILO. KG. .... Kilogram
APPROX. ..... Approximately	L. LIT. ...... Liter
ART. .... Artificial	LB. .... Pound
AV. .... Avoirdupois	M. P. .... Melting Point
BÉ. .... Baumé	NEUT. .... Neutral
C. .... Centigrade	POT. .... Potassium
C.C. .... Cubic Centimeter	POWD. .... Powdered
C. D. .... Current Density	PPTE. .... Precipitate
C. P. .... Chemically Pure	PULV. .... Pulverized
CARB. .... Carbonate	Q. S. .... Quantity Sufficient
COM. .... Commercial	R. P. M. ..... Revolutions per **Minute**
COMPN. ...... Composition	R. S. .... Regular Soluble
CONCN. ...... Concentration	S. D. .... Semi Denatured
CONTG. ...... Containing	SAPON. ...... Saponified
D. .... Density	SATD. .... Saturated
DEN. .... Denatured	SOD. .... Sodium
DIL. .... Dilute	SOLN. .... Solution
ESS. .... Essence	SP. GR. ..... Specific Gravity
EXT. .... Extract	TINCT. ...... Tincture
F. .... Fahrenheit	U. S. P. ...... United States
F. F. C. ..... Free from Chlorine	Pharmacopœia
F. F. P. A. ... Free from Prussic Acid	V. .... Voltage
FL. .... Fluid	VISC. .... Viscosity

READ THIS BEFORE YOU BEGIN

Why should you pay $2.00 for an article that you can make for 11 cents—or pay 35 cents for one that you can duplicate for 3 cents?

There is no reason in the world why you shouldn't make your own tooth-paste, shaving cream, hair tonic, mouthwash, ink, paste, sun-tan-oil, weed-killer, paint remover, or any of a thousand other homely articles that national advertising and the cost of merchandising has sky-rocketed to a retail price out of all proportion to the cost of manufacture.

This Book Will Not Only Save You Many Dollars, but Will Also Give You a Comforting Knowledge of Exactly What's in the Various Preparations in Daily Use in Your Home.

In the following preliminary pages are listed the recipes for articles most prominently in daily use. They are typical of the hundreds of valuable formulas that this book contains. They are in their simplest form as being most practical for beginners without experience or training. No expensive equipment is needed for their making. It should prove fun to do. And it will not only save you money, but, if you take intensive advantage of the opportunity it affords, and mix quantities for your friends as well as yourself, it may even put you in the way of *making money!*

See how simply many household necessities may be compounded. Then read the introduction carefully and you'll be ready to start.

VANISHING CREAM

Stearic Acid	**18 oz.**

Melt above in an aluminum or enameled double boiler (the water in the outer pot must be boiling.) To the above add, in a thin stream, while stirring vigorously with a fork, the following boiling solution made in an aluminum or enameled pot:

Potassium Carbonate	¼ oz.
Glycerin	6½ oz.
Water	5 lbs.

Continue stirring until the temperature falls to 135° F., then stir in a little perfume oil and stir from time to time until cold. Allow to stand over night and stir again the next day. Pack into jars which should be closed tightly.

BRUSHLESS SHAVING CREAM

White Mineral Oil	10 oz.
Glycosterin or Glyceryl Monostearate	10 oz.
Water	50 oz.

Heat the first two ingredients together in a pyrex or enamel dish to 150° F. and into this run slowly, while stirring with a fork the water which has been heating to boiling. Allow to cool to 105° F. and while stirring add a few drops of perfume oil. Continue stirring until cold.

MOUTH WASH

Benzoic Acid	⅝ oz.
Tincture of Rhatany	3 oz.
Alcohol	20 oz.
Oil of Peppermint	⅛ oz.

Just shake together in a dry bottle until it is dissolved and it is ready. A teaspoonful is used to a small wine-glassful of water.

MOSQUITO REPELLING OIL

Cedar Oil	2 oz. liquid
Citronella Oil	4 oz. liquid
Spirits of Camphor	8 oz. liquid

Just shake together in a dry bottle and it is ready for use. This preparation may be smeared on the skin as often as is necessary to repel mosquitoes and other insects.

FLY SPRAY

Deodorized Kerosene	89 oz. fluid
Methyl Salicylate	1 oz. fluid
Pyrethrum Powder	10 oz. fluid

Mix thoroughly by stirring from time to time; allow to stand covered over night and then filter through muslin.

Caution! This spray is inflammable and should not be used near open flames.

HAND LOTION (MILKY LIQUID)

Lanolin	¼ teaspoonful
Glycosterin or Glyceryl Monostearate	1 oz.
Tincture of Benzoin	2 oz.
Witch Hazel	25 oz.

Melt the first two items together in an aluminum or enameled double boiler. If no double boiler is at hand improvise one by standing the dish in a small pot containing boiling water. When the mixture becomes clear remove from the double boiler and add slowly, while stirring vigorously with a fork or stick, the Tincture of Benzoin and then the Witch Hazel. Continue stirring until cool and then put in one or two large bottles and shake vigorously. The finished lotion is a beautiful milky liquid comparable to the best hand lotions on the market sold at high prices.

COLD CREAM

Liquid Petrolatum (white mineral oil)	4 lbs.
White Beeswax	1 lb.

Heat the above in an aluminum or enameled double boiler (the water in the outer pot should be brought to a boil.) In a separate aluminum or enameled pot dissolve

Borax	1 oz.
Water	2 pints

and bring this to a boil. Then add this in a thin stream, while stirring vigorously in one direction only, to the melted wax mixture. Use a large fork for stirring. When the mixture turns to a smooth thin cream, immerse the bottom of the thermometer in it from time to time, stirring continuously. When the temperature drops to 140° F add ⅛ ounce of perfume oil and continue stirring until the temperature drops to 120° F. At this point pour into jars where the cream will "set" after a while. If a harder cream is desired, reduce the amount of liquid petrolatum. If a softer cream is wanted increase it.

CLEANSING CREAM (LIQUEFYING)

Liquid Petrolatum (white mineral oil)	5½ oz.
Paraffin Wax	2½ oz.
Petrolatum	2 oz.

Melt together with stirring in an aluminum or enameled dish and allow to cool. Then stir in a dash of perfume oil. Allow to stand until a haziness appears and then pour into jars, which should be allowed to stand *undisturbed* over night.

ANT POISON

Sugar	1 lb.
Water	1 qt.
*Arsenate of soda	125 grams

Boil and stir until uniform; strain through muslin; add a teaspoonful of honey.

*Poison.

LINIMENT—SORE MUSCLE

Olive Oil	6 oz. fluid
Methyl Salicylate	3 oz. fluid

Shake together and keep in a well-stoppered bottle. Apply externally but do not apply to chafed or cut skin.

TOOTH POWDER

Magnesium Carbonate	420 grams
Precipitated Chalk	565 grams
Sodium Perborate	55 grams
Sodium Bicarbonate	45 grams
Soap, Powdered White	50 grams
Sugar, Powdered	90 grams
Oil of Wintergreen	8 c. c.
Oil of Cinnamon	2 c. c.
Menthol	1 gram

Dissolve the last three ingredients together and then rub well into the sugar. Add the soap and perborate mixing in well. Add the chalk with good mixing and then the sodium bicarbonate and magnesium carbonate. Mix thoroughly and sift through a fine wire screen. Keep dry.

"CHEST RUB" SALVE

Yellow Petrolatum	1 lb.
Paraffin Wax	1 oz.
Oil of Eucalyptus	2 oz. fluid
Menthol	1/2 oz.
Oil of Cassia	1/8 oz. fluid
Turpentine	1/2 oz. fluid
Carbolic Acid	1/8 oz. fluid

Melt the vaseline and paraffin wax together in a double boiler and then add the menthol. Remove from the heat and stir and cool a little; then stir in the oils, turpentine and acid. When it begins to thicken pour into tins and cover.

MALTED MILK POWDER

Powdered Malt Extract	5 oz.
Powdered Skimmed Milk	2 oz.
Powdered Sugar	3 oz.

Mix thoroughly by shaking and rolling in a dry can. Pack in an airtight container.

RUST PREVENTION COMPOUND

Lanolin	1 oz.
*Naphtha	2 oz.

Mix until dissolved.

The metal to be protected is cleaned with a dry cloth and then coated with the above composition.

*Inflammable—Keep away from flames.

ARTIFICIAL VANILLA FLAVOR

Vanillin	¾ oz.
Coumarin	¼ oz.
Alcohol	2 pints

Stir the above in a glass or china pitcher until dissolved. Then stir in the following solution which has been made by stirring in another pitcher.

Sugar	12 oz.
Water	5¼ oz.
Glycerin	1 pint

Color brown by adding sufficient "burnt" sugar coloring.

WRITING INK (BLUE-BLACK)

Napthol Blue-Black	1 oz.
Gum Arabic Powdered	½ oz.
Carbolic Acid	¼ oz.
Water	1 gal.

Stir together in a glass or enameled vessel until dissolved.

LEATHER PRESERVATIVES

Neatsfoot Oil (cold pressed)	10 oz.
Castor Oil	10 oz.

Just shake together.

This is an excellent preservative for leather book bindings, luggage or other leather goods.

WHITE SHOE DRESSING

Lithopone	19 oz.
Titanium dioxide	1 oz.
Shellac (Bleached)	3 oz.
Ammonium Hydroxide	¼ oz. fluid
Water	25 oz.
Alcohol	25 oz.
Glycerin	1 oz.

Dissolve the last four ingredients by mixing in a porcelain vessel. When dissolved stir in the first two pigments. Keep in stoppered bottles and shake before using.

AUTO POLISH (CLEAR OIL TYPE)

Paraffin (Mineral Oil)	5 pints
Raw Linseed Oil	2 pints
China Wood Oil	½ pint
*Benzol	¼ pint
*Kerosene	¼ pint
Amyl Acetate	1 tblspfl.

Shake together in a glass jug and keep stoppered.

*Inflammable—Keep away from flames.

FLOOR WAX (PASTE TYPE)

Yellow Beeswax	1 oz.
Ceresin	2½ oz.
Carnauba Wax	4½ oz.
Montan Wax	1¼ oz.
*Naphtha or Mineral Spirits	1 pint
*Turpentine	2 oz.
Pine Oil	½ oz.

Melt the waxes together in a double boiler. Turn off the heat and run in the last three ingredients in a thin stream and stir with a fork. Pour into cans; cover and allow to stand undisturbed over night.
*Inflammable

PATCHING PLASTER

Plaster of Paris	32 oz.
Dextrin	4 oz.
Pumice Powder	4 oz.

Mix thoroughly by shaking and rolling in a dry container. Keep away from moisture.

CEMENT FLOOR HARDENER

Magnesium Fluosilicate	1 lb.
Water	15 lbs.

Mix until dissolved.
In using this, the cement should first be washed with clean water and then drenched with the above solution.

WOOD FLOOR BLEACH

Sodium Metasilicate	90 oz.
Raw Linseed Oil	16 oz.

Mix thoroughly and keep dry in a closed can. Use one pound to a gallon of boiling water. Mop or brush on the floor, allow to stand ½ hour, then rub off and rinse well with water.

*PAINT REMOVER

Benzol	5 pints
Ethyl Acetate	3 pints
Butyl Acetate	2 pints
Paraffin Wax	½ lb.

*Inflammable

FURNITURE POLISH (OIL AND WAX TYPE)

Thin Paraffin (Mineral) Oil	1 pint
Carnauba Wax, Powdered	¼ oz.
Ceresin Wax	⅛ oz.

Heat together until all of the wax is melted. Allow to cool and pour into bottles before mixture turns cloudy.

METAL POLISH

Naphtha	62 oz.
Oleic Acid	⅓ oz.
Abrasive	7 oz.
Triethanolamine Oleate	1 oz.
Ammonia (26°)	1 oz.
Water	128 oz.

In one container mix together the naphtha and oleic acid to a clear solution. Dissolve the Triethanolamine in water separately, stir in the abrasive, if it is a clay type, and then add the naphtha solution. Stir the resulting mixture at a high speed until a uniform creamy emulsion results. Then add the ammonia and mix well, but do not agitate as vigorously as before.

LIQUID SOAP (CONCENTRATED)

Water	11 oz.
*Caustic Potash (Solid)	1 oz.
Glycerin	4 oz.
Red Oil (Oleic Acid)	4 oz.

Dissolve the caustic in water, add the glycerin and bring to a boil in an enamel pot. Remove from heat; add the red oil slowly while stirring. If a more neutral soap is wanted, use a little more Red Oil.

*Do not get on skin as it is corrosive.

MECHANICS HAND SOAP PASTE

Water	1.8 qt.
White Soap Chips	1.5 lb.
Glycerin	2.4 oz.
Borax	6 oz.
Dry Sodium carbonate	3 oz.
Coarse Pumice Powder	2.2 lb.
Safrol	enough to scent

Dissolve the soap in ⅔ of the water by heating. Dissolve the last three in the rest of the water. Pour the 2 solutions together and stir well. When it begins to thicken, sift in the pumice, stirring constantly till thick, then pour into cans. Vary amount of water, for heavier or softer paste (Water cannot be added to the finished soap.)

LIBRARY PASTE

White Dextrin	6 oz.
Dilute Acetic Acid	1 oz.
Oil of Cloves	10 drops
Glycerin	1 oz.
Water to make	16 oz. fluid

First make a paste of the dextrin with 6 oz. of cold water and then add 8 oz. of boiling water; boil 5 minutes, in a double boiler, keep stirring and add 14 fluid ounces of hot water. Allow to cool a little and then stir in other ingredients (above).

DRY CLEANING FLUID

Glycol Oleate	2 parts
Carbon Tetrachloride	60 parts
Varnoline (Naphtha)	20 parts
Benzine	18 parts

An excellent cleaner that will not injure the finest fabrics.

GREASE, OIL, PAINT AND LACQUER SPOT REMOVER

Alcohol	1 oz.
Ethyl Acetate	2 oz.
Butyl Acetate	2 oz.
Toluol	2 oz.
Carbon Tetrachloride	3 oz.

Place garment with spot over a piece of clean paper or cloth and wet with the above fluid; rub with clean cloth toward centre of spot. Use a clean section of cloth for rubbing and clean paper or cloth for each application of the fluid. The above product is inflammable and should be kept away from flames. Use of cleaners of this type should be out-of-doors or in well-ventilated rooms as the fumes are toxic.

FIRE EXTINGUISHING LIQUID

Carbon Tetrachloride	95 parts
Solvent Naphtha	4 parts
Ammonium Oleate	1 parts

Stir until dissolved.

The inclusion of the naphtha minimizes production of toxic fumes when extinguishing fires.

SILK STOCKING SOLUTION

To lengthen the life of silk stockings and also to keep them from bagging at the knees, soak them in a warm solution of alum.

SEALING WAX

Shellac	8½ oz.
Venice Turpentine	6¾ oz.
Resin	2 oz.
Vermilion	⅛ oz.

Melt together in a double boiler, stirring until uniform.

CANARY BIRD FOOD

Yolk of eggs, dried and chopped	2.oz.
Poppy heads, coarse powder	1 oz.
Cuttlefish bone, coarse powder	1 oz.
Granulated sugar	2 oz.
Soda crackers, powdered	8 oz.

Mix well together.

INTRODUCTION

THIS book is written in a simple way so that anyone, regardless of technical education or experience, can start making simple products without any complicated or expensive machinery. For commercial productions, however, suitable equipment is necessary.

Oils, greases, fats, waxes, emulsifying agents, water, chemicals of great diversity, dyestuffs, and perfumes are required in the composition of these various formulas. To assemble some of these ingredients with some of the others requires certain definite and well-studied procedure, any departure from which will inevitably result in failure. The successful steps are given with the formulas. Follow them explicitly. If the directions require that A should be added to B, carry this out literally, and not in reverse fashion. In making an emulsion (and many cosmetics and polishes, particularly creams, are emulsions), the job is often quite as tricky as the making of mayonnaise. In making mayonnaise, you add the oil to the egg, *slowly,* with constant and even and regular stirring. If you do it correctly, you get mayonnaise. If you depart from any of these details: if you add the egg to the oil, or pour the oil in too quickly, or fail to stir regularly, the result is a complete disappointment. The same disappointment might be expected if the prescribed procedure of any formula is violated.

The next point in importance is the scrupulous use of the proper ingredients. Substitutions are sure to result in inferior quality, if not in complete failure. Use what the formula calls for. If a cheaper product is desired, do not obtain it by substituting a cheaper material for the one prescribed: resort to a different formula. Not infrequently a formula will call for some ingredient which is difficult to obtain: in such cases, either reject the formula or substitute a similar material only after preliminary experiment demonstrates its usability. There is a limit to which this rule may reasonably be extended. In some instances the substitution of an equivalent ingredient may legitimately be made. For example: when the formula calls for *white wax* (beeswax), yellow wax can be used, if the color of the finished product is a matter of secondary importance. Yellow beeswax will replace white beeswax, making due allowance for color: but paraffin will *not* replace beeswax, even though its light color recommends it above yellow beeswax.

And this leads to the third point: the use of good quality ingredients, and ingredients of the correct quality. Lanolin is not the same thing as *anhydrous* lanolin: the replacement of one for the other, weight for weight, will give discouragingly different results. Use exactly what the formula calls for: if you are unacquainted with the material and a doubt arises as to just what is meant, discard the formula and use one that you understand. Buy your materials from reliable sources. Many of the ingredients in cosmetics are obtainable in a number of different grades: if the formula does not designate the grade, it is understood that the best grade is to be used. Remember that a formula and the directions can tell you only a part of the story. Some skill is often required to attain success. Practice with a small batch in such cases until you are sure of your technique. Many instances can be cited. If the formula calls for soaking quince seed for 30 minutes in cold water, your duplication of this procedure may produce a mucilage of too thin a consistency. The originator of the formula may have used a fresher grade of seed, or his conception of what "cold" water means may be different from yours. You should have a feeling for the right degree of mucilaginousness, and if steeping the seed for 30 minutes fails to produce it, steep them longer until you get the right kind of mucilage. If you do not know what the right kind is, you will have to experiment until you find out. Hence the recommendation to make small experimental batches until successful results are arrived at. Another case is the use of dyestuffs for coloring lotions, and the like. Dyes vary in strength: they are all very powerful in tinting value: it is not always easy to state in quantitative terms how much to use. You must establish the quantity by carefully adding minute quantities until you have the desired tint. Gum tragacanth is one of those products which can give much trouble. It varies widely in solubility and bodying power: the quantity prescribed in the formula may be entirely unsuitable for *your* grade of tragacanth. Hence a correction is necessary which can only be made after experiments to determine *how much* to correct.

In short, if you are completely inexperienced, you can profit greatly by gaining some experience through recourse to experiment. Such products as mouth washes, hair tonics, astringent lotions, need little or no experience, because they are as a rule merely mixtures of simple liquid and solid ingredients, the latter dissolving without difficulty and the whole being a clear solution that is ready for use when mixed. On the other hand, face creams, tooth pastes, shaving creams, which require relatively elaborate procedure and which depend for their usability on a definite final viscosity, must be made with the exercise of some skill, and not infrequently some experience.

Figuring

Some prefer proportions expressed by weight, volume or in terms of percentages. In different industries and foreign countries various systems of weights and measures are used. For this reason no one set of units could be satisfactory for everyone. Thus divers formulae appear with different units in accordance with their sources of origin. In some cases, parts instead of percentages or weight or volume is designated. On the pages preceding the index, tables of weights and measures are given. These are of use in changing from one system to another. The following examples illustrate typical units:

Example No. 1

Permanent Wave Lotion

Ammonium Sulphite	50
Water	1000
Glycerin	30
Turkey Red Oil	5

Here no units are mentioned. When such is the case it is standard practice to use parts by weight, using the same system throughout. Thus we may use ounces or grams as desired. But if ounces are used for one item, then ounces must be the unit for all the other items in the particular formula.

Example No. 2

Cream Rouge

Stearic Acid	25 %
Water	61½%
Glycerin	10 %
Potassium Hydroxide	1 %
Oil Soluble Dye	2½%
Perfume	to suit %

When no units of weight or volume but percentages are given, then forget the percentages and use the same instructions as given under Example No. 1.

Example No. 3

Antiseptic Ointment

Petrolatum	16 parts
Coconut Oil	12 parts
Salicylic Acid	1 part
Benzoic Acid	1 part
Chlorthymol	1 part

The same instructions as given under Example No. 1 apply to Example No. 3.

It is not wise in many cases to make up too large a quantity of material until one has first made a number of small batches to first master the necessary technique and also to see whether it is suitable for the particular purpose for which it is intended. Since, in many cases, a formula may be given in proportions as made up on a commercial factory scale, it is advisable to reduce the proportions accordingly. Thus, taking the following formula:

Example No. 4

Neutral Cleansing Cream

Mineral Oil, White	80 lb.
Spermaceti	30 lb.
Glyceryl Monostearate	24 lb.
Water	90 lb.
Glycerin	10 lb.
Perfume	to suit

Here, instead of pounds, grams may be used. Thus this formula would then read:

Mineral Oil, White	80 g.
Spermaceti	30 g.
Glyceryl Monostearate	24 g.
Water	90 g.
Glycerin	10 g.
Perfume	to suit

Reduction in bulk may also be obtained by taking the same fractional part or portion of each ingredient in a formula. Thus in the following formula:

Example No. 5

Vinegar Face Lotion

Acetic Acid (80%)	20
Glycerin	20
Perfume	20
Alcohol	440
Water	500

We can divide each amount by ten and the finished bulk is only 1/10th of the original formula. Thus it becomes:

Acetic Acid (80%)	2
Glycerin	2
Perfume	2
Alcohol	44
Water	50

Apparatus

For most preparations pots, pans, china and glassware, such as is used in every household, will be satisfactory. For making fine mixtures and emulsions a "malted-milk" mixer or egg-beater is necessary. For weighing, a small, low priced scale should be purchased from a laboratory supply house.

For measuring of fluids, glass graduates may be purchased from your local druggist. Where a thermometer is necessary a chemical thermometer graduated in both Centigrade and Fahrenheit degrees should be obtained from a druggist or chemical supply house.

Methods

Better to understand the products which you intend making, it is advisable that you read the complete section covering such products. Very often an important idea is thus obtained. You may learn different methods that may be used and also avoid errors which many beginners are prone to make.

Containers for Compounding

Where discoloration or contamination is to be avoided (as in light colored preparations or in food and drug products) it is best to use enamelled or earthenware vessels. Aluminum is highly desirable in such cases but it should not be used with alkalies as the latter dissolve and corrode this metal.

Heating

To avoid overheating, it is advisable to use a double boiler when temperatures below 212°F (temperature of boiling water) will suffice. If a double boiler is not at hand, any pot may be filled with water and the vessel containing the ingredients to be heated is placed therein. The pot may then be heated by any flame without fear of overheating. The water in the pot, however, should be replenished from time to time as necessary—it must not be allowed to "go dry." To get uniform and higher temperatures, oil, grease or wax is used in the outer container in place of water. Here of course care must be taken to stop heating when thick fumes are given off from the oil, grease or wax mixture as these are inflammable. When higher uniform temperatures are necessary, molten lead may be used as a heating medium. Of course, where materials melt uniformly and stirring is possible, direct heating over an open flame is possible, but not often recommended.

Where instructions indicate working at a certain temperature, it is important that the proper temperature be attained—not by guess work, but by the use of a thermometer. Deviations from indicated temperatures will usually result in spoiled preparations.

Temperature Measurements

In Great Britain and the United States, the Fahrenheit scale of temperature measurement is used. The temperature of boiling water is 212° Fahrenheit (212°F); the temperature of melting ice is 32 degrees Fahrenheit (32°F).

In scientific work and in most foreign countries the Centigrade scale is used. On this scale of temperature measurement, the temperature of boiling water is 100 degrees Centigrade (100°C) and the temperature of melting ice is 0 degrees Centigrade (0°C).

The temperature of liquids is measured by a glass thermometer. The latter is inserted as deeply as possible in the liquid and is moved about until the temperature remains steady. It takes a little time for the glass of the thermometer to come to the temperature of the liquid. The thermometer should not be placed against the bottom or side of the container, but near the center of the liquid in the vessel. Since the glass of the bulb of the thermometer is very thin, it can be broken easily by striking it against any hard surface. A cold thermometer should be warmed gradually (by holding over the surface of a hot liquid) before immersion. Similarly the hot thermometer when taken out should not be put into cold water suddenly. A sharp change in temperature will often crack the glass.

Mixing and Dissolving

Ordinary solution (e.g., sugar in water) is hastened by stirring and warming. Where the ingredients are not corrosive, a clean stick, bone or composition fork or spoon is used as a mixing device. These may also be used for mixing thick creams or pastes. In cases where most efficient stirring is necessary (as in making mayonnaise, milky polishes, etc.) an egg beater or a malted milk mixer is necessary.

Filtering and Clarification

When dirt or undissolved particles are present in a liquid, they are removed by settling or filtering. In the former the solution is allowed to stand and if the particles are heavier than the liquid they will gradually sink to the bottom. The upper liquid may be poured or siphoned off carefully and in some cases is then of sufficient clarity to be used. If, however, the particles do not settle out, then they must be filtered off. If the particles are coarse they may be filtered or strained through muslin or other cloth.

If they are very small particles then filter paper is used. Filter papers may be obtained in various degrees of fineness. Coarse filter paper filters rapidly but will not take out extremely fine particles. For the latter, it is necessary to use a very fine grade of filter paper. In extreme cases even this paper may not be fine enough. Here it will be necessary to add to the liquid ⅓ % of infusorial earth or magnesium carbonate. The latter clog up the pores of the filter paper and thus reduce their size and hold back undissolved material of extreme fineness. In all such filtering, it is advisable to take the first portions of the filtered liquid and pour them through the filter again as they may develop cloudiness in standing.

Decolorizing

The most commonly used decolorizer is decolorizing carbon. The latter is added to the liquid to the extent of ⅕ % and heated with stirring for ½ hour to as high a temperature as is feasible. It is then allowed to stand for a while and filtered. In some cases bleaching must be resorted to.

Pulverizing and Grinding

Large masses or lumps are first broken up by wrapping in a clean cloth and placing between two boards and pounding with a hammer. The smaller pieces are then pounded again to reduce their size. Finer grinding is done in a mortar with a pestle.

Spoilage and Loss

All containers should be closed when not in use to prevent evaporation or contamination by dust; also because, in some cases, air affects the material adversely. Many materials attack or corrode the metal containers in which they are received. This is particularly true of liquids. The latter, therefore, should be transferred to glass bottles which should be as full as possible. Corks should be covered with aluminum foil (or dipped in melted paraffin wax when alkalies are present).

Materials such as glue, gums, olive oil or other vegetable or animal products may ferment or become rancid. This produces discoloration or unpleasant odors. To avoid this, suitable antiseptics or preservatives must be used. Too great stress cannot be placed on cleanliness. All containers must be cleaned thoroughly before use to avoid various complications.

Weighing and Measuring

Since, in most cases, small quantities are to be weighed, it is necessary to get a light scale. Heavy scales should not be used for weighing small amounts as they are not accurate for this type of weighing.

For measuring volume (liquids) measuring glasses or cylinders (graduates) should be used. Since this glassware cracks when heated or cooled suddenly it should not be subjected to sudden changes of temperature.

Caution

Some chemicals are corrosive and poisonous. In many cases they are labeled as such. As a precautionary measure, it is advised not to smell bottles directly, but only to sniff a few inches from the cork or stopper. Always work in a well ventilated room when handling poisonous or unknown chemicals. If anything is spilled, it should be wiped off and washed away at once.

Where to Buy Chemicals and Apparatus

Many chemicals and most glassware can be purchased from your druggist. Notices of suppliers of all products will be found at the end of this book.

ADVICE

This book is the result of co-operation of many chemists and engineers who have given freely of their time and knowledge. It is their business to act as consultants and, for a fee, to give advice on technical matters. As publishers, we do not maintain a laboratory or consulting service to compete with them.

Please, therefore, do not ask us for advice or opinions, but confer with a chemist in your vicinity.

CONTENTS

CONTENTS

CONTENTS

CHAPTER I

ADHESIVES

ADHESIVES are gummy or gelatinous substances which are used to unite two surfaces. They usually are substances which swell in water and harden when the water evaporates or is absorbed by the materials being joined, e.g., glue. Another class of adhesives softens or melts on heating and "sets" on cooling to form a bond. This is exemplified by shellac. Another class is exemplified by materials which swell in organic solvents and "set" when the solvent evaporates after application, e.g., rubber-cement. The number of formulae used commercially are many and varied—depending on the type of material, degree of adhesion desired, conditions of application, use of the finished product, cost considerations, etc. For this reason a formula may have to be modified to meet specific conditions. Thus an adhesive which may work perfectly for labels on shoe boxes will be ruled out for cracker boxes because of its odor. One which works perfectly when applied hot cannot be used because heating in another case is not desirable.

In uniting two surfaces the latter should be cleaned thoroly and the adhesive should be applied thinly to both surfaces; then uniform pressure should be applied. The following method is used for dissolving casein, one of the most important bases for industrial adhesives.

Dissolving Casein

3 to 4 parts of cold water by weight to each pound of dry Casein.

1 ounce 26° Ammonia to each pound of dry Casein.

If a heavy solution is required, use 3 to 1 proportion; if a thinner solution is desirable, use 4 to 1.

Pour water into a double boiler and add the Casein. Stir well to break down any lumps that may form and then add Ammonia. Stir the mixture after adding the Ammonia and immediately turn on the heat. Heat, while stirring, to about 160° F. Turn off the heat when this temperature is reached and continue to stir, preferably with a mechanical agitator, until the Casein is completely dissolved, which will take about half an hour.

If the temperature exceeds 160° during the heating, it is not serious, although it is advisable not to apply excessive heat, particularly when Ammonia is used, as

there is a tendency to somewhat weaken the Casein and to darken it in color.

When the Casein is completely dissolved it may be diluted, if necessary, by the addition of warm water and used, as dissolved, either hot or cold, in the same manner as ordinary glue.

10 pounds Casein

1½ lb. Powdered Borax

40 to 60 pounds *cold* water

Stir cold for about 15 minutes or until the Casein commences to swell.

Then heat in a double boiler for 40 to 60 minutes at a temperature not higher than 160° F. stirring constantly.

Ammonia 26° can be used in place of Borax.

To make a thin solution use equal parts of Ammonia 26° and Trisodium Phosphate or Borax and Trisodium Phosphate.

If a preservative is desired you can use

1

about 2% of Benzoate of Soda or ¼ of 1% Carbolic Acid.

Note—do not dissolve Casein in a copper kettle as this tends to discolor the Casein particularly if the solvent is Ammonia.

Handling of Glue

Special precaution should be used in all cases to insure a soaking of the glue in the required amount of cold water for at least 4 hours. In order to effect solution of glue the temperature should be increased to about 160° F. Prolonged heating and excessive heating should be avoided, because this has been shown to result in extensive loss due to the hydrolyzing action of the water. In applying the heat, the most advantageous method is to apply heat (e.g., steam or electricity) to a water jacket in which glue container is placed.

To employ glue such that the greatest benefit may be derived from its physical and chemical characteristics, the surface should be made so warm that the melted glue will not be chilled before it has time to effect a thorough adhesion.

For high-class joint work only the better grades of hide glue should be employed.

For Veneer work the medium grades are indicated. In this case a high viscosity is desirable on account of the tendency of a thin liquid to penetrate the pores of the thin sheet of wood and show itself on the opposite surface.

Liquid Glue

Animal Glue	46.7%
Water	46.7%
Sodium Nitrate	6.6%

Dissolve sodium nitrate in cool water, stir glue into solution, allow to soak 2 hours, melt in water bath at temperature between 140–160° F. Heat a couple of hours or until mixture remains fluid at room temperature. Glue may be preserved by adding phenol or other common preservative.

Glue Composition for Plaster Casting Molds

Powdered Hide Glue	1	part
Glycerin	1½	parts
Water	1	part
Sugar	½	part
Finely Powdered Silica	1	part

Tungstic Glue

(Substitute for Hard India Rubber)

Mix thick solution of glue with tungstate of soda and Hydrochloric Acid, by means of which a compound of tungstic acid and glue is precipitated which at a temperature of 86–104° F. is sufficiently elastic to admit of being drawn out into thin sheets. On cooling, this mass becomes solid and brittle and on heating is again soft and plastic. It can be used for all purposes to which hard rubber is adapted.

Starch Paste

The strength of starch paste is increased by the addition of a small quantity of ammonium hydroxide. Paste may be rendered flexible by the addition of glycerine. The following formula produces satisfactory results:

Water	100 grams
Ammonium Hydroxide	4 grams
Paste Starch	8 grams
Glycerin	1 gram

Starch Paste

Corn or Tapioca Starch	4
Cold Water	8
Boiling Water	64

Make a paste of starch and cold water then pour into boiling water and stir until translucent.

Putty

Whiting	800
Corn Oil	20
Crude Cottonseed Oil	10
Thin Mineral Oil	69
Sod Oil	3

Elastic Putty

Turpentine	5
Rosin Oil	8
Linseed Oil and drier	5.5
Barytes	8.5
Whiting	73.0

Non-Shrinking Putty

White Lead	150 lb.
Raw Linseed Oil	16 gal.
Whiting	505 lb.
Silica	41 lb.
Flour Paste	41 lb.

Sculptors' Putty

Linseed Oil (Boiled)	15%
Fullers' Earth	15%
Calcium Carbonate	70%

Mix all ingredients thoroughly.

◆

Tape, Coating for Adhesive

Heat 10 parts Castor Oil to 270° C. and to it add slowly with stirring 6 parts shellac and 1 part rosin. The addition of glycerol or glycols produces more sticky products.

◆

Tape, Masking

As above except that 9 parts of shellac are used.

◆

Glue for Cellophane

1.

Gum Arabic	17½	parts
Water	52½	parts
Glycerin	30	parts
Formaldehyde	.05	part

2.

Animal Glue	40%
Water	40%
Glycol Bori-borate	20%

Use grade of glue common to paper box work; soak glue in cool water for around one hour, melt in water bath at 140° F. and stir in Glycol bori-borate. Add sufficient water to produce the proper working consistency at 130–140° F.

◆

Jeweler's Cement

Dissolve over a water bath 25 parts of fish glue in a small quantity of alcohol-water mixture 40%, add 2 parts of gum ammoniac. Separately dissolve 1 part of mastic gum in 5 parts alcohol-water solution. Mix the two solutions and keep in well-stoppered bottles.

◆

Stratena—Household Cement

Dissolve 12 parts of white glue in 16 acetic acid, and then add this solution to 2 parts gelatine in 16 of water. After mixing add 2 parts shellac varnish.

◆

Aquarium Cement

To 10 lbs. of glazier's putty add 1 lb. dry litharge, 1 lb. dry red lead, and 1 gill of asphaltum. Mix to a stiff consistency with boiled linseed oil and add sufficient lampblack to give a slate color.

Another well-known formula consists of 10 parts by bulk of plaster of Paris, 10 of fine sand, 10 of litharge, 1 part of powdered rosin, and sufficient boiled linseed oil to make a stiff putty. A third formula is as follows: Red lead 3 parts, litharge 7, fine sand 10, powdered rosin 1 part, and spar varnish sufficient to make a stiff cement.

In each case add the linseed oil or varnish little by little and mix the ingredients very thoroughly. If the putty should become too soft, merely add more of the dry materials as the exact proportions are not especially important.

◆

Iron Cement (for castings)

Iron filings	128
Plaster of Paris	20
Whiting	8
Gum Arabic	8
Carbon Black	1
Portland Cement	4

Make into a paste with water directly before using.

◆

Linoleum Cement

Clay	20
Red Oxide of Iron	20
Dextrin	60

The powders are thoroughly mixed and made into a paste of desired consistency with water.

◆

Case Making Machine Glue

Glue	175 lb.
Glycerin	10 lb.
Water	175 lb.
Betanaphthol	½ lb.
Terpineol	½ lb.

◆

Floor Crack Filler

Plaster of Paris	32 lb.
Silica	200 lb.
Dextrin Yellow	33 lb.

Make into a stiff dough with water before use.

◆

Albumin Glue

Blood albumin (90 per cent solubility)	100 parts
Water	170 parts
Ammonium hydroxide (specific gravity 0.90)	4 parts
Hydrated lime	3 parts
Water	10 parts

Pour the larger amount of water over the blood albumin and allow the mixture to stand undisturbed for an hour or two. Stir the soaked albumin until it is in solution and then add the ammonia while the mixture is being stirred slowly. Slow stirring is necessary to prevent foamy glue. Combine the smaller amount of water and the hydrated lime to form milk of lime. Add the milk of lime, and continue to agitate the mixture for a few minutes. Care should be exercised in the use of the lime, inasmuch as a small excess will cause the mixture to thicken and become a jellylike mass. The glue should be of moderate consistency when mixed and should remain suitable for use for several hours. The exact proportions of albumin and water may be varied as required to produce a glue of greater or less consistency or to suit an albumin of different solubility from that specified.

Blood albumin (90 per cent solubility)	100	parts
Water	140–200	parts
Ammonium hydroxide (specific gravity, 0.90)	5½	parts
Paraformaldehyde	15	parts

The blood albumin is covered with the water and the mixture is allowed to stand for an hour or two, then stirred slowly. The ammonium hydroxide is next added with more stirring. Then the paraformaldehyde is sifted in, and the mixture is stirred constantly at a fairly high speed. Paraformaldehyde should not be poured in so rapidly as to form lumps nor so slowly that the mixture will thicken and coagulate before the required amount has been added.

The mixture thickens considerably and usually reaches a consistency where stirring is difficult or impossible. However, the thickened mass will become fluid again in a short time at ordinary temperatures and will return to a good working consistency in about an hour. It will remain in this condition for 6 or 8 hours, but when the liquid finally sets and dries, as in a glue joint, it forms a hard and insoluble film.

This glue may be used in either hot or cold presses. When cold pressed, however, it has only moderate strength, and for that reason is not to be depended upon in aircraft construction where maximum strength is required. If hot pressed, it is high in strength and very water resistant.

Banknote or Mouth Glue

Dissolve gelatine with about ¼ to ½ of its weight of brown sugar in as small a quantity of water as possible. When liquid, cast mixture in thin cakes and when cold cut to size. When required for use moisten one end.

◆

Tablet Binding Glue

Glue	120
Glycerin	113
Water	113
Zinc Oxide	5
Betanaphthol	½
Terpineol	½

◆

Paste for Cardboard

Dissolve 14 oz. of high-grade glue in 26 oz. water. Add 1 oz. of a solution composed of 1 part shellac in 7 parts alcohol and stir as long as solution is warm. Next dissolve ½ oz. of dextrine in 7 oz. of alcohol and 3½ oz. of water, stir and place vessel in warm water until solution is complete. Mix two solutions and allow to cool. When wanted for use cut off a small piece and liquefy by warming.

◆

Flexible Paper Box Glue

Animal Glue	45%
Glycerine	15%
Water	39%
Preservative	1%

Soak the animal glue (bone glue suitable for paper box work) in cool water for approximately two hours and melt at 140° F. Stir the glycerine into the glue solution after the temperature has reached 140° F. If the glue is kept for a period of time, some effective preservative should be added.

◆

Pastes for Paper and Fine Fancy Articles

Dissolve 100 parts glue in 200 water and add a solution of 2 parts of bleached shellac in 10 of alcohol. Stir constantly while adding. Keep temperature below 50° C.

◆

Glue for Hectograph

One part glue, 1 part glycerine and smallest amount of water possible is used as a hectograph mass for the transfer of matter, when with concentrated solution of aniline color.

Label Paste

Soak glue in 15% Acetic Acid solution and heat to boiling and add flour.

Sausage Casing Glue

Add to 1 quart of hide glue 20% solution, ¾ to 1 oz. bichromate of potash. Warm slightly when about to use it and before application moisten paper, latter must be dried rapidly and then exposed to light until yellow glue becomes brownish, boiled in sufficient quantity of water to which 2 to 3% alum is added until chromate is dissolved out.

Metal Cap Seal

Rubber Factice	20
Gutta Percha	20
Asbestos Flour	60
Dark Red Iron Oxide	1.5

Decorators' Paste

Rye Meal	4
Fine Whiting	2
Casein	1
Powdered Alum	½

Mix the above ingredients together and rub to a fine powder. Use 2 lb. of the mixture to one quart of water either hot or cold.

Flour Paste

Wheat Flour	4 lb.
Cold Water	2 qt.
Boiling Water	3 gal.

Make smooth paste of flour and cold water and then pour into boiling water. Stir and boil for 5 minutes.

Glue—Starch Paste

Starch (Cassava)	30%
Glue (Bone Glue)	10%
Water	60%

The starch and glue are put into solution separately and mixed hot. Any additional water necessary to produce the desired consistency is incorporated later.

Paste for Fixing Labels (Machine)

Make 10% solution of glue and add to this 25% of dextrin on weight of glue. Mix while warm and add to every pound thereof ½ oz. each of boiled linseed oil and turpentine. This paste resists dampness and thus prevents printed labels from falling from metallic surfaces.

Sealing Wax

1.		
	Shellac (Button)	14
	Rosin	24
	Vermilion	1¼
	Barytes	14
	French White	4
	Turpentine	1

Melt shellac and rosin; keep hot and work in pigment and finally the turpentine. Cast in sticks.

2.		
	Shellac	84
	Venice Turpentine	60
	Rosin	21

3.		
	Limed Rosin	3
	Tallow	6
	Turpentine	3
	Precipitated Chalk	4
	Red Lead	4

4.		
	Orange Shellac	39
	Rosin	78
	Turpentine	14
	Whiting	56
	Silex	36
	Pale Vermilion	5¼

5.		
	Orange Shellac	26
	Rosin—H grade	83
	Turpentine	7¼
	Whiting	32
	Silex	31
	Burnt Umber	4

Mucilage

To 30 gallons water add 75 lb. gum arabic, clean sorts. Mix at 160° F. until completely dissolved; add 6 lb. carbolic acid, 1 lb. oil of cloves. Strain and fill.

Envelope Mucilage

Gum arabic	1 part
Starch	1 part
Sugar	4 parts

Water, sufficient to produce the desired consistency.

The gum arabic is first dissolved in water, the sugar added, then the starch, breaking up all lumps, after which the mixture is boiled for a few minutes in order to dissolve the starch, after which it is thinned down to the desired consistency with more water.

Mucilage, Stick Form

Powdered White Glue	10 parts
Powdered Gum Arabic	2 parts

| Sugar | 5 parts |
| Water | Sufficient |

Mix the glue and gum, stir in enough cold water to make the solution the consistency of thick syrup. Soak overnight to allow the glue and gum to absorb the water, then add enough water to again bring it to a thick syrup. Pour into a flat bottom pan that has been chilled and cut into sticks of desired size when almost solid. If poured into molds the molds should first be well greased and then chilled by setting upon cracked ice. The addition of 0.1% of Moldex in the water used will prevent spoilage.

Mounting Paste

White dextrin	1	lb.
Gum arabic	1	oz.
Water	1½	pt.
Acetic acid	1	oz.
Oil of wintergreen	20	drops
Oil of cinnamon	20	drops
Salicylic acid	20	gr.

The dextrin and the gum, which should be pulverized, are dissolved in the water, and then the salicylic acid added and dissolved. This liquid is heated with the dextrine, and when the whole has become pasty, which should require a quarter of an hour, the acetic acid is added, stirring in slowly. The heating is continued, taking care not to boil the mass. The paste will soon become pearly, and should then be removed from the fire and the perfume oils added while it is cooling. It should be stirred thoroughly while the oils are being added.

Library Paste

1.

Tragacanth (powdered)	20
White Dextrin	10
Wheat Flour	60
Glycerin	10
Cold Water	40
Salicylic Acid	3
Boiling Water	400

Mix the tragacanth with 160 parts of boiling water, stir well and set aside. Mix the dextrin and the flour with the cold water, stir well and add to the tragacanth mucilage. Pour into the resulting mixture the rest of the boiling water stirring constantly. Rub up the salicylic acid with the glycerin, add to the mucilage and boil for 5 to 6 minutes with constant stirring.

2.

White Dextrin	6 oz.
Diluted Acetic Acid	1 oz.
Oil of Clove	10 drops
Glycerin	1 oz.
Water to make	16 fl. oz.

Make a paste of the dextrin with 6 ounces of cold water, add 8 ounces of boiling water, boil 5 minutes with constant stirring, then add enough hot water to make 14 fluid ounces. Let cool then add the other ingredients.

3.

Flour	16
Gum Acacia	12
Gum Tragacanth	3
Salicylic Acid	0.5
Clovel	0.6
Water	160

Use part of water to make a paste of flour. Heat another part of water with gums until dispersed. Mix these two well and other ingredients and bring to a boil while stirring.

4.

To 30 gallons cold water, add 75 lb. white potato dextrin. Break up all lumps then heat to 180° F. Add 6 lb. carbolic acid and 1 lb. oil of wintergreen. Strain and fill into jars while hot. Allow to stand for three days.

Photo Mounting Paste

White Potato Dextrine	15	lb.
Water	15	lb.
Glycerin	1	lb. 15 oz.
Formaldehyde	2½	oz.
Oil of Sassafras	2½	oz.

Adhesive for Celluloid to Celluloid

| Gum Camphor | 1 part |
| Alcohol | 4 parts |

Dissolve the camphor in the alcohol and then add 1 part Shellac. Warm to dissolve. This cement is applied warm, and the parts united must not be disturbed until the cement is hard.

Adhesives for Cellulose Acetate and Pyroxylin

1.

15 parts nitrocotton.
6 parts camphor.
79 parts acetone.
10 parts filler.

2.

20 parts scrap film.
60 parts ethyl acetate.
20 parts ethyl alcohol.
10 parts aluminium powder.

3.

16 parts nitrocotton.
10 parts ethyl acetanilide.
74 parts acetone.
15 parts starch.

4.

12 parts cellulose acetate.
 8 parts tricresyl phosphate.
20 parts methyl alcohol.
30 parts ethyl acetate.
30 parts methyl acetate.
25 parts filler.

5.

12 parts nitrocotton.
 4 parts ethyl acetanilide.
 2 parts castor oil.
20 parts ethyl acetate.
20 parts methyl acetate.
17 parts methyl alcohol.
25 parts starch.

6.

14 parts scrap film.
 2 parts ethyl acetanilide.
 2 parts castor oil.
 3 parts tricresyl phosphate.
13 parts ethyl acetate.
13 parts methyl acetate.
 6 parts methyl alcohol.
21 parts acetone.
 6 parts benzine.
20 parts starch.

7.

10 parts nitrocotton.
 4 parts camphor.
 2 parts tricresyl phosphate.
50 parts acetone.
20 parts butyl acetate.
14 parts filler.

___

Adhesive Wax

Rosin	100
Paraffin Wax	10
Thin Mineral Oil	88

___

Marble and Onyx Adhesive

Carnauba Wax	28
Rosin	8
Silica	64

The wax and rosin are heated until molten and the Silica is mixed in by stirring. For use object and adhesive are heated.

A more liquid adhesive than the above can be made as follows.

Carnauba Wax	63
Dammar Gum	37

The wax and gum are mixed when they are heated until liquid. This adhesive is used the same way as the previous one.

___

Glass to Brass Adhesive

Caustic Soda	1
Rosin	3
Plaster of Paris	3
Water	5

Boil together until all lumps disappear and cool before using. This sets in about 20 min.

___

Glue for Cementing Glass

(To be exposed to boiling water)

Five parts hide glue, one part dissolved acid chromate of lime; the glue prepared becomes, after exposed to light, insoluble in water in consequence of a partial reduction of chromic acid.

___

Cement, Safety Glass

Pyroxylin	12
Camphor	2
Ethyl Methyl Ketone	30
Alcohol	15
Gum Benzoin	2
Triacetin	5
Benzyl Alcohol	2.5

___

Waterproof Glass and Metal Cement

This cement will also stand fairly high temperatures.

Cement and litharge in equal parts are thoroughly mixed. Then glycerin in an amount equal in volume to half the volume of the mixed powder is added and the whole thoroughly mixed with a spatula. This cement will set under water.

To repair leaks in pipes, fill the hole with the cement and bind it in place with cheese cloth. Then daub a quantity of the cement on the cloth and wrap the whole tightly together with iron wire.

The powders may be mixed ready for use, but the glycerin must only be added as needed.

___

Quicksetting Insulating Adhesive

Modified Alkyd Resin	11–20
Pyroxylin Solution (35%)	64–73

| Tricresyl Phosphate | 4–8 |
| Lacquer Thinner | 11–21 |

This is useful on coils and radio parts.

Sticky Wax

Rosin	100
Talc	16
Lanolin	60
Paraffin	8
Saponified Wax	2

Melt together and while stirring rapidly add slowly a boiling caustic soda solution (10° Bé.); stir until uniform.

Adhesive for Wigs

Damar	20
Rosin	20
Beeswax	40
Venice Turpentine	20

Heat to 90° C. and stir until uniform; cast in sticks.

Waterproof Adhesive for Wood

Light gasoline	0.5 gal.
Acetone	0.5 gal.
Soft cumarone	10.0 lb.
Pine oil	0.5 lb.
Tricresyl phosphate	0.25 lb.

Adhesives for Fixing Wood, Tin, etc. to Celluloid

Shellac	2 gm.
Spirits of Camphor	3 gm.
Alcohol	4 cc.

Warm together until dissolved.

Cork and Wood Flour, Binders for

A. Rosin	100
Dibutyl Phthallate	35
Sodium Silicate	4
Nitrocellulose	4
Castor Oil	2
B. Ester Gum	50
Coumarone Resin	50
Linseed Oil bodied	10
Dibutyl tartrate	35
C. Urea formaldehyde resin	50
Coumarone Resin	25
Rosin	25
Tricresyl phosphate	20
Dibutyl phthallate	20

Adhesive Cement (For Fine Furniture)

| Casein (fine ground) | 12 lb. |
| Lime (powdered, unslaked) | 13 lb. |

| Mica (dry, ground) | 15 lb. |
| Barium sulphate (barytes) | 60 lb. |

Mix all ingredients. Keep in dry container. To use, mix with water until pasty. Hardens in about 24 hours.

Furniture Glue

Animal glue	10 lb.
Powdered white lead	2½ lb.
Powdered Chalk	5 oz.
Sodium salicylate	2 lb.
Wood alcohol	1¼ pt.
Water	19 lb.

Dissolve sodium salicylate in water. Dissolve animal glue in the same water. Mix lead and chalk; add to the sodium salicylate water and glue. Add wood alcohol to the batch.

Cabinet Makers' Glue

Glue	175 lb.
Glycerin	10 lb.
Water	175 lb.
Betanaphthol	½ lb.
Terpineol	½ lb.

In the above formula the glue is soaked in cold water over night and heated not over 150° F. and stirred until dissolved. The other ingredients are then dissolved in it and the liquid is then poured into molds where it sets on cooling.

Hard Wax Stopping for Filling Screw Holes in Wood

Carnauba wax	16 lb.
Paraffin wax	8 lb.
Rosin	8 lb.
Asphaltum	1 lb.

Melt the above together and apply hot.

Adhesives for Hard Rubber

1. Carefully melt together 1 part gutta percha and 2 parts coal tar pitch. Immediately apply the fluid, homogeneous hot mass to the parts to be joined, these first having been degreased. Allow the repair to cool under pressure.

2. Broken hard rubber can be repaired by applying to the 2 surfaces to be joined, concentrated silicate of potassium and subjecting them to strong pressure.

3. Marine glue is made of 10 parts rubber dissolved in 120 parts benzol or turpentine. Add 20 parts asphalt or 18 parts gum lac and allow to digest until the mass is homogeneous. The solid

glue, when it is to be used, is liquefied by careful heating; while the surfaces to be joined are first heated.

4. Melt together equal parts of pitch and gutta percha. Apply hot.

5. Dissolve 20 parts of rubber in 160 parts benzol or naphtha and mix with a solution of 20 parts gum lac and 50 parts mastic in the smallest possible amount of 90% alcohol.

When the surfaces to be adhered are smooth, it is always necessary to roughen them first by filing them lightly.

Raincoat Rubber Cement

Hevea Rubber	50
Litharge	20
Whiting	26.5
Rosin	2
Sulfur	1.5

Grind and mix thoroughly.

Cement, Rubber Tire

Crude Rubber	2 lb.
Rosin	2 lb.
Carbon Bisulfide (inflammable)	1 gal.

Allow to stand, covered until dissolved.

Rubber Cement (For Use on Leather Shoes)

Naphtha (62° Bé.)	9.8 pt.
Carbon Tetrachloride	5.4 lb.
Crepe Rubber	0.33 lb.

Makes 1 gal. cement on allowing to swell.

Paste for Joining Leather to Pasteboard

Dissolve 50 parts of glue with 50 parts water, add 1% Venice turpentine and next a thick paste made with 100 parts starch in water.

Airplane Propeller Glue

1. Black Blood		Mix at 15°
Albumen	1 g.	C., stop mix-
Water	6 g.	ing for two
		hours

Add:

Slaked Lime	0.06 g.	Mix until
Water	1 g.	thick

Leather to Metal Giue

Digest a quantity of nutgalls (approx. 1 part), reduced to powder, in 8 parts distilled water for 6 hours and filter. If tannic acid is available use 5% solution instead. Dissolve 1 part by weight of glue in same quantity of water. Leather moistened with decoction of nutgalls or acid solution, and glue applied to metal previously roughened and heated. Dry under pressure.

Leather Sole Glue

Rosin	60
Crepe Rubber	40
Varnish	20

Digest on a water-bath and when dissolved cool and add

Naphtha	30

Adhesive, Leather Shoe

Good leather adhesives for use by the shoe industry are based on nitrocellulose, rubber or casein. A nitrocellulose compn. contains nitrocellulose 200, amyl acetate 15, amyl alcohol 15, rosin 10, camphor 5, Venice turpentine 15 and linseed oil 20 parts. Soft leather is made to adhere especially well by the following compn.: gutta percha 85, rosin 25, asphalt 26, petroleum 130 and carbon bisulfide 300–350 parts.

Glue for Joints in Leather Driving Belts

Soak 1 part domestic isinglass and 25 parts glue in 75 parts water until thoroughly soft. Heat until solution has been effected. Add 0.2% Beta Naphthol and 0.1% Venice Turpentine. Surfaces to be cemented should be free from grease, slightly roughened and glue applied at a temperature of 150° F.

Mordant for Handles of Kitchen Knives

a. Potassium Bichromate	15 g.
Water	1000 cc.
b. Ammonia (25%)	150-200 g.

Dissolve the chromate a, and add b. Treat wood with solution, dry, rub over with a hard brush (horse-hair), optionally a thin polish.

CHAPTER II

AGRICULTURAL AND GARDEN SPECIALTIES

THE specialties of greatest interest to the farmer and gardener are fertilizers, insecticides and weed-killers. All of these must be adapted to the particular conditions present. The fertilizer will vary with the type of soil. The insecticide will vary with the type of pest being combated; the degree of infestation and the effect on plant or the harvested material. Here, too, cognizance must be taken of the possibility of a poisonous residue being left on fruits and vegetables. Animal remedies as well as vermin destroying formulae are likewise given in this chapter.

Brown Patch in Lawns and Golf Greens: Control of

Mercuric chloride	1 oz.
Mercurous chloride	2 oz.

Three ounces of the mixture are sufficient for 1000 sq. ft. of lawn. Application may be made by thoroughly mixing 3 oz. of the finely powdered chemicals to 2–3 pounds of pure sand or sifted composition. The treated spots should be well watered afterward. Treatment may also be made by adding 3 oz. of the mercury salt mixture to 5 gallons of water and applying with a sprinkling can or sprayer. This treatment is also a reliable control for foot-root, another disease sometimes found in lawns.

------◆------

Care of Lawns

As a diluent for artificial fertilizer sharp river or sea sand is the best, and sand should make up about 50 per cent of the total constituents. Of the components, iron sulphate will kill weeds and also increase the greenness of the new grass; sulphate of ammonia will kill weeds; sulphate of potash is useful if the sand is for spring use, but bone meal should replace it in the autumn.

Sulphate of ammonia	6
Superphosphate	1½
Sulphate of potash	½
Sulphate of iron	2

This mixture is applied when the grass is dry and the soil is moist, at the rate of 1 oz. per square yard. It can be applied at monthly intervals up to the end of June or later. If it is mixed with sand the directions are modified accordingly. Lawn sand should be sprinkled over the parts of the lawn where weeds are growing, and the operation should be carried out during a dry spell of sunshine. If rain comes within a few hours of application it will wash the sand off the leaves into the soil and the product will act as a fertilizer only, not as a weed-killer. If the fertilizing action of the mixture is required it should be watered in or applied when rain is expected. Sulphate of ammonia, 1 oz. to the square yard, applied every fourteen days, is used on lawns which are troubled with clover or daisies to stimulate the growth of grass. Sulphate of ammonia reduces the lime content of the soil; limed land encourages the growth of clover, and this is often a result of applying lime to prevent the growth of moss. The latter trouble is due usually to faulty drainage, shade, and drip from trees. Lime should be sprinkled liberally over the lawn, brushed in, and the application repeated in a month. Equal parts of crushed sulphate of iron and charcoal (or soot) can also be applied to destroy moss.

------◆------

Fertilizers

Commercial fertilizers are compounded from various raw materials which con-

10

tain one or more of the three necessary ingredients: Nitrogen, Phosphoric acid and Potash.

Different crops need different proportions of these chemicals and in general it is better to have the Nitrogen present in two or more forms such as Ammonium Sulfate, Sodium Nitrate, Organic (such as tankage, blood, cottonseed or other meals, etc.) The phosphoric acid is de-

rived from super-phosphate or animal bone: the Potash from mineral salts such as Muriate, Sulfate or mixtures such as Kainit or Manure Salt, and in special cases, Carbonate. Typical formulae follow.

In a formula the first figure represents the percentage of Nitrogen, the second, Available Phosphoric acid and the third, Potash.

A simple formula   4–8–4

Ammonium sulfate (contains 20% Nitrogen)........	400 lb.	equal	80 lb.	N
Super-phosphate (contains 16% Available P₂O₅).....	1000 lb.	"	160 lb.	P₂O₅
Muriate Potash (contains 50% K₂O)...............	160 lb.	"	80 lb.	K₂O
Earth (to make up one ton)....................	440 lb.			

4–8–7 Potato Fertilizer

Am. Sulfate (20% N).........................	100 lb.	contain	20 lb.	N
Sodium Nitrate (16% N)......................	100 lb.	"	16	"
Blood (13% N).............................	340 lb.	"	44	"
Super-phosphate (16% P₂O₅)..................	1000 lb.	"	160	"
Muriate Potash (50%K₂O)....................	280 lb.	"	140	"
Earth....................................	180 lb.			

Tobacco Fertilizer

	Pounds
Sulfate Ammonia (20.50% N)	293
Tankage (7% N)	286
Cottonseed Meal (5.50 N)	351
Superphosphate (18% P₂O₅)	778
Sulfate Potash (48% K₂O)	292
	2000

General Garden Fertilizer

	Pounds
Sulfate Ammonia (20.50% N)	293
Nitrate Soda (16% N)	125
Tankage (7% N)	286
Superphosphate (18% P₂O₅)	889
Muriate Potash (50% K₂O)	200
Filler	207
	2000

Corn Fertilizer

	Pounds
Sulfate Ammonia (20.50% N)	341
Tankage (6% N)	166
Superphosphate (18% P₂O₅)	1333
Muriate Potash (50% K₂O)	160
	2000

Grass Fertilizer

	Pounds
Sulfate Ammonia (20.50% N)	585
Castor Pomace (4.50% N)	440
Superphosphate (18% P₂O₅)	667
Muriate Potash (50% K₂O)	80
Filler	228
	2000

Pine Oil Cattle Sprays

The axiom ''contented cows produce more milk'' has been the basis for considerable research work on pine oil cattle sprays.

Various cattle sprays are being marketed, differing in ingredient content, but producing comparable results in combating warble and horse flies. There are also a few pine oil cattle sprays on the market that have outstanding merit. These sprays could be materially improved by the addition of more pine oil as evidenced by the subsequent data.

A series of four sprays were subjected to identical conditions for a period of time at an agricultural college and a city sanitation department.

The sprays were composed of the following ingredients, all figures computed on a volume basis:

	Formula No. 1	Formula No. 2A	Formula No. 2B	Formula No. 3
(a) Heavy-bodied Paraffin Oil....	15%	20%	.....	30%
(b) Kerosene Ext. of Pyrethrum....	5%	8%	8%	8%
(c) Pine Oil.........	25%	30%	30%	50%
(d) Long-time Burning Oil........	55%	42%	.....	12%
(e) Petroleum Distillate.........	.....	.....	62%	.....
	100%	100%	100%	100%

The product is prepared by simple mixing of the ingredients. Care must be taken that the ingredients are not allowed to absorb water as this may produce a cloudy product. The cloudiness is easily removed, however, by filtration through kieselguhr or like material.

◆

Formulae No. 2A and No. 2B

	Down in 10 min.	Dead after 24 hrs.
Test No. 1	100	70
2	98	72
3	98	75
4	100	66
5	98	55
6	99	62
7	100	49
8	97	47
9	100	71
	—	—
	99	63

These sprays were later tested on a practical scale at an agricultural college and a city sanitation department. The comments are indicative of what to expect when they are applied in the field. Formulae No. 2A and No. 2B received the unanimous vote as being the most effective and most presentable products of the four. They possessed the following characteristics:

1. Burning or blistering of hides— negative
2. Odor—mild odor of the pine forest
3. Tainting of milk—negative if sprayed 30 min. before milking time and usual care exercised
4. Clarity — free from suspended matter
5. Color—dark amber
6. Repellency—three to six hours
7. Volatility—relatively slow drying
8. Kill—63%
9. Knock-down—99%

10. Matting of hair—negative
11. Healing properties—the pine oil content promotes healing of open wounds and cuts.

Results of field tests may be duplicated provided no deviations are made in raw materials specified.

◆

Cattle Spray

Kerosene Extract of
Pyrethrum Flowers   8 parts
Steam-distilled Pine Oil 10 to 15 parts
Petroleum Oil (40 to
65 secs. viscosity)
    to make 100 parts by volume

The kerosene extract is made at the rate of five pounds of flowers to a gallon of oil. The kerosene used should be highly refined so as to be as nearly non-irritant as possible. Pine Oil is the repellent in the formula. *Steam-distilled pine oil is more repellent to flies and less irritating to the skin than the cheaper destructively distilled pine oil.* If necessary the latter may be used at the rate of 20 to 25 parts per hundred.

◆

Cattle Louse Insecticide

Dust with
Sodium Fluoride   1
Diatomaceous Earth   1

◆

Cattle Parasiticide

Precipitated Chalk   40
Rock Salt   60
Pine Tar   2
Copper Sulfate   1
Make into plastic mass with water; cast into blocks and dry.

◆

Apples, Removing Arsenic Spray Residue from

Removal of arsenic to within tolerance limits is effected by washing with 0.33% hydrochloric acid, provided no oil-spray has been used on the fruit. Accumulations of oil or wax may necessitate the use of 0.66–1.33% hydrochloric acid. Apples were injured by 2% hydrochloric acid. Oils having viscosity >65–75 or lighter oils applied very late in the season rendered arsenic removal very difficult. Storage of apples at ordinary temp. prior to washing also increased the difficulty of cleaning, but cold storage had little effect. Kerosene emulsion, prepared with kaolin and used in conjunction with hot hydrochloric acid, facilitated oil and wax removal. Heating the acid (35–40°) improved washing efficiency more than did increasing the concentrate of hydrochloric acid used.

Banana Plants, Combating "Panama Disease"

Best results were gotten by treating roots and surrounding soil of each plant with 1½ pints heavy gas oil (sp. gr. up to 0.8869).

Prevention of Black Rot in Delphinium

Mercuric Chloride	1
Sodium Nitrate	1
Water	1280

Dissolve the above and saturate soil around roots.

Potato Flake Fodder

Potato flakes contain all the solid constituents of the tubers and are an easily digested fodder material. The potatoes are washed, cooked or steamed under pressure, and then mashed to a pulp, which is dried as a film on steam-heated rollers, scraped off, broken up and stored. 400 kg. of potatoes contg. 18% starch yield 100 kg. of flakes contg. 12–15% water, 6–7% protein, 0.3–0.5% fat, 1.2–1.5% cellulose and 72–77% nitrogen-free exts.

Seed Potato Disinfectant

The dip is prepared by adding to 25 gals. of water, a mixture of 6 oz. of mercuric chloride dissolved in 1 qt. of hydrochloric acid. Forty bu. of potatoes can be treated with 25 gals. of the dip. The soaking period is 5–40 min. according to the severity of *Rhizoctonia* and scab infection.

Seed Disinfectant

Hydrated Lime	95
Water	500

Stir well and add while agitating

Mercuric Chloride	5
Water	100

Filter and dry precipitate.

Seed Disinfection

Formalin vapor is generated by boiling a formalin solution containing 1 part of 40% in 100 parts water and the seed is exposed 1–10 min. Tests in 4 widely separated areas for 4 yrs. have given efficient control of oat smut (*Ustilago avenae*) and wheat bunt (*Tilletia caries* and *T. foetens*) in every case and the cost is extremely low. The germinability of the seed grain is not impaired.

Weed Killer for Seed Beds

Zinc Sulfate	8 gm.
Water	250 c.c.

Dissolve and apply above equally to every square foot of seed bed. Careless application will damage root tips. The second dose for a succeeding crop should be half of above strength.

Earthworm Poison

Corrosive Sublimate	1 oz.
Water	75 gal.

Sprinkle ground with this solution which is unharmful to plant life; vegetation should be sprinkled with water after this treatment.

Snail Killer

Ferrous Sulfate	20
Ferric Sulfate	20
Copper Sulfate	45

Grass Killer

Grass between the bricks or stones of a walk may be killed by adding a strong solution of calcium chloride in water.

Quack-Grass Killer

Sodium Chlorate	1 lb.
Water	1 gal.

Spraying two or three times yearly is efficacious.

Lettuce Bottom Rot, Control of

Ethyl Mercury Phosphate	1
Powdered Bentonite	2

Potato Blight Control

Dusting with following gives good results

Anhydrous Copper Sulfate	1
Slaked Lime	8

Agricultural Spray

Nicotine	1.20
Soap	20.20
Water	75.20

Agricultural Spray

Anthracene Oil	75
Fish Oil Soap	3
Water	22

Spray, Horticultural

0.84 pound of casein is slowly poured into about 2 gallons of cold water, and stirred until thoroughly wet and soaked, then 0.63 pound of dehydrated sodium carbonate is added, stirring until all the casein is well in solution. Six gallons of denatured alcohol is then added, and 1.67 pounds of powdered gamboge. The gamboge is added slowly with constant stirring. Enough water is then added to make up a total of 20 gallons. With this composition, the oil to be emulsified is incorporated, preferably by slow additions, with agitation. Most oils emulsify therein readily. Heat may be applied if quicker emulsification is desired. For petroleum oil for example, with a specific gravity of 0.891 a proportion of 1 part by volume of the foregoing composition to 5 parts of the oil affords a satisfactory product. Such emulsion will contain about 83.3% of oil, making up to a consistency about that of lard at the same temperature. Such a product, even after standing in a warm place for months is free from separation.

For horticultural spraying, a petroleum oil emulsion as indicated, would ordinarily be used at a spraying strength of 2% oil. This would be obtained from the preparation referred to in the above example by diluting 2.4 gallons of the emulsion to 100 gallons with water. When sprayed, a highly satisfactory coating on the vegetation is had, with a minimum loss from run-off or drip, and at the same time the oil is well protected against damaging tender foliage.

Bordeaux Mixture

The following is the method of making Bordeaux Mixture for horticultural spraying. The customary wash is known as "4-4-50," and the official formula and instructions are as follows:

Copper Sulfate (98 per cent)	4 lb.
Best Quicklime (in lump form)	4 lb.
Water	50 gal.

The copper sulfate should be dissolved in a small wooden vessel at the rate of 1 gal. of water per lb. of sulfate (iron or tin vessels must not be used). The lime should be slaked to a fine paste with a little water in another vessel, and water added gradually to make a milk, and finally diluted in a large barrel to the requisite amount (46 gal.). The 4 gals. of copper sulphate may now be poured slowly into the diluted milk of lime and the mixture stirred thoroughly during the process. The two components of the mixture may be kept separately for a long time, but, after mixing, the spray fluid should be used as soon as possible—at all events, within 24 hours. When used on a large scale it may be convenient to make up a stock of each ingredient which may be diluted down and mixed as required. For this purpose, 50 lb. of copper sulphate may be dissolved in 50 gals. of water and 50 lb. of lime, slaked and diluted to 50 gals. of milk of lime. Each gallon will then represent 1 lb. of copper sulphate and 1 lb. of lime. When required for use, the contents of the barrels should be thoroughly stirred and the requisite number of gallons taken out and diluted according to the above formula. For a 50-gallon barrel, for instance, 4 gals. of lime-milk should be removed and diluted with 42 gals. of water, and when thoroughly stirred and strained the 4 gals. of copper solution may be added slowly. The addition of refined sugar (2 oz. to 50 gals.) is useful in delaying flocculation.

Beet Fly, Spray for

Eggs and pupae are not greatly harmed by contact insecticides. The larvae may be killed by 5–6% barium chloride soln. or 0.15% nicotine spray (40 gallons per acre, min.), but it is more advisable to destroy the flies with a spray contg. 0.3–0.4% Sodium Fluoride and 2% sugar.

Nematodes, Spray for Combating

Carbon Bisulfide	68
Rosin Soap	8
Water	26

Agitate violently and dilute 1 : 50 with water before use. Formaldehyde may be added to control fungus pests.

Warble-Fly, Control of

Good results are gotten by spraying with

Soft Soap	¼ lb.
Water	1 gal.
Derris Powder	½ lb.

Bracken, Eradication of

Spray with 1% solution of sodium chlorate.

◆

Peach-Borer (lesser), Control of

Paradichlor Benzol	1 lb.
Crude Cottonseed Oil	2 qt.

Other oils are not as satisfactory as cottonseed oil.

◆

Tree-Bands, Insect

Rolls of corrugated paper are saturated with following and wrapped around trees

Mineral Oil	1½ lb.
Alpha Naphthylamine	1 lb.
Paraffin Wax	4 oz.

◆

Treeband Composition

Sulfur Flowers	6
Linseed Oil	75

Heat 1½ hrs. with stirring until uniform. Cool and thin with cottonseed oil.

◆

Sprout Killer

Sprouts or shoots of young trees can be killed by injecting into them a twenty per cent solution in water of sodium arsenite. Since this material is very poisonous it must be handled with the utmost care.

◆

Pine Oil Insecticides

Steam-distilled Pine Oil is rapidly displacing such ingredients as methyl salicylate, citronella, lemon oil, safrol and oil of wintergreen in household insecticides for it possesses a pronounced germicidal value, aside from its pleasant perfume odor.

1. Formulae

A. Pyrethrum Extract	1 qt.
Gasolene-kerosene	5 qt.
Citronella	1 oz.
Pine Oil	6 oz.
Paradichlorbenzene	6 oz.
B. Pyrethrum Extract	1 qt.
Gasolene-kerosene	5 qt.
Paradichlorbenzene	4 oz.
Cedarwood Oil	3 oz.
Pine Oil	3 oz.
Methyl Salicylate	2 oz.
C. Pyrethrum Extract	1 qt.
Gasolene-kerosene	5 qt.
Pine Oil	5 qt.

Fungicide

A composition consisting of 95 per cent dusting sulfur and 5 per cent by weight of either of the following dry and finely ground substances: aluminum hydroxide, zinc oxide, or hydroxide, aluminum sulfate or zinc sulfate, was found to be much superior to straight sulfur dusts, and at least equal to the most efficient lime-sulfur liquid sprays without having any of the drawbacks of the latter.

◆

Fungus Killer

Copper Carbonate	36
Copper Sulfate	3
Sulfur	58

◆

Insect Exterminator

Kerosene, Refined Grade	1 gal.
Pyrethrum Powder, Best Grade	½ lb.
Paradichlorbenzene	1 lb.
Perfume	sufficient

◆

Insecticide Spray

(Agricultural Quick-Breaking)

Diglycol Oleate	2 lb.
Pyrethrum Extract (Mineral Oil or Kerosene)	50 lb.

Mixing the above together gives a concentrated spray base free from alkalies. The active principle of pyrethreum is thus unaffected. Burning due to alkali is also eliminated.

The above concentrate emulsifies readily on stirring in water with a pump. It is "quick-breaking" when sprayed on the foliage.

◆

Insecticide

Naphthalene	2 lb.
Oleo-resin Pyrethrum	2 oz.
Methyl Salicylate	2½ pt.
Deodorized Kerosene	6¼ gal.

Dissolve the first two ingredients in the kerosene by mixing or shaking and add the methyl salicylate.

◆

Insecticidal Dust

Sulfur	60.00
Nicotine	1.90
Lead Arsenate	10.00
Arsenic	2.00
Talc	28.00

Insecticide, Cabbage Maggot

Calomel	4
Gypsum Powder	96

Vegetable Weevil, Insecticide for

Sodium silicofluoride when used as a dust (about 30–40 lbs. per acre) gives good results.

Weevils, Killing Corn

Fumigation with carbon bisulfide is recommended. Approx. 1 lb. of carbon bisulfide is used to 100 cu. ft. of space to be fumigated. If the contact period exceeds 36 hrs., germination is injured. Optimum results were obtained at temps. of 75–90° F. in closed bins.

Field Mouse Poison

Whole Wheat	125	lb.
Thallium Sulphate	1½	lb.
Hot Water	6	qt.
Starch, Dry	½	lb.
Glycerin	½	pt.

The thallium sulphate is dissolved in the hot water, and to this is added the starch, previously mixed with a little cold water. The clear starch paste thus made is boiled for 2 to 3 minutes, the glycerin is added and the mixture boiled for a short time and then incorporated with the wheat.

A simple rat poison consists of a tapioca flour paste, containing 2½% of thallium sulphate, and spread on slices of bread. Another bait which has been used successfully is made as follows: ¼ oz. of thallium sulphate is dissolved in a large tea cup of boiling water and half a cupful of corn syrup, and 12 oz. of peanut butter are added. Thin slices of bread from two loaves are well covered with this mixture and cut into small squares. Tablespoonful doses of these squares are placed in the tracks of the vermin.

Lice and Mite Tablets (Poultry)

Calcium Sulfide	16.13
Silica Sand	7.52
Gypsum	6.48
Sugar	57.80
Starch	11.64

Poultry Louse Powder

Nicotine	0.28
Naphthalene	9.98
Sulfur	19.80
Sodium Fluoride	0.54

Mixture for Fowl Ailments

A useful mixture of sulphate of copper and iron for tonic and worm deterrent purposes is: Copper sulphate, 3 oz.; iron sulphate, 1 oz.; vinegar, 1 quart. Dose: 1–2 ounces added to each gallon of the drinking water. This is suitable for all ages of poultry, as the dose automatically becomes regulated by the amount of water taken. It is commonly employed in cases of catarrh, roup, and fowl-pox, also in worm infestations, and in anæmic conditions.

Veterinary Gall Salve

Tribromphenol	5.75
Petrolatum	67.15
Beeswax	9.2
Lard Compound (Paraffin added in summer)	29.9
Alum	13.8
Sulfur	27.6
Indigo	2.25

Melt the wax; add the other ingredients, and rub thoroughly through ointment mill.

Cough Electuary for Horses

Powdered camphor	½	oz.
Powdered myrrh	½	oz.
Potassium chlorate	1	oz.
Honey	4	oz.
Glycerin	4	oz.

Dose: One tablespoonful three times a day.

Horse Conditioning Powder

Gentian Root Powder	4
Sulfur Powder	4
Potassium Nitrate	1
Glaubers Salts	2
Fenugreek Powder	1
Licorice Root Powder	4

A tablespoonful at meals is a usual dose.

Mange Ointment

Mercurous Iodide Yellow	10	gr.
Salicylic Acid	½	oz.
Sulfur Sublimed	3	oz.
Coal Tar (neutral)	½	oz.
Pine Tar	3	oz.
Fish Oil	24	oz.
Diglycol Oleate	1	oz.

Shake well before using: apply at night and wash off next day.

Mange Cure

Potassium Carbonate	8 gr.
Flowers of Sulphur	64 gr.
Oil of Picis	12 c.c.
Oil of Cade	12 c.c.
Linseed Oil	to make 11 liters

Animal Worm Expeller

Oil of Chenopodium	6 oz.
Oil of Cloves	1 oz.
Oil of Anise	2 oz.
Chloroform	4 oz.
Castor Oil	115 oz.

Dog Vermifuge

Oil of chenopodium	16 minims
Turpentine	2 minims
Oil of aniseed	10 minims
Castor oil	3½ fl. drachms
Olive oil	3 fl. drachms

Mix.

For a full-size or medium-size puppy under six weeks old give ½ drachm in a teaspoonful of milk. Between six and eight weeks the dose is 1 drachm, and at eight weeks 1 drachm, to be repeated in an hour. If the bowels do not act within an hour, give ½ to 1 teaspoonful of castor oil. For small puppies reduce the doses to one-half, and for the toy breeds to one-quarter. If no worms be expelled, the mixture may be repeated in a few days. The only really satisfactory method of treating animals is to dose each one separately, otherwise there is no way of ensuring that every animal obtains a correct dose.

Distemper Cure for Dogs

Fluid Extract of Buckthorn	1 oz.
Fluid Extract of Ginger	⅛ oz.
Syrup of Poppies	2 oz.
Simple Syrup	1 oz.
Cod Liver Oil	4 oz.

Shake well.

Dose—A tablespoonful is given twice daily.

Worm Expeller

Magnesium Sulfate	12.04
Calcium Sulfate	9.05
Calcium Silicate	6.85
Venetian Red	7.34
Sand	2.11
Nicotine	0.22

Animal Condition Powder

Sulfur	5
Rosin	5
Fenugreek Seed	5
Flaxseed Meal	5
Magnesium Sulfate	5
Ginger African	4
Gentian Root	4
Copperas	4
Sodium Bicarbonate	4
Antimony	2
Salt	2
Potassium Nitrate	1

All of above materials should be powdered and then mixed thoroughly.

Argentine Ant Poison

This poison consists of a syrup, attractive to the insects, containing from one to two tenths of one per cent of the chemical element arsenic in the form of sodium arsenite. In view of the uncertain purity of commercial sodium arsenite, it is advisable to prepare the chemical in solution from arsenious oxide, a stable, standard compound universally obtainable and of known poison strength. The poisoned syrup prepared from this material is not immediately fatal to the worker ants, but instead is carried by the insects to the nests, where the queen and brood are killed.

Inasmuch as the syrup does not keep very well without a preservative, it is perhaps better to make up a small supply each time it is used. In order that such a plan may be convenient, a "stock solution" of sodium arsenite is made up. This does not ferment and if kept in a well-stoppered bottle will not deteriorate appreciably. The stock solution is mixed as desired with thin syrup.

One ounce arsenious oxide (common "white arsenic")

¾ ounce sal soda crystals (if the soda has crumbled down into a fine white powder, use only ⅜ ounce)

Boil the above ingredients together with about one pint of water in a granite-ware pan. Do not use aluminum or galvanized vessels. After the arsenic is practically all dissolved, add enough water to make the total volume of the solution *one quart*. Sometimes the arsenic is not quite pure, and leaves a little cloudiness which will settle over night, and which does no harm anyway. Mix thoroughly, bottle and label POISON. At the time the syrup is desired for use, mix the Stock Solution as above prepared with honey according to the following figures:

Stock Arsenic Solution	1 fl. oz.
Thin Honey	1 pt.

Method of Use.—Soak pieces of excelsior in the syrup, place in cans; cover with loose-fitting lids, and place outfit in path of ants.

Note. Ants seem to like straight honey best. If economy is desired, cane syrup may be substituted for a part of the honey ingredient.

Ant Repellent

1 lb. sugar in 1 qt. of water
125 grams arsenate of soda

Boil and strain.

Add spoonful of honey.

Ant Destroyer

Tartar Emetic	1 lb.
Sugar Powd.	1 lb.
Arsenic Sulfide Powd.	½ oz.

Ants, Carpenter, Destroying

Bore sloping hole at top of infested wood and pour in a mixture of equal parts of carbon disulfide and carbon tetrachloride. The heavy liquid and its vapor will sink down and permeate crevices.

Another method is to dissolve one pound paradichlorbenzene in two quarts of kerosene and spray this solution.

Ants, Preventing Entry of

Sprinkle Clovel or Oil of Sassafras at entrances. Ants do not like these odors and will not enter.

Ant Powder

Sodium Fluoride	78
Pyrethrum Powd.	8
Starch	14

Fire Ant, Insecticide for

Thallium Sulfate	2	oz.
Sugar	5	lb.
Honey	½	lb.
Water	4½	pt.

Ant Poison

Thallium sulphate has been found effective in exterminating in 3 or 4 weeks small red ants in houses, where arsenic compounds had previously failed. The following mixture was used:

Water	1 pt.
Sugar	1 lb.
Thallium Sulphate	27 gr.
Honey	3 oz.

The whole is brought to the boil and well stirred.

Beetle Powders

Formula No. 1

Barium Carbonate	10 oz.
Borax	20 oz.
Sugar	5 oz.

No. 2

Sodium Fluoride	10 oz.
Kaolin	10 oz.

No. 3

Kieselguhr	22 oz.
Sodium Fluoride	40 oz.
Sodium Chloride	10 oz.

No. 4

Powdered Borax	4 oz.
Flour	2 oz.
Chocolate Powder	1 oz.

No. 5

Powdered Borax	10 oz.
Insect Powder	1 oz.
Starch	1 oz.

Lime Sulphur Spray

Directions for making 50 gal. of lime sulphur spray are as follows:

Sulphur	8 lb.
Spent Carbide Residue	3 gal.
Calcium Arsenate	8 oz.

Heat about ⅓ of the total amount of water, adding the sulphur slowly to make a thick paste. When the water is hot, add all the carbide residue, thoroughly stirred. Mix and add another third of water and continue to cook and stir for about 45 to 60 minutes until a clear, orange-colored solution is obtained. Then add the rest of the water and the calcium arsenate. Let the mixture settle and run it through a fine sieve as it is poured into the spray tank. This should be diluted in a ratio of about six parts water to one part of the solution.

CHAPTER III

COATINGS, PROTECTIVE AND DECORATIVE

COATINGS such as paint, varnish, lacquer, etc., are applied to myriad products for protections against corrosion and to preserve their original appearance. Since some coatings or finishes need be but temporary whereas others must last indefinitely, a coating that is satisfactory in one case is unsuitable for another. Thus an indoor wall paint need not be resistant to water and salt whereas a ship paint must be. A lacquer for automobiles should be durable under the conditions to which it is exposed (sun, rain, cold and heat) whereas a lacquer for furniture requires an entirely different formulation. This chapter deals exhaustively with the different types of finishes.

Nitrocellulose Lacquers

These lacquers may be divided into two parts:—volatile and non-volatile constituents. Under the former may be classed the liquids used to carry the solids into solution. The non-volatile matter consists of nitrocellulose, gums or resins, and a plasticizer.

A film of nitrocellulose alone is not satisfactory for most uses, as it lacks adhesion, is stiff and brittle, lacks flexibility and elasticity; and as a result of this, it will split or peel off the surface. Nitrocellulose has a high viscosity, and a lacquer solution will not contain as much solids for the coating as a paint or varnish of like viscosity.

Resins are used to give a lacquer more solids without increased viscosity, greater adhesion, more gloss and sometimes greater hardness. The resins used are both natural and synthetic. The former class contains such well known materials as rosin, shellac, dammar, kauri, copals, sandarac, mastic, and elemi. A legion of names may be mentioned in the latter class. But we will confine ourselves to the most representative and popular members of each kind. In this class are found ester gum, bakelite, beckacite, amberols, lewisols, and the rezyls and teglacs.

Lacquer films become hard and brittle with age. To overcome the cracking and peeling of a brittle film due to the ex-

pansion, contraction, or bending of the coated surface, a plasticizer is incorporated into the lacquer. These materials may be oils, such as castor oil, blown castor oil, blown rape seed oil, OKO oil, and lacquer linseed oil. A very important class is the high boiling esters which are solvents for the cotton and many times for the resins. In this class will be found the ethyl, butyl and amyl esters of the phthalates, tricresyl phosphate, triphenyl phosphate; just to mention a few of the most common ones in use. These plasticizers are non-volatile and will remain in the film for a very long time. They tend to form solid solutions with the nitrocellulose. A very important class and coming to the fore are the resin-plasticizers. In this class will be found ethyl or methyl abietate, beckolac 1308, paraplex 5B as those most popular to-day.

By the use of the term solvents, we mean those liquids that are used to dissolve the nitrocellulose. Solvents are classified as low boilers and high boilers. Each class performs a certain function. Low boilers are used to carry the cotton into solution, provide volatility for the lacquer, and also give the initial set for the film. Usually the low boiler is a faster solvent for the cotton than the high boiler. The most popular member of this class is ethyl acetate. The high boilers provide smooth flow, prevent blush, orange peel and give homogeneity to the film. In this class are found butyl acetate, amyl acetate, butyl proprionate, ethyl lactate, butyl lactate, and the cellosolves.

Latent solvents are compounds or liquids that are not solvents for cotton by themselves. But they become so, by the mere addition of a solvent. In this class are the methyl, ethyl, butyl, propyl, and amyl alcohols.

In the making of a solvent mixture or thinner for a lacquer, other liquids are used, such as benzol, toluol, xylol, solvent naphtha, and also special petroleum naphthas. These do not dissolve cotton, and also lower the solvent power of a solvent when mixed with them. This class of liquids is called diluents, and though they are excellent solvents for a great many of the resins, we will call them diluents as they are not solvents for the nitrocellulose. They give bulk to the mixture, aid in keeping the resins in solution, help balance the formula, and also lower the price.

In the compounding of lacquers, certain standards or stock solutions are used; nitrocellulose or cotton solutions, and the resin solutions. They are blended in various proportions, a plasticizer and the solvents added to bring it to the desired viscosity or concentration.

The nitrocellulose solutions are usually made to contain a definite amount of ounces to the gallon, or to hold a certain amount of cotton in the gallon of solution. Or else it may be cut according to the percentage formula, as a 20%, 25%, or 35% solution.

◆

"Cotton" Solution No. 1

Dry ½sec Cotton	25 %
Den. Alcohol	10.7%
Butyl Acetate	16.1%
Toluol	32.1%
Ethyl Acetate	16.1%

This solution contains 2 pounds of dry cotton in the gallon of solution. The solution weighs 8.3 pounds per gallon.

◆

"Cotton" Solution No. 2

Dry ½sec Cotton	35.8 lb.
Ethyl Acetate	24.8 lb.
Toluol	24.2 lb.
Ethyl Alcohol	15.2 lb.

This solution is a 36% cut, and contains approximately 59.5 ounces of dry cotton in the gallon.

"Cotton" Solution No. 3

Dry 70sec Cotton	1.13 lb.
Alcohol	.51 lb.
Benzol	3.10 lb.
Ethyl Acetate	3.00 lb.

This yields one gallon of solution of a high viscosity cotton.

◆

Cotton Solution No. 4

To 24 ounces of film scrap add one gallon of solution of 25% Ethyl Acetate, 25% Alcohol, 16% Toluol, and 34% Bayway Solvent No. 55 (Petroleum Fraction).

Resin solutions are cut from 4 to 14 pounds of resin to the gallon of solvent, or else as a 50/50 cut of resin and the solvent. The solvents used are generally benzol, toluol, xylol, alcohol, and ethyl acetate. In general, different resins will require different solvents. Some manufacturers cut their resins in a thinner to insure greater compatibility with the cotton solutions. Ester gum, Lewisol, Beckacite, Amberol are dissolved in one gallon of toluol or thinner. The proportions are 8 pounds of the resin to one gallon of the solvent. Elemi gum is dissolved in an equal weight of solvent. For Kauri gum, dissolve 40 pounds of the resin in 60 pounds of a solution of 85% denatured alcohol and 15% ethyl acetate. Dammar Solution is made by dissolving 80 pounds of dammar in a mixture of 20 pounds of ethyl acetate and 40 pounds of petroleum naphtha of boiling range between 80 and 130° C. When completely dissolved add 100 pounds of ethyl alcohol, agitate for a while and allow to settle overnight for a thorough dewaxing. The shellac solution may be the ordinary 4 or 5 pound cut of shellac in alcohol.

A good solvent should possess high solvent power, offer excellent blush resistance, give good flow, make for excellent compatibility and a thoroughly homogeneous film, and should be fast in its action. The formulae listed below may be used for solvents and reducers to thin the various stock solutions, when incorporating them with the other ingredients for a lacquer for sale or use.

Solvent	No. 1	No. 2	No. 3	No. 4	No. 5
Toluol	65%	60%	50%	50%	70%
Ethyl Acetate	10%	15%	15%	15%	15%
Den. Alcohol	....	....	15%	10%	5%
Butyl Alcohol	15%	15%	....	5%	....
Butyl Acetate	....	....	....	20%	....
Amyl Acetate	....	....	13%	....	....
Amyl Alcohol	....	....	7%	....	....

	No. 1	No. 2	No. 3	No. 4	No. 5
Cellosolve.....	5%	5%	....	....	5%
Butyl Cellosolve	5%	5%	....	....	5%

The following formulae contain the main elements of a good thinner for general use, namely

1—Good solvent power.
2—Good blush resistance.
3—Proper speed of evaporation.
4—Low cost.

Solvent (Thinner)....	No. 1	No. 2	No. 3	No. 4	No. 5
Petroleum Naphtha....	44%	....	....	20%	30%
Toluol.........	....	50%	70%	40%	32%
Ethyl Acetate..	22%	18%	15%	10%	....

	No. 1	No. 2	No. 3	No. 4	No. 5
Ethyl Alcohol..	12%	12%	10%	10%	10%
Butyl Acetate..	....	....	....	10%	23%
Butyl Alcohol..	....	....	....	10%	5%
Amyl Acetate..	22%	20%	....	....	....
Amyl Alcohol..	....	....	.....	....	....
Butyl Cellosolve	....	....	5%	....	....

Wood Lacquers

In a general run of wood lacquers, one will be called upon to supply a sanding sealer, high gloss clear, flat lacquers, rubbing or polishing lacquers, and various specialties as required by the trade such as alcohol proof lacquer, and rubbed effect lacquer.

SANDING SEALER:	No. 1	Non-Volatile Dry Basis	No. 2	Non-Volatile
"Cotton" Solution No. 1.......	1 qt. or 2 lb.	½ lb.	4 lb.	1 lb.
"Cotton" Solution No. 4.......	1 qt. or 2 lb.	5 oz.		
Resin Solution................	1 pt. or 1 lb.	½ lb.	2 lb.	1 lb.
Dibutyl Phthalate ....⅛ lb.		⅛ lb.	½ lb.	½ lb.
Blown Castor Oil.............⅛ lb.		⅛ lb.		
Zinc Stearate (R. B. H.) Paste..	1 lb. paste	32½% solids	1 lb.	32½% solids
Solvent No. 3................	1 qt.		1 qt.	

The resin in No. 1 is amberol No. 801 and in No. 2 is Lewisol No. 2. Each solution is made by cutting 8 pounds of the respective resin in 1 gal. of a cheap thinner.

Below we will give a table of various wood lacquers. In this table will be found the non-volatiles. By the use of the standard solutions of cotton, resin and solvents as given above these formulae may be compounded. The addition of solvent and amount will be left to the individual, to meet his specific problem of price and quality.

Clear Lacquers

½ Sec. Nitrocellulose.	2	2	2	2	2	2
Dammar Solution....	5	2	..	..	..	..
Ester Gum Solution..	..	3	1	3	2	4
Kauri Solution......	..	..	1½	..	..	..
Amberol Solution....	..	..	...	2	..	2
Lewisol Solution.....	..	..	...	..	3	..
Blown Castor Oil....	½	¼	½	¼	¼	¼
Dibutyl Phthalate...	½	..	¼	½	..	¼
Tricresyl Phosphate..	..	½	...	..	½	..

Flat Lacquer

"Cotton" Solution No. 1	2	lb.
"Cotton" Solution No. 3	½	lb.
Amberol Solution	1	lb.
Zinc Stearate (RBH) Paste	1	lb.
Tricresyl Phosphate	¼	lb.

Solvent No. 4 to one gallon.

Flat Lacquer Paste
(All by Weight)

½" RS "Cotton"—dry basis	4	oz.
Aluminum Stearate	16	oz.
Dibutylphthalate	1	oz.
Ethyl Alcohol, including alcohol in cotton	10	oz.
Ethyl Acetate	13½	oz.
Butyl Acetate	3	oz.
Butyl Alcohol	4	oz.
Toluol	13½	oz.

Grind 18 hours in a one-gallon porcelain mill with stone pebbles. The above gives proper size batch for such a mill. The mill should be one-half full of one-inch flint pebbles.

Rubber or Polishing Lacquer

"Cotton" Solution No. 1	4	lb.
"Cotton" Solution No. 3	1	·lb.
Lewisol Solution	1	lb.
Dibutyl Phthalate	½	lb.

Solvent No. 4 to one gallon.

Alcohol Proof Lacquers

"Cotton" Solution No. 1	1	gal.
Amberol Solution	1	qt.
Paraplex 5B	2	lb.

Solvent No. 3 to spraying consistency.

"Cotton" Solution No. 1   4 lb.
Lewisol Solution   2 lb.
Dibutyl Phthalate   12 oz.
Solvent No. 3   1 qt.

By combining the flat and gloss lacquers in varying proportions, any desired effect of semi-gloss, satin finish or rubbed effect may be obtained.

Wood Enamels (Pyroxylin)

In a discussion of the pigmented enamels two factors must be considered. The ability to grind the pigment in the plant, or must the ground pigment be bought from an outside source. For the former we will list below some represented grinds in a plasticizer and gum solution. These will be explained in detail and the difference from the mill ground product shown.

	Pigment, Lbs.	Blown Oil, Lbs.	D.B.P., Lbs.	Ester Gum, Sol., Lbs.	Lewisol Sol., Lbs.
Black......	10	16	8	12	12
White......	60	8	4	12	12
Red........	40	26½	13½	18	..
Blue.......	45	22	11	9	6
Orange.....	80	14	6	..	..
Yellow.....	67	15	5	9	..
Green......	58	13½	6½	3	6
Indian Red.	68	14	6	9	..

To make these all equal to 100 pound basis add enough toluol to make 100 pounds. This will also thin the mixture to the proper grinding consistency for a roller mill. For a ball or pebble mill slightly more thinning will be required. The R.B.H. pigments are dispersed in a medium consisting of ½ second nitrocellulose in a solvent mixture. These lacquer pigments will be found to be of a uniform dispersion, excellent covering power, smooth, and may be obtained in any quantity from a gallon can to a fifty gallon drum. In the use of the R.B.H. pigments additional plasticizer must be added to compensate for the added cotton and pigment. It will also be found necessary to carefully watch the resin content for gloss lacquers as these pigments have a tendency to flatten a lacquer.

In the formulation of a wood enamel, a good clear lacquer is usually taken as the base and the pigment grind added to this to meet the required specification for covering power. Sometimes more resin is added to bring up the gloss. If flattening is desired a zinc stearate mixture is added. The base clear used will depend on the price of the enamel.

If a cheap enamel is being formulated, a base clear high in ester gum will be indicated. Also the viscosity may be increased by the use of high viscosity cotton or the film solution. For the better grade enamels, the lower viscosity cotton is used to give more solids, and the better resins increased, such as amberol, lewisol, beckacite, and the rezyls. These resins will also give the tougher and more flexible film.

Metal Lacquers

These lacquers are used as a protective and ornamental coating on all class of metal objects, such as, brass goods, plated ware, and even iron and steel, and some of the newer alloys. When the purpose is to protect the highly polished surface against tarnishing, the lacquer is made of a rather high viscosity cotton, as this type will give a tougher film than ½ second cotton. The film is thin and almost imperceptible. The resin used is usually low in acid number and of a very pale color. The low acid number being required so as not to attack the metal coated. The resin will add to the adhesion of the lacquer.

High Viscosity "Cotton"	4	4	4	4
Elemi Solution	–	–	2	–
Dammar Solution	–	1½	–	1
Lindol	1	–	–	–
Dibutyl Phthalate	–	1	1	1
Blown Castor Oil	–	–	–	1
Ester Gum Solution	–	–	–	1

Clear finishing lacquers for metal and automobile work may be included in this class.

Dry Pyroxylin	10 parts
Rezyl 19	20 parts
Dibutyl Phthalate	5 parts

Dry Pyroxylin	10 parts
Rezyl 113	30 parts
Dibutyl Phthalate	3 parts

Dry Pyroxylin	6
Ester Gum	1¼
Blown Castor Oil	1½
Dibutyl Phthalate	1½

For the enamels for metal, we again refer to the grinds given under wood enamels and follow the same system of incorporating the pigment. That is, take a clear base, and add sufficient pigment to reach the requirements for good covering power. In this class of material it is advisable to increase the plasticizer, for better flexibility and better adhesion.

Automobile Lacquers

Primer Surfacer.—This type of material should possess excellent adhesion, extreme flexibility and toughness, dry quickly, high filling power, and be easily sanded by the dry or wet paper in either water or naphtha.

To 2 pounds of dry ½ sec. cotton add 12 lb. of grind of

40	lb.	Keystone Filler
20	lb.	Lithopone
10	lb.	Talc or Barytes
40	lb.	Beckolac No. 1308
6½	lb.	Blown Castor Oil
3½	lb.	Dibutyl Phthalate

in 1 gal. Butyl Acetate

Polishing Black.—High solids, good covering power, good color, excellent flow, easy rubbing and must come to a high polish with the least amount of Rubbing. To

1 lb. dry ½ sec. ''cotton''
½ lb. dry 30 sec. ''cotton''

add 2 lb. of the following pigment grind

10	lb.	Super Spectra Black
15	lb.	Blown Castor Oil
15	lb.	Tricresyl Phosphate
2½	lb.	Butyl Stearate
15	lb.	Lewisol Solution
42½	lb.	Toluol

make up to two gallons with an extremely good solvent.

High Gloss Black.—This lacquer should possess high gloss of a lasting quality, good coverage, good color, excellent flow and smoothness and be able to stand the wear of the sun's rays.

Dry ½ sec. ''cotton''	5 lb.
Dry 15 sec. ''cotton''	3 lb.
Ester Gum	3 lb.
Lewisol	9 lb.
Lindol	2 lb.
Blown Castor Oil	2 lb.
Black Grind (above)	10 lb.
Solvent q.s.	10 gal.

Leather Lacquers

Leather lacquers or leather dopes are used in the manufacture of artificial leather and split leather. The solvents are quick drying. These lacquers are usually made from a medium to a high viscosity cotton. They contain castor oil and other oils as plasticizers and no resins. The resins are not used as they tend to detract from the flexibility. The usual starting point in this work is to begin with the plasticizer equalling the dry cotton. The plasticizers that may be recommended for this work are numerous. The old favorites are blown castor, raw castor oil, blown rapeseed oil and treated linseed oil. The newer ones are ADM 100, butyl acetyl ricinoleate, beckolac 1308 and hydroresin.

Bronzing Lacquer

A special grade of nitrocellulose is usually used for this type of material. It is called bronzing cotton and has a viscosity of from 30 to 40 seconds. Resins are not used as the free acid may cause the powder to turn. A formula that has been tested and used is:

Dry Pyroxylin	4	parts
Dibutyl Phthalate	1¼	parts
Bronze Powder	5	lb.
Solvent	5	gal.

Specialty Lacquers

A lacquer in vogue today for decorating purposes is the crystal lacquer. This material depends on the action of naphthalene to crystallize out of a cotton solution and at the same time not affect the strength of the film.

''Cotton'' Solution No. 1	15	lb.
''Cotton'' Solution No. 3	5½	lb.
Naphthalene Flakes	4	lb.
Cyclohexanone	6½	lb.
Amberol Solution	2	lb.
Tricresylphosphate	½	lb.
Amyl Acetate	5	lb.

Fill to 10 gal. with solvent.

A ''matt'' lacquer for the furniture trade may be made by taking:

5 lb. Cut White Shellac	2½	lb.
A. S. Solution ''Cotton''	½	lb.
Raw Linseed Oil	2	oz.
Blown Castor Oil	2	oz.
Acetone	1	pt.
Toluol	1	pt.

Fill to gal. with denatured alcohol.

Nail Polish Lacquer (Clear)

''Cotton'' Solution No. 1	32 oz.
''Cotton'' Solution No. 3	16 oz.
Dammar Solution	16 oz.
Tricresyl Phosphate	16 oz.
Butyl Cellosolve	16 oz.

C.P. Acetone to one gallon (1 qt.).
The above may be colored to suit.

All the formulae given above though having proved their practical use by standing the test of sale and resale to consumers are only offered as a starting basis for one's problem. In each trade there are individual requirements, and it is up to the skill and ingenuity of the compounder to adapt or change his formulae to meet these requirements.

Pearl Wood Lacquer

18   oz. ½ second Nitrocellulose
8   oz. High Viscosity Nitrocellulose
6   oz. Dammar Gum-Pale
6   oz. Shellac
2   pt. Butyl Acetate
1   pt. Butyl Alcohol
¼   pt. Amyl Acetate
4   pt. Toluol
3   oz. Dibutyl Phthalate
4   oz. Pearl Essence

White Lacquer Enamels

(1) Nitro-cotton Solution:
    10 parts Nitro-cotton No. 6—dry
    30 parts Butyl Acetate
    10 parts Toluol
    10 parts Ethyl Acetate
The ingredients are mixed and the cotton dissolved.

(2) Pigment Paste:
    10 parts Alftalate 222 A 100 per cent
    10 parts Toluol
    20 parts Titanium D i o x i d e (100 per cent Titanium White)
The paste is ground finely on a mill.
(3) 60 parts nitro-cotton solution are mixed thoroughly with 40 parts of pigment paste, and the enamel then diluted with the above-mentioned solvent mixture to brushing, spraying or dipping consistency.

Lacquer Thinners

A

Butyl Acetate	20
Ethyl Acetate	10
Denatured Alcohol	10
Toluol	60

B

Butyl Acetate	25
Ethyl Acetate	15

Butyl Alcohol	10
Toluol	50

Imitation Chinese Lacquer

Alcohol	1 gal.
Shellac	4 lb.
Sealing Wax	4–16 oz.

Different colored sealing waxes produce different colored lacquers.

Artificial Flower Pearl Lacquer

40   oz. High Viscosity Nitrocellulose
1½   pt. ''Cellosolve'' Acetate
½   pt. Dibutyl Phthalate
1   qt. Butyl Acetate
1.2 lb. Glyptal
2½   gal. Toluol
1½   gal. Ethyl Acetate
32   oz. Pearl Essence

Pearl Dipping Solution

3   lb. High Viscosity Nitrocellulose
4½   gal. Amyl Acetate
8   oz. Pearl Essence

Pearl Enamels

1 pt. Lacquer Enamel (Black, Blue, Red, etc.)
7 pt. Outdoor Durable Clear Lacquer
8 oz. Pearl Essence

Pyroxylin and Rubber Lacquer

Pyroxylin	10
Rubber	5
Ethyl Crotonate	100

Paper Lacquer

Dry nitrocellulose, 100 lb.; rezyl No. 11, 250 to 300 lb.; tricresyl phosphate, 50 to 100 lb.; and paraffin wax, 4 to 8 lb. Extra wrappings in cardboard containers are sometimes rendered unnecessary by coating one or both surfaces of the container with the foregoing type of coating. Rezyl lacquer coatings are suggested also for washable and other wallpapers.

Lacquer, Airplane Dope, Non-Inflammable

The following formula is used by the U. S. Government for coating the fabric of the wings of airplanes. While it may be applied by brush, it is preferably sprayed on.

	By Weight
Cellulose Acetate	7.5%
Tri-phenyl Phosphate	2.5%
Acetone	30.0%
Benzol	30.0%
Methanol	20.0%
Di Acetone Alcohol	10.0%

The solvents are mixed in the above proportions and the cellulose acetate added with continued stirring to avoid formation of lumps. When solution is complete, the triphenyl phosphate is added and again stirred. The mass is allowed to settle at room temperature and then decanted, or if need for immediate use may be readily filtered by passing same through a piece of finely woven fabric.

Airship Fabric Dope

The rubberized fabric composing the gas bags of airships is also treated with Pyroxylin dope as follows:

Amyl Acetate	21%
Butyl Acetate	36%
62° Gasoline	28%
Denatured Alcohol	2%
Castor Oil	8%
Pyroxylin	5%

Pigmented Airplane Wing Coating

The addition of pigments, oxide of zinc for instance, still further decreases inflammability. Metallic salts applied to the cloth as the first step would act as retarding agents, but they are not used as the dope would adhere less firmly to the cloth. In this connection, it must be noted that the presence of a non-saponifiable substance, such as petrol, in the cloth completely prevents the adherence of dope.

The aeroplane wings are brushed with the acetate of cellulose solution. Pads or other machines are not much used for the cloth, as the solution is so volatile. After drying a second and even a third coat is given.

The dry dope should stick tightly on the tissue, like the skin of a drum, and should resist changes of temperature, wet weather and sunlight. It is recommendable to protect it by means of a varnish, generally with a base of nitrocellulose, to which pigments are added to decrease very considerably its inflam-mability. This protecting varnish can be prepared as follows:

Viscous Solution of Nitro-	
cellulose	118 kgs.
Castor Oil	23 kgs.
Acetone	90 l.
Amyl Acetate	67 kgs.
Methylated Spirits	67 kgs.

Cellulose Coatings

After treatment with a dilute mineral acid at a moderate temperature, cellulose (in the form of cotton fibre, rags, or waste) can be disintegrated and reduced to a fine powder. In the latter condition it is capable of even dispersion in a dilute adhesive medium, such as nitrocellulose solution, drying oil or starch. A paint for metal or wooden surfaces can be obtained, for example, by incorporating ·twenty parts of the powdered disintegrated cellulose with a clear solution of nitrocellulose plasticized with tricresyl phosphate. Similarly, the new material can be mixed with viscose solution to form a paste-like product, which can be applied as a paper coating.

Clear Gloss Lacquer
(By Weight)

½" RS ''Cotton''—dry basis	7½%
Pale Dewaxed Dammar—	
solid basis	4½%
Dibutylphthalate	3 %
Blown Castor Oil	1½%
Methyl Alcohol	4 %
Ethyl Alcohol, including	
that in cotton	7½%
Butanol	6 %
Ethyl Acetate	8 %
Butyl Acetate	18 %
Toluol	40 %

Gloss Furniture Lacquer

Gals.	Pts.	Lbs.	Material	Wt. %
28	7.63	213.64	Cotton Solution	48.76
10	8.45	84.55	Lewisol No. 3 Solution	19.30
2	8.65	16.30	Dibutyl Phthalate	3.72
6	7.51	45.06	Butyl Cellosolve	10.28
3	7.29	21.87	Butyl Acetate	4.99
3	6.76	20.28	Butyl Alcohol	4.63
6	6.07	36.42	Lactol Spirits A	8.32
		438.12		100.00

This lacquer, to quote a finisher, "flows like a varnish." It, therefore will rub down with a minimum of labor, which leaves more lacquer on the work. It is very tough and three months of exposure facing south at 45° to the horizontal did not damage it.

"Cotton" Solution

Gals.	Pts.	Lbs.	Material	Wt. %
		193.00	Wet Cotton	28.09
22	7.36	161.92	Ethyl Acetate	23.57
46	7.22	332.12	Toluol	48.34
		687.04		100.00

YIELD 90 GALLONS OF SOLUTION

WEIGHT, 7.63 LBS. PER GAL.

This solution contains 1½ lbs. of dry cotton in each gallon of solution (or 19.66% by wt.). The 193 lbs. of wet cotton is a standard weight drum and is composed of 135 lbs. of dry cotton and 58 lbs. of alcohol.

Lewisol No. 3 Solution

Gals.	Pts.	Lbs.	Material	Wt. %
		8.00	Lewisol No. 3	52.56
1		7.22	Toluol	47.44
		15.22		100.00

YIELD, 1.8 GALS.   WEIGHT, 8.45 LBS. PER GAL.

Each gallon of solution contains 4.4 lbs. of gum.

Sanding Sealer

Gals.	Pts.	Lbs.	Material	Wt. %
8		61.04	"Cotton" Solution	41.70
3		25.35	Lewisol Solution	17.32
	6	5.83	Zinc Stearate Base	3.98
	2	2.16	Dibutyl Phthalate	1.47
1	4	10.83	Butyl Acetate	7.40
1	4	10.14	Butyl Alcohol	4.93
1		6.76	Denatured Alcohol	4.62
4		24.28	Lactol Spirits A	16.58
20		146.39		100.00

SAND IN TEN MINUTES

Tinting Lacquers, Shellacs, Etc.

(Light Yellow to a Ruby Red Color)

Resublimed iodine added in the proportion of 2 grams of iodine to 1 gallon of lacquer or shellac will produce a clear golden yellow color that is fast.

This yellow color can be deepened by the addition of more iodine to a point when it begins to take on a clear ruby red color at about 50 grams per gallon. This color is also fast.

Paints

Paints are surface coatings consisting essentially of pigments ground in vehicles of drying oils and varnishes. The quantity and type of pigments determine the color, hiding value and to a large extent the body or consistency of the material. They may also influence the drying time as well as the life of the paint.

The vehicle portion, both as to quantity and type, influences essentially the life, gloss, flexibility and drying time of the material. It consists of drying oils, gums, varnishes, dryers and volatile matter.

Dryers are metallic soaps of fatty acids, such as Co, Pb, and Mn, compounds of linoleic and abietic acids, known as linoleates and resinates. These are the important metals used for dryers. More recently, other organic acids have been used in place of the fatty acids, particularly naphthenic acid. The naphthenates are quite commonly used at present.

Volatiles, such as turpentine, solvent naphtha, varnolene, benzine, etc., are used merely to give fluidity in order to permit application by spraying, brushing and dipping.

Typical paint formulas follow:

For exterior use where surfaces are exposed to atmospheric conditions.

House Paint

1. White

White Lead	210 lb.
Zinc Oxide	60 lb.
Asbestine	30 lb.
Refined Linseed Oil	12 gal.

Grind and add

Varnolene	1 gal.
Linseed Oil	7 gal.
Liquid Dryer (containing 5% Manganese and 5% Lead metal)	1 gal.
Yield	27 gal.

2. Black

Lamp Black	30 lb.
Litharge	8 lb.
Whiting	52 lb.
Asbestine	60 lb.
Raw Linseed Oil	25 gal.

Grind and add

Mixed Dryer (containing about 5% each of Lead and Manganese and 1% Cobalt)	3 gal.
Linseed	11 gal.
Yield	53¼ gal.

3. Green

Chrome Green	75 lb.
Barytex	75 lb.
Silica	75 lb.
Asbestine	75 lb.
Linseed Oil	22 gal.

Grind and add

Dryers Mixed	1¾	gal.
Varnolene	1½	gal.
Linseed	11	gal.
Yield	47¼	gal.

In grinding the pastes above add the oils first into the mixer and while mixing follow with the pigments. After the grind, the remaining vehicles are added.

Other Colors:

For light tints such as ivory, cream, buff, gray, light brown, light green, and light blue, use the white house paint formula and add small quantities of colors in oil to the finished product to obtain the required shades. The colors in oil most generally used for ivory, cream, buff, gray, and light brown are raw and burnt umbers, lamp black, chrome yellows, ochers, and red oxides.

For light blue, use either prussian or ultramarine blue and lamp black, chrome yellow and red oxide, depending upon shade required. These are the most usual combinations but others may be used. It depends entirely upon the shades required.

Bright red or vermilion, use Formula as the above black or green, substituting Toluidine red for the colored pigments, leaving the rest of the formulas the same. Because of the price, toluidine is little used. Para Toner is generally substituted.

Red Lead

Red Lead	1,000	lb.
Linseed Oil	10	gal.

Grind and add

Linseed Oil	5	gal.
Kettlebodied Linseed Oil	10	gal.
Varnolene	1¼	gal.
Lead, Manganese Dryer	1¼	gal.
Yield	41½	gal.

Metal Protective Paint
Zinc Dust Paint

Zinc Oxide	250	lb.
Zinc Dust	750	lb.
Linseed Oil	10	gal.

Grind and add

Linseed Oil	5	gal.
Kettlebodied Linseed Oil	10	gal.
Varnolene	1¼	gal.
Lead, Manganese Dryer	1¼	gal.

Outside House Paints are also made in paste form and sold as such. The user reduces them gallon to gallon with linseed oil and adds about 1 pint of Pb-Mn Dryer.

Roof Coating

Asphalt	10	lb.
Varnolene	3	gal.
Short Fibered Asbestos	5	lb.

Size for Paints and Calcimine

In the painter's trade glue is employed both as a size for the treatment of walls prior to the application of paint, merely to fill up the pores of the wall, for which bone glue is satisfactory; or it may be mixed with a little paint, an insoluble base, and water, in the preparation of a calcimine. In the higher grades of these calcimines which must be used with hot water, the better grades of hide glue are used.

PASTE PAINTS
Zinc Oxide

Zinc Oxide	415	lb.
Refined Linseed	11	gal.
Yield	500	lb.

Red Lead

Red Lead	465	lb.
Raw Linseed Oil	4½	gal.
Yield	500	lb.

White Lead

Corroded White Lead	430	lb.
Refined Linseed Oil	6½	gal.
Yield	500	lb.

Both the white ready mixed and paste paints are made also by combining White Lead, Zinc Oxide, and Titanox with inerts in various proportions.

INTERIOR PAINTS
White Flat Wall Paint

Lithopone (high oil absorption)	400	lb.
Asbestine	100	lb.
Refined Linseed Oil	7½	gal.
3 Hour Kettle Bodied Linseed Oil	2½	gal.
60% Limed Rosin Soln. in Varnolene	2½	gal.
Varnolene	5	gal.

Grind and add		
Varnolene	15	gal.
Pb-Mn Dryer	⅝	gal.
Yield	45	gal.

Eggshell

Low Oil Lithopone	400	lb.
Asbestine	50	lb.
Whiting	50	lb.
50 Gal. Ester Wood Oil Varnish	17½	gal.
3 Hr. Kettle Body Linseed Oil	7½	gal.
60% Limed Rosin Soln.	5	gal.

Grind and add		
Varnolene	5	gol.
Mixed Dryer	1⅜	gal.
Yield	51⅞	gal.

Gloss

Low Oil Lithopone	375	lb.
Zinc Oxide	125	lb.
Refined Linseed Oil	12½	gal.
3 Hr. Kettle Bodied Oil	10	gal.

Grind and add		
3 Hr. Kettle Bodied Oil	5	gal.
60% Pale Ester Gum Soln.	16¼	gal.
Mixed Dryer	1⅞	gal.
Varnolene	11¼	gal.
Yield	70	gal.

Tint as above under house Paints, before painting.

Wall Sealer

Silica	20	lb.
Asbestine	10	lb.
50 Gal. Ester Wood Oil Varnish	3	gal.

Grind and add		
50 Gal. Ester Wood Oil Varnish	7	gal.
Blown Linseed Oil	2	gal.
Varnolene	2	gal.
Mixed Dryer	½	gal.
Yield	16	gal.

used on walls for reducing porosity.

Wall Wash for Neutralizing Free Lime on Fresh Plaster Walls

Zinc Sulphate	1	lb.
Water	1	gal.

Floor Paint

Lithopone	150	lb.
Zinc Oxide	50	lb.
22 Gal. Varnish *	8	gal.

Grind and add		
No. 22 Gal. Varnish *	16	gal.
Varnolene	3	gal.
Mixed Dryer	½	gal.
	32¾	gal.

* Ester gum-wood oil varnish may be used. Preferably however use a partial phenol-formaldehyde condensation gum variety such as paranol or amberol.
Tint to required color with colors ground in Varnish. For colors note discussion under outside paints.

Enamel for Walls and Wood Work

Low Oil Lithopone	350	lb.
Zinc Oxide	25	lb.
3 Hr. Kettlebodied Linseed Oil	12	gal.
Light Ester Wood Oil	5	gal.

Grind and add		
Light Ester Wood Oil 50% Varnish	15	gal.
Dammar Soln. in Varnolene	1	gal.
Varnolene	10	gal.

Dammar Cut

Dammar	500	lb.
Varnolene	25	lb.

Dissolve as under Ester Cut above.

Varnolene	25	gal.
	100	gal.

There are many other gums that may be used, particularly the innumerable synthetics, but the above illustrate the general type of formula. Many of the modern synthetic gums are really complete varnishes and need be merely dissolved and driers added in order to make a finished product.

The main type of synthetics may be divided into two parts: 1 Phthalic anhydride glycerin condensation products and 2, Phenol-formaldehyde condensation products. Fatty acids are always incorporated with these materials and thus the gums really contain oils and the finished product in many cases are in reality varnishes and may be so used.

The first type, the phthalics, are best used when light color is required, but they do not dry hard through unless applied in a very thin film. They tend to remain soft underneath. They are particularly good in white baking enamels where discoloration is not permissible.

The phenolics are excellent for fast drying and give excellent dry, hard films. They however discolor badly, particularly on baking.

For exterior purposes (spar varnishes) the long oil 50 gal. type is used. For interior the shorter 25 gallon type. A 25 to 30 gal. ester varnish is generally sold as a general purpose varnish for floors, furniture, etc.

Up to a certain point drying of all varnishes can be hastened by adding driers, cobalt being a top or surface drier while manganese and lead are through driers. Excessive driers, however, hasten the deterioration of the film and may cause wrinkling, particularly in baking. A proper balance should always be sought. The quantity of metal should be determined empirically. Based on solid content, lead is used up to about .1%, Manganese .05%, Cobalt up to .05%. Of course these ratios can vary greatly with individual requirements.

Speed of drying depends largely also on the length of the varnish, the shorter drying faster.

Baking Enamels, primer and undercoats can be formulated after the manner of floor paint and 4 hour enamels. Each particular problem requires its own special formula and must be made up largely empirically. Certain fundamental facts of course should be known such as increase in pigment content increases the flatness of the finish; increase in non-volatile oil and gums increase the gloss; the longer the varnish the more flexible the film and also the softer; Phenolics give harder films than phthalics and in general less gloss; certain pigments such as toners do not stand excessive baking, that is high temperature and long baking. Also dryers must be used in much smaller amounts with the latter than in air drying paints.

Interior Enamel I

Pigment	40%
Vehicle	60%
	100%

Pigment	
Zinc Oxide, French Process	100%

Vehicle

Heat Bodied Linseed Oil	60%
Mineral Spirits	12%
Turpentine	25%
Lead-Cobalt Liquid Drier	3%
	100%

Interior Enamel II

Pigment	47%
Vehicle	53%
	100%

Pigment	
Lithopone	80%
Zinc Oxide, French Process	20%
	100%

Vehicle	
Heat Bodied Linseed Oil	50%
Dammar	10%
Turpentine	8%
Mineral Spirits	30%
Cobalt Liquid Drier	2%
	100%

Interior Enamel III

Pigment	34%
Vehicle	66%
	100%

Pigment	
Lithopone	100%

Vehicle	
Limed Rosin	20%
China Wood Oil	35%
Linseed Oil	10%

Above cooked together and reduced with

Mineral Spirits	33%
Cobalt Liquid Drier	2%
	100%

Interior Flat Paint I

Pigment	65%
Vehicle	35%
	100%

Pigment	
Lithopone	85%
Extenders *	15%
	100%

* Extenders for interior flat paints include asbestine, talc, silica, whiting, china clay, barytes.

Vehicle
Limed Rosin	8%
Linseed Oil	7%
China Wood Oil	25%

The above cooked together
and reduced with

Mineral Spirits	58%
Lead-Cobalt-Manganese	
Liquid Drier	2%
	100%

Interior Flat Paint II

Pigment	65%
Vehicle	35%
	100%

Pigment
Lithopone	80%
Zinc Oxide	5%
Extenders	15%
	100%

Vehicle
Refined Linseel Oil	30%
Blown Linseed Oil	6%
Limed Rosin †	4%
Mineral Spirits	57%
Lead-Cobalt-Manganese	
Liquid Drier	3%
	100%

† Limed Rosin dissolved in part of Mineral Spirits.

Interior Gloss Paint I

Pigment	60%
Vehicle	40%
	100%

Pigment
Lithopone	65%
Zinc Oxide	20%
Extenders *	15%
	100%

* Extenders for interior gloss paints include whiting, barytes, china clay, asbestine.

Vehicle
Heat Bodied Linseed Oil	65%
Mineral Spirits	32%
Lead-Cobalt Liquid Drier	3%
	100%

Interior Gloss Paint II

Pigment	55%
Vehicle	45%
	100%

Pigment
Lithopone	80%
Extenders	20%
	100%

Vehicle
Limed Rosin †	20%
China Wood Oil †	25%
Refined Linseed Oil	25%
Mineral Spirits	27%
Cobalt Liquid Drier	3%
	100%

† Limed Rosin and China Wood Oil cooked together and reduced with Mineral Spirits.

Interior Gloss Paint III

Pigment	52%
Vehicle	48%
	100%

Pigment
Lithopone	90%
Asbestine	10%
	100%

Vehicle
Refined Linseed Oil	45%
Blown Linseed Oil	10%
Limed Rosin ‡	7%
Mineral Spirits	35%
Lead-Cobalt Liquid Drier	3%
	100%

‡ Limed Rosin dissolved in part of Mineral Spirits.

Exterior House Paint I

Pigment	67%
Vehicle	33%
	100%

Pigment
White Lead	70%
Zinc Oxide	20%
(Amer. Process)	
Extenders *	10%
	100%

* Extenders for exterior paints include barytes, asbestine, silica.

Vehicle
Raw Linseed Oil	80%
Kettle Bodied Linseed Oil	5%
Turpentine	11%
Lead-Manganese Liquid Drier	4%
	100%

Exterior House Paint II

Pigment	64%
Vehicle	36%
	100%

Pigment

Lithopone	40%
Zinc Oxide, 35% Leaded	45%
Extenders	15%
	100%

Vehicle

Raw Linseed Oil	83%
Kettle Bodied Linseed Oil	7%
Mineral Spirits	5%
Lead-Manganese Liquid Drier	5%
	100%

Exterior House Paint III

Pigment	65%
Vehicle	35%
	100%

Pigment

White Lead	40%
Titanox B	20%
Zinc Oxide, Amer. Process	25%
Extenders	15%
	100%

Vehicle

Raw Linseed Oil	80%
Kettle Bodied Linseed Oil	5%
Mineral Spirits	11%
Lead-Manganese Liquid Drier	4%
	100%

Exterior House Paint IV

Pigment	63%
Vehicle	37%
	100%

Pigment

Zinc Sulphide	25%
White Lead	15%
Zinc Oxide, 35% Leaded	40%
Silica	10%
Asbestine	10%
	100%

Vehicle

Raw Linseed Oil	80%
Kettle Bodied Linseed Oil	5%
Turpentine	5%
Mineral Spirits	6%
Lead-Manganese-Cobalt Liquid Drier	4%
	100%

Black Stoving Enamels or Baking Japans

These are applied by dipping, brushing or spraying and are stoved at 150° F. to 400° F. from 1 to 4 hours according to the nature of the japan. Egg shell gloss or flats are made by adding vegetable black in sufficient quantity to give the desired result and thinned down with volatile thinner.

General Method of Procedure

The japans are made by cooking linseed oil with litharge, red lead and black oxide of manganese (or burnt umber) for about five hours at 450° F. to 475° F. The driers are gradually taken up and the oil oxidized to an almost solid mass. This is known as lead oil. Stearine pitch, together with a bone pitch, to increase blackness, are added to the hot mass and thoroughly cooked for two to three hours until they are all completely amalgamated. It is then thinned down with kerosene and tar spirits, strained and tanked until impurities have settled out. Sometimes a half to one ounce of Prussian blue to the gallon is added during heating. This increases opacity and in parts increased hardness and drying to the oil. These japans are used for the cycle and bedstead trade, also as insulating varnish for impregnating armature and field coils of motors and dynamos, also transformer and magnet coils.

Black Stoving Enamel

Gilsonite Selects	100 lb.
Manjak	10 lb.
Linseed Oil	10 gal.
Burnt Umber	5 lb.
Kerosene	16 gal.
Tar Spirits	16 gal.

Stove at 300° F. for four hours.

Black Stoving Enamel

Stearine Pitch	100 lb.
Rosin	20 lb.
Raw Linseed Oil	50 gal.
Flake Litharge	24 lb.
Manganese Dioxide	2 lb.
Kerosene	20 gal.
Tar Spirits	40 gal.

Stove at 300° F. for four hours.

Black Varnish (Cycles)

Prepared Pitch	37.5 parts
Boiled Linseed Oil	31.5 parts

Petroleum	12.5 parts
White Spirit	18.5 parts
Stove at 180° C.	

Black Stoving Enamel

Stearine Pitch	34 parts
Asphaltum	11 parts
Boiled Linseed Oil	22 parts
Turpentine	13 parts
White Spirit	20 parts
Stove at 120° C.	

Air Drying Black Enamels and Varnishes

Formula A

Asphaltum	100 lb.
Boiled Linseed Oil	4 gal.
Red Lead	2 lb.
Manganese Dioxide	1 lb.
White Spirit	20 gal.

The White Spirit is added to the mixture of the other materials.

Formula B

Asphaltum	100 lb.
Boiled Linseed Oil	2 gal.
White Spirit	14 gal.

Brunswick Black A

Asphaltum	100 lb.
Dark Rosin	80 lb.
Litharge	2 lb.
Manganese Dioxide	1 lb.
White Spirit	18 gal.

Brunswick Black B

Asphaltum	30 lb.
Dark Rosin	100 lb.
Slaked Lime	4 lb.
Boiled Linseed Oil	3 gal.
Litharge	2 lb.
Manganese Dioxide	1 lb.
White Spirit	30 gal.

Brunswick Blacks are only for indoor use such as for coating iron work and are too brittle for outdoor use.

Berlin Black

Berlin Blacks are air drying enamels which give a mat or eggshell finish.

Brunswick Black	12 gal.
Vegetable Black	20 lb.
Turpentine	6 gal.

Wood Paints

No. 1 Paint. Weight per gallon 14.8 lb.

Pigment	62%	
Lithopone		50%
35% Leaded Zinc Oxide		40%
Silica		5%
Asbestine		5%
Vehicle	38%	
Raw Linseed Oil		80%
Kettle Bodied Oil		8%
Naphtha		7%
Turp, Drier		5%

The above paint was reduced for primer by the addition of one quart of raw linseed oil and one quart of turpentine to one gallon of paint.

No. 2 Paint. Weight per gallon 11½ lb.

Pigment	44%	
Titanox B		70%
Titanium Dioxide		15%
Zinc Oxide		15%
Vehicle	56%	
* Phenol Rosin Varnish		75%
Boiled Linseed Oil		12%
Turpentine		6%
Xylol		3.4%
Solution		2.6%
Drier		1.0%

* The Phenol Rosin Varnish is made up (by weight) as follows:

Phenol Rosin	13.0%
Wood Oil	45.0%
Heavy Naphtha	42.0%

This paint is reduced for priming purposes by the addition of one-half gallon raw linseed oil and one-half pint of turpentine to one gallon of paint.

No. 3 Paint. Weight per gallon 11½ lb.

Pigment	43%	
Titanox B		70%
Titanium Dioxide		15%
Zinc Oxide		15%
Vehicle	57%	
* Phenol Ester Varnish		77%
Boiled Linseed Oil		12%
Turpentine		5.4%
Solution		2.5%
Drier		3.1%

* The Phenol Ester Varnish consists of:

100% Phenol Formaldehyde

		By Weight
Type—Resinoid	25%	
Ester Gum	71%	19.1%
Rosin	4%	
Wood Oil	67%	
Bodied Linseed Oil		35.0%
(Body Q Oil	33%	
Heavy Naphtha		37.1%
Xylol		2.8%
Turpentine		6.0%

Reduction of the No. 3 paint for priming purposes is accomplished by adding one-half gallon raw linseed oil and one-half pint of xylol to one gallon of paint.

No. 4 Paint. Weight per gallon 11½ lb.

Pigment	43%	
Titanox B		70%
Titanium Dioxide		15%
Zinc Oxide		15%
Vehicle	57%	
* Phthalic Anhydride Varnish		83.5%
Boiled Linseed Oil		11.4%
Drier		2.4%
Solution		2.7%

* Phthalic anhydride varnish percentages by weight:

Glycerol Phthalate Linseed Acid Resin		42.5%
Heavy Naphtha	90%	57.5%
Pine Oil	10%	

Reduction of this paint for priming purposes is effected by the addition of one-half gallon of raw linseed oil to one gallon of paint.

Plastic Paint Powder

Whiting	1000
Clay	520
Glue	60
Gypsum	80

Water Paint

Trihydroxyethylamine Linoleate	0.6
Glue	10
Water	32
Varnish	16
Naphtha	4
Sodium Ortho Phenyl Phenate	0.1

Silicate Water Paint

Sodium Silicate	40
Potassium Silicate	25
Asbestine	15
Pigment (High Density)	20

Dilute with sufficient water before use.

A paint similar to this, but containing much less pigment, may be used for coating electric light bulbs, which should first be cleaned with care or trouble will be experienced with adhesion. The following modification works more smoothly and gives a better coating, but is not so durable or waterproof.

Sodium Silicate (S.G.1.4)	20%
Rice Starch	5%
Pigment	20%
Water	55%

Cement Water Paint

White Portland Cement	50	lb.
Gypsum	5	lb.
Calcium Chloride	4½	lb.
Hydrated Lime	½	lb.
	60	lb.

Mix intimately in pebble mill. Stir about 7 to 8 lb. of the above into 1 gal. of water and paint over wet surface. When paint sets up, wet down with ordinary tap water.

Cold Water Paint, Outside

Whiting	55 lb.
Clay	15 lb.
Dextrine	2 lb.
Casein	12 lb.
Lime	15 lb.
Trisodium Phosphate	1 lb.
Corrosive Sublimate	1 oz.
China Wood Oil	10–15
Linseed Oil	5–10
Turpentine	10–20
Manila Copal	5–10
Alcohol	50–70
Ethyl Acetate	30–50

Paint, Cold Water

Casein	10
Lime	10
Chalk	60
Clay	10
Pigment	10

To the above dry mixture water is added just before use.

Water Soluble Shellac Solution

(1) To 5 parts of sulfonated rape oil add 1 part of sodium hydroxide. Warm in a water bath until the excess water has been evaporated.

(2) Dissolve 3 parts of No. 1 in 36 parts of water.

(3) Add 5 parts of a 20% ammonia solution to the 39 parts of No. 2.

(4) To 44 parts of No. 3 add 25 parts of flaked orange shellac and agitate in a mechanical churn until solution is complete. Under normal conditions this will require about 6 hours.

The resultant heat should dissipate about 22 parts of the water so that the completed mixture will contain approximately 4½ lb. of shellac per gallon of mixture.

Fireproof Paint

Aluminum Powder	1 lb.
Sodium Silicate 22° Bé.	1 gal.

White Distemper

Casein	1,000 parts
Urea	340 parts
Hexamethylenetetramine	210 parts
Lithopone	6,950 parts
Zinc Oxide	1,000 parts
Lime	500 parts

Paint, Oil Emulsion

Trihydroxyethylamine Linoleate	0.3
Glue	5
Water	16
Linseed Oil Varnish	8
Phenol	0.2

Procedure for the above oil emulsion paints is to dissolve the water soluble materials and heat together with stirring until free from lumps. The oil, varnish or other water insoluble material is run in slowly while stirring vigorously with a high-speed mixer. Best results are obtained by not too long mixing and occasional rest periods.

Traffic Line Paint

	Pounds	
Neville Hard Resin.	550	
Light Naphtha....	350	(56 gal. approx.)
(Gasoline or V.M.P.)		
Neville Commercial Xylol.........	100	(14 gal. approx.)
	1000	(130 gal. 125 gal. yield)

DIRECTIONS

Cold Cut in tumbling drum or agitator tank.

The above results in a thin bodied vehicle, fast drying (Ten Minutes) in which a high percentage of pigment may be incorporated without losing gloss. A small amount (2 to 5%) of bodied Linseed Oil is sometimes added. Any desired pigment combination may be used dependent on the color desired.

Wall Sealers

The following formula may be used by the paint and varnish manufacturer in developing a good wall sealer.

Varnish II

China Wood Oil	30 gal.
Kettle Bodied Linseed or Perilla Oil *	3 gal.
Cumar W 1	88 lb.
N Rosin	12 lb.

Candy Glaze

1.	Shellac (arsenic free)	4 lb.
	Alcohol	6.5 lb.
	Isopropyl Acetate	2.4 lb.

2.	Copal Bold Chips	6 lb.
	Isopropyl Alcohol (98–99%)	12 lb.
	Isopropyl Acetate	2 lb.

Antifouling Paint

a.	Rosin	2 lb.
	Lithopone	1 lb.
	Naphtha	160 lb.

b.	Chrome Green	1 lb.
	Lithopone	2 lb.
	Rosin	3 lb.
	Naphtha	160 lb.

c.	Chrome Green	21 lb.
	Rosin	12 lb.
	Naphtha	160 lb.

First apply a coat of (a) and when dry apply a coat of (b). When this has dried apply (c).

Blackboard Paint

Carbon Black	15 lb.
Shellac	14 lb.
Prussian Blue	1 lb.
Lithopone	1 lb.
Powdered Carborundum	7 lb.

Drier Liquid	16 lb.
Alcohol	130 lb.
Linseed Oil Boiled	7½ lb.

Flexible Paint for Marking or Stencil Work

Adheres well to rubber goods. Can be hot pressed into fabrics.

Gutta Percha	60
Colored Pigment	40

The colored pigment is milled into the Gutta Percha on a roll mill. Pigments such as vermilion, cadmium sulphides, ultramarine, etc., may be used. Organic color lakes are also satisfactory. On account of the smaller quantity of lake needed, the difference should be made up with blanc-fixe.

The mixed compound is dissolved in solvent naphtha with slight warming. A 20% solution gives good coverage and may be sprayed easily.

Galvanized Iron, Treatment before Painting

Some people, before painting it, wash the galvanized metal with vinegar. This is said to be good. Others scrub it well with burlap wet with benzine. Scrubbing the surface with soap and sand can be recommended. The best method seems to be, however, to leave the galvanized metal exposed to the weather for a few months.

Still others report good results from washing the well-cleaned surface with a one per cent solution of copper chloride, acetate or sulphate. The solution is left on for a time and then brushed off before painting is attempted. A few months of exposure is probably better, however, even than this treatment.

Light sand-blasting is also said to have been used for cleaning galvanized iron and putting it in condition to take paint. No doubt this would accomplish the purpose.

Even in the case of perfectly clean zinc, it is not easy to get paint to stick always. No paint yet invented adheres to it as well as in the case of iron or wood. What chemists call "the surface tension" is different. Not that any good paint invariably all comes off. Generally most of it stays on but that is not very satisfactory.

If galvanized iron is weathered and then well cleaned, there is seldom any trouble encountered when the paint is red-lead. Probably most of the difficulties in painting galvanized surfaces are traceable to improper preparation done by not too expensive labor. This is why weathering, which does not skip anything, is best.

Paint Grinding

A small percentage of Oleic Acid materially helps the grinding of Carbon Black.

Heat Resisting Paint

Powdered Graphite	1	lb.
Lampblack	1	lb.
Black Oxide of Manganese	0.33	lb.
Japan Gold Size	0.33	pt.
Turpentine	0.50	pt.
Boiled Linseed Oil	0.33	pt.

Mix together until a uniform consistency is obtained.

Heat Sensitive Paints
The Double Iodide of Silver and Mercury

Silver Iodide	5 parts
Mercuric Iodide	1 part

This compound mixed with shellac and painted on thin strips of steel changes from a very bright yellow to a deep red as the temperature increases.

High Light Reflecting Paint

The following formulae are suggested for obtaining proper illumination in interiors and providing desirable paints that can be washed repeatedly:

100 lb. Pure White Lead (heavy paste)
2 gal. Flatting Oil

or

100 lb. Pure White Lead (heavy paste)
2 gal. Pure Turpentine
1 pt. Floor Varnish
½ pt. Pure Drier

They may be tinted as follows (Quantities are per 100 lb. white lead):

Ivory White—	3 oz.	French Ochre
Cream —	¾ lb.	French Ochre
Light Buff —	3 lb.	French Ochre

Ultra Violet Paints

The following formulae produce a strong fluorescence when exposed to ultra violet light.

Blue-violet.—Dissolve five parts of vaseline and 12 parts of white paraffin wax of melting point 140° F. in 175 parts of benzene and add, by stirring in, five parts of finely powdered calcium salicylate. Alternately, five parts of aesculin may be substituted for the calcium salicylate.

Apple-green.—Repeat the above mixture and substitute anthracene for the calcium salicylate.

Brilliant Green.—Dissolve 20 parts of cellulose acetate in 300 parts by weight of chloroform into which one part of vaseline to 15 to 37 parts of chloroform are dissolved. Mix the solutions thoroughly and introduce 10 to 30 parts of finely powdered potassium uranyl sulphate.

Orange Yellow.—Substitute zinc sulphide containing about one part of 1000 of manganese for the potassium uranyl sulphate in the above formula.

Red.—Mix 100 parts of zinc sulphide with 200 parts of cadmium sulphate and incorporate in gum. The consistency of the mixture determines the brilliance of resulting hue.

Violet Luminous Paint

Twenty parts, by weight, of calcium oxide (burnt lime) free from iron; 6 parts by weight of sulphur; 2 parts by weight of starch; 1 part by weight of a 0.5 per cent solution of bismuth nitrate; 0.15 part by weight of potassium chloride; 0.15 part by weight of sodium chloride. The materials are mixed, dried and heated to 1300° C. (2373° F.). The product gives a violet light. To make this effective, it is exposed for a time to direct sunlight, or a mercury lamp may be used. Powerful incandescent gaslight also does well, but requires more time.

Phosphorescent Pigments

A method for producing phosphorescent pigments from alkaline earth sulphides has been developed. This is stated to have advantages over the usual method which employs a zinc sulphide base in that usual commercially available materials can be used. The following formulas are offered:

Greenish-blue Phosphorescence. A mixture of 20.7 gms. strontium hydrate, 8.0 gms. sulfur, 1.0 gms. lithium sulfate, and 6 cc. of 0.3 per cent aqueous colloidal bismuth is heated in a porcelain crucible for 40 minutes to a point of incandescence. The mass is then allowed to cool slowly.

Red Phosphorescence. 40.0 gms. barium oxide, 9.0 gms. sulfur, 0.7 gm. lithium phosphate, 3.5 cc. of 0.4 per cent alcoholic copper nitrate. If the lithium phosphate cannot be obtained, it can be replaced by a mixture of magnesium phosphate and sulfate or carbonate of lithium.

Red Phosphorescence. 40.0 gms. magnesium phosphate, 0.7–1.0 gm. lithium sulfate, and 3.5 cc. of 0.4 per cent alcoholic copper nitrate.

Luminous Paint

Barium Sulfate	34 lb.
Indian Lake	22 lb.
Madder Lake	23 lb.
Luminous Calcium Sulfide	76 lb.
Varnish	73 lb.

Luminous Paint

The following is for luminous paint giving a yellow glow:

	I	II
Strontium Carbonate	100	100
Sulphur	100	30
Potassium Chloride	0.5	—
Sodium Carbonate	—	2
Sodium Chloride	0.5	0.5
Manganese Chloride	0.4	0.2

The mixture is heated in a crucible for three-quarters of an hour at about 1,300° C. The more permanent variety of luminous paint used for watch hands consists of zinc sulphide activated with radium bromide.

Olive Drab Paint

White Lead (ground in raw linseed oil)	6 lb.
Raw Umber (ground in raw linseed oil)	3 lb.
Chrome Yellow (ground in raw linseed oil)	½ lb.
Linseed Oil Raw	1 pt.
Turpentine	½ pt.
Japan Drier	¼ pt.

Outside White Paint

Material	Pounds
Carbonate White Lead	41.0
Zinc Oxide	20.5
Asbestine	7.3
Linseed Oil	25.8
Turpentine and Driers	5.4
	100.0

Pounds
21.8 Titanox B.
21.8 Basic Carbonate White Lead
12.4 Zinc Oxide
 6.0 Asbestine
31.9 Linseed Oil
 6.1 Turpentine, Varnolene and
       Driers

Pounds
24.0 Lithopone (Albalith)
24.0 Zinc Oxide (American Proc-
       ess) XX
 6.0 Asbestine
 6.0 Silica
30.9 Alkali refined or mechanically
       refined Linseed Oil
 2.5 Kettle Bodied Linseed Oil
 6.6 Turpentine, Varnolene and
       Driers

Pounds
25.5 AX1 Lithopone
28.7 35 per cent Leaded Zinc Oxide
 4.8 Asbestine
 4.8 Silica
29.9 Refined Linseed Oil
 2.6 Kettle Bodied Linseed•Oil
 1.8 Drier
 1.9 Thinners

Cheap Outside White Paint

Lithopone	300 lb.
Paris White	200 lb.
Asbestine	130 lb.
Refined Linseed Oil	7 gal.
Refined Fish Oil	7 gal.
Limed Gloss Oil	11¾ gal.
Varnolene (Naphtha)	5¼ gal.
Kerosene	5¾ gal.
Liquid Japan Drier	2¼ gal.
Spar Varnish	3 gal.
Water	4¾ gal.

Where colored paint is desired raw oils and dark gloss oil may be used with suitable pigments replacing all or part of the above pigments.

Tire Paint

Precipitated Chalk	40 lb.
Spanish White	20 lb.
Gilder's Whiting	15 lb.
Gum Tragacanth	10 lb.
Phenol Crude	10 oz.
Amyl Acetate	10 oz.

Allow gum to soak overnight in 7 gal. water; add phenol and pigments while stirring; if too thick add more water and then stir in the Amyl Acetate.

Ship Paint

The experts in charge of dry-dock work on the Atlantic coast have found satisfaction in repainting work done with the following formula:

Paste Red-Lead	100 lb.
Raw Linseed Oil	1½ gal.
Japan Drier	1 qt.
Turpentine or Mineral Spirits	1½ qt.
	4¼ gal.

Liquid Paint Drier

1. Rosin W. W.	200 lb.
2. Calcium Hydroxide	16 lb.
3. Lead Acetate (Powd.)	16 lb.
4. Chinawood Oil	8 gal.
5. Manganese Borate	2 lb.
6. Benzine	98 gal.
7. Kerosene	9 gal.

Melt (1) and (2) and strew (3) over surface. Heat slowly raising temperature to 450° F. and heat until odor of acetic acid is gone. Mix (4) and (5) and stir into above and mix thoroughly while heating. Raise temperature to 540° F. stirring and beating down foam. Cool to 460° F. and add Kerosene while stirring. When cooled to 240° F. add benzine with stirring.

This gives a practically colorless quick drier.

Wood Paint Primer

Pigment	65.6%	
Basic Carbonate of Lead		60%
Zinc Oxide		20%
Titanox B		19%
Aluminum Bronze Pwd.		1%
Vehicle	34.4%	
Raw Linseed Oil		40%
Boiled Linseed Oil		30%
Turpentine		16½%
Solvent Naphtha		10%
Drier (Lead-Manganese)		3½%

Weight per Gallon 16.7 lb.

Whitewash

The following will give good results. Dissolve six pounds of trisodium phosphate in two gallons of water. Soak ten pounds of casein in four gallons of water for two hours, or until soft, add to the first solution and dissolve. Stir to smoothness twenty-five pounds of whiting and fifty pounds of hydrated lime in seven gallons of water. When the mixtures are cold, slowly add the first solution to the lime, stirring continuously. Dissolve five pints of formaldehyde in three gallons of water and just before use add it slowly to the whitewash, stirring hard. Do not make more than can be used in one day.

Whitewash (Without Glue)

Dissolve 15 lb. salt in 7½ gallons water and add slowly with stirring 50 lb. hydrated lime.

High Grade Whitewash

Slack a bushel of a good grade Quicklime and remove the lumps by pouring through a screen. Add sufficient water to produce a good whitewash, which is usually forty gallons.

Then add 20 lbs. of whiting, 17 lbs. of Rock Salt, and 12 lbs. Brown Sugar in the order started. Stir thoroughly and if necessary add some more water until the desired fluidity is attained.

Apply with a brush. Two coats are needed on wood, and three on brick or stone surfaces. This produces a coat that does well in preserving the surface, looks well, and does not readily rub or wash off. The sugar serves as a binder.

Hot Water Calcimine

Water-floated chalk whiting	86%
China clay	10%
Hot water animal glue (high grade)	4%
	100%

Zinc sulphate (preservative) 1 part to 300.

Paint Remover

Benzol	5	gal.
Ethyl Acetate	3	gal.
Butyl Acetate	2	gal.
122° M. P. Paraffin	3½	lb.
Nitrocellulose	½	oz.

Dissolve nitrocellulose in acetates. Dissolve paraffin in benzol. Mix two.

Paint and Varnish Remover

1.

Benzol (90%)	3 gal.
Denatured Alcohol	2 gal.
Paraffin Wax	1 lb.

2.

Benzol	50
Methanol	25
Acetone	15
Gasoline	10
Paraffin Wax	2½

3.

Gasoline	50
Benzol	15
Acetone	35
Paraffin	3

4.

Trisodium phosphate and sodium metasilicate will quickly and easily remove varnish. They will also work on paint if not too old or too thick. Use 1 lb. to 1 gallon of boiling water. Mop or brush on, and let stand 20 to 30 minutes. Then rub off and rinse well with water.

5.

Acetone	1/3
Methanol	1/3
Ethylene dichloride	1/6
Naphtha 140–220° F.	1/6

Add a small amount of paraffin wax to prevent too rapid evaporation.

Artificial Snow Coating

Permanently flexible artificial snow which will stand exposure to the elements and remain sufficiently plastic to prevent film breakage by contraction and expansion of wood, metal or tar roofing may be made by melting rosin, 40%, and adding boiled linseed oil, 5%, and Rezinel No. 2, 5%. When the batch is completely mixed work in 50% of oil free white lead. This plastic is painted on hot and swept over with whiting to quickly remove the fresh tack. Ground mica may be mixed with the chalk sweeping compound to give sheen. Coatings of this type can be washed with water and reswept with whiting when they become dirty.

Varnishes

Varnish is a gum cooked in a drying oil and thinned with volatile solvents. Driers are added in the form of metallic compounds during the heat process or they are added as metallic linoleates and resinates after the varnish is made. (Other organic compounds of these metals are also used such as the napthenates.)

The presence of lead, manganese, and cobalt in solution accelerates the drying of varnishes very materially. They act as oxygen carriers, absorbing oxygen from the air and surrendering it to the oils, which combine with it to form a hard rubbery material.

Gums impart hardness to a varnish film, and oils impart flexibility. The "longer" a varnish the more flexible it is. This length is measured by the number of gallons of oil used per 100 lb. of gum, 50 gal., 25 gal., 10 gal., etc., denoting the addition of the corresponding gallons of, say, combined linseed and china wood oils to 100 lb. of gum.

The most common gum used is ester gum, the glyceryl compound of abietic acid or rosin. Limed rosin is also used extensively but gives more discoloration and is not as neutral as the ester. Neutrality is important, particularly when used in paint formulation when such basic pigments are used as White Lead and Zinc Oxide. An acid varnish may result in coagulation or "livering" of the paint caused by metallic soap formation.

◆

Gloss Oil

W. W. Rosin	100 lb.

Melt and heat to 450° F. and add slowly

Hydrated Lime (stir when adding)	7 lb.

Raise temp. to 550° F. continue stirring for about 15 minutes. Draw from fire, let temp. drop to 400° and add slowly while stirring

Varnolene	10 gal.

Centrifuge while hot.

Yield	20 gal.

◆

25 Gal. Rosin Varnish

W. W. Rosin	100	lb.
Hydrated Lime	7	lb.
China Wood Oil	21	gal.
Litharge	3	lb.
3 Hour Linseed Oil	4	gal.
Manganese Resinate	1½	gal.
Varnolene	35	gal.
	70	gal.

50 Gal. Rosin Varnish

W. W. Rosin	100 lb.

Melt and heat to 450° F. and add slowly while stirring

Hydrated Lime	7 lb.

Raise to 550 and hold 10 minutes, add slowly

China Wood Oil	43 gal.

Heat to 520 and add

Litharge	5 lb.

Raise to 570 and let cool to 550. Hold 20 min. and add

3 Hr. Linseed Oil	7 gal.

Heat to 535 and add

Manganese Resinates	3 lb.

Draw from fire and add

Varnolene	60 gal.
Centrifuge while hot	120 gal.

◆

25-Gallon Rosin Varnish Formula

I Wood Rosin	50	lb.
Raw China Wood Oil	25	gal.
Hydrated Lime	2	lb.

heat to 550° F. (to 570° F. off fire). Check with

I Wood Rosin	50 lb.

add

Sublimed Litharge	6 lb.

allow to cook at 500° F. for 1½ hours, cool and reduce with

Turpentine	20 gal.
Varsol	20 gal.
Cobalt Linoleate Paste Drier	4 lb.

◆

50-Gallon Rosin Varnish Formula

I Wood Rosin	100	lb.

run to 450° F. and add

Hydrated Lime	6 lb.

run to 560° F. and add slowly with constant stirring

Raw China Wood Oil	37½	gal.
Raw Linseed Oil	10	gal.

heat to 550° F. (to make 575° F. off fire). Check with

Linseed Oil	2½ gal.

Sprinkle on top of batch

Sublimed Litharge	4	lb.

allow to cook down to 450° F. and reduce with

Turpentine	30	gal.
Varsol	30	gal.

in which has been dissolved

Cobalt Linoleate Paste Drier	6	lb.

75-Gallon Rosin Varnish Formula

I Wood Rosin   100   lb.
run to 450° F. and add
Hydrated Lime   7   lb.
run to 560° F. and add in a slow stream
with stirring
Raw China Wood Oil   37½ gal.
and
Raw Linseed Oil   9½ gal.
run to 590° F. and add
Raw Linseed Oil   28   gal.
Sublimed Litharge   8   lb.
run to 510° F. and cook at this temperature until proper body is obtained (about 4 hours). Reduce with
Turpentine   40 gal.
Varsol   45 gal.
in which is dissolved
Cobalt Linoleate Paste Drier 4 lb.

---

4 Hour Varnish (Partial Phenol-formaldehyde Type of Resin)

China Wood Oil   21 gal.
25%   Phenol-formaldehyde
condensation gum like paranol or amberol   100 lb.
Heat to 500° F. and add
PbO. Heat to 550 and hold
for about 20 min.   3 lb.
Add
3 Hour Bodied Linseed Oil   4 gal.
Manganese Resinate. Heat
to 530 and draw from fire   2 lb.
Add
Xylol   10 gal.
Varnolene   25 gal.
                           ——
                           70 gal.

---

40 Gal. Phenol-formaldehyde Type of Gum

Resin (Durez 500 Gum
Plastic)   100 lb.
China Wood Oil   32 gal.
Heat gum and oil to 460 F.
in 20 min. Add
3 Hour Linseed Oil   8 gal.
Hold for body for about
20–30 minutes and add
Cobalt   Linoleate (5¾%
metal)   1 lb.
Lose heat to 425, add
Xylol   24 gal.
Varnolene   30 gal.

Ester Cut

Ester Gum   500 lb.
Varnolene   25 gal.
Heat to about 400 care-
fully in 200 gal. kettle. Draw
from fire and add
Varnolene   25 gal.
                           ——
Yield   100 gal.
Similarly varnishes of any length can be made.

---

50 Gal. Ester Varnish

Ester Gum   100 lb.
China Wood Oil   42 gal.
Melt and heat to 520
Lead Oxide and heat to 570 5 lb.
Drop to 550 hold for ½ hour.
Add
3 Hr. Bodied Linseed Oil   8 gal.
Raise to 540 and add
Manganese Resinate, draw
from fire   3 lb.
Varnolene   60 gal.
                           ——
                          120 gal.

25 Gal. Ester Varnish

Ester   100   lb.
China Wood   21   gal.
Litharge   3   lb.
3 Hour Linseed Oil   4   gal.
Manganese Resinate   1½ gal.
Varnolene   35   gal
                           ——
                           70   gal.

25-Gallon Ester Gum Varnish Formula

Ester Gum   40 lb.
China Wood Oil   9 gal.
Bodied Linseed Oil   1 gal.
Litharge   1 lb.
Manganese Acetate   4 oz.
Cobalt Acetate   1 oz.
Turpentine   5 gal.
Mineral Spirits   10 gal.
Heat 9 gal. Wood Oil and 35 lb. Ester Gum to 400° F. Add 1 lb. Litharge. Raise quickly to 580° F., gain 590 off fire; hold for light string from stirring rod. Add immediately 5 lb. Ester Gum and 1 gal. of Bodied Linseed Oil (3 hrs.). At 440° F. add driers. Then thin.

50-Gallon Ester Gum Varnish Formula

Ester Gum   36   lb.
China Wood Oil   13   gal.

Perilla Oil	1½ gal.
Bodied Linseed Oil	3½ gal.
Litharge	1 lb. 11 oz.
Manganese Acetate	6 oz.
Cobalt Acetate	1½ oz.
Turpentine	13½ gal.
Mineral Spirits	12 gal.

Heat Wood Oil, Perilla Oil and Ester Gum to 400° F. Add Litharge. Quickly raise to 580–590° F. off fire. Hold for light string. Add Bodied Linseed Oil. At 440° F. add driers and reduce.

75-Gallon Ester Gum Varnish Formula

Ester Gum	40	lb.
China Wood Oil	15	gal.
Bodied Linseed Oil	15	gal.
Litharge	3	lb.
Manganese Acetate	½	lb.
Cobalt	⅛	lb.
Turpentine	22	gal.
Mineral Spirits	10	gal.

Heat Wood Oil and Ester Gum and 5 gal. Linseed Oil to 400° F. Add Litharge. Raise quickly to 580°. Gain 590. Hold for light string. Add balance of Linseed Oil. Reheat to 500° F. until 2" to 3" string established from stirring rod. Cool to 440° and thin.

Ester Gum Mixing Varnish

China Wood Oil	22½ gal.
Imperial Ester Gum No. 8	22½ lb.

Heat to 525° F. and hold for string and add 45 lb. Imperial Ester Gum No. 8, 2½ lb. Red Lead, 3¾ lb. Ground Litharge, and gain to 550° F. and add 6 gal. LV-150 Oil.
Stir well, and, if necessary, hose to 500 and let cool to 425°.
Add Oleum (Solvent).

Rubbing Varnish

Lewisol No. 2	100	lb.
Hardened Rosin (800 lb. Rosin, 64 lb. Lead Acetate, 40 lb. Lime)	20	lb.
China Wood Oil	10	gal.
Powdered Litharge	5	lb.
Zinc Sulfate	2½	lb.
Dipentene	8	gal.
Benzine	30	gal.
No. 49 Drier	4	lb.

Directions:

CW Oil and H Rosin run to 510° F.	4 gal.
More China Wood Oil added and run to 540° F.	4 gal.

China Wood Oil, Litharge, Zinc Sulfate and the	2 gal.
Lewisol No. 2 added and run to 500° F.	100 lb.

Hold for 20 minutes to hard pill. Cool and reduce.

Four Hour Varnish

The following formula using Nevindene is suggested where rapid drying is desired in a medium oil varnish. The Limed Rosin is used to assist kettle manipulation, to prevent drier precipitation and to keep the Nevindene completely dissolved. To obtain maximum speed of drying no Linseed Oil is used.

Medium Oil Varnish

Nevindene	81	lb.
Limed Rosin (5%)	13	lb.
No. 1 Fused Lead Resinate	6	lb.
China Wood Oil	25	gal.
No. 1 Cobalt Drier	1	gal.
No. 1 Manganese Drier	⅜	gal.
Mineral Spirits	44	gal.

0.60% Lead Metal based on weight of China Wood Oil.
0.03% Cobalt Metal based on weight of China Wood Oil.
0.011% Manganese Metal based on weight of China Wood Oil.

Procedure

Heat the Wood Oil to 400° F. and add 13 lb. of Limed Rosin and 40 lb. of Nevindene. Run the batch so as to get to the top heat of 565° F. in approximately 30 minutes from the start of the cook. Hold at 565° F. until a few drops "spun" on glass "pick up" 12 to 15 inches before "breaking." Chill with the Lead Resinate and balance of 41 lb. of Nevindene to cool around 495° F. Hold here for a syrupy body but do not "string" the varnish. As soon as the desired body is obtained add enough Mineral Spirits to completely "check" the batch. Add the liquid driers at 350° F.

Remarks

This varnish is a so-called "four hour" varnish. It is highly water and alkali resistant. Samples have been maintained at a temperature of 30° F. for 7 days without showing precipitation.

Cobalt Drier

W. W. Rosin	100 lb.
Refined Linseed Oil	100 lb.
Cobalt Acetate	16 lb.
Mineral Spirits	35 gal.

Heat Rosin and Linseed Oil to 350° F. and add Cobalt Acetate slowly. Keep the temperature rising. When nearly all the Acetate has been added, the mixture may crystallize but in raising the temperature to 500° F. it will again become liquid. Add the balance of Acetate if not already added and hold at 500° F. until all acetic acid fumes have been eliminated. Cool to 390° F. and add Mineral Spirits.

This drier contains one ounce of Cobalt Metal per gallon.

Manganese Drier

W. W. Rosin	100	lb.
Refined Linseed Oil	100	lb.
Manganese Acetate	15½	lb.
Mineral Spirits	35	gal.

The procedure in making this drier is the same as that described for the Cobalt drier.

This likewise contains one ounce of Manganese Metal per gallon.

Liquid Drier

Rosin	60	lb.
Cobalt Acetate	40	lb.
Mineral Spirits	100	gal.

Baking Varnish for Wrinkle-Finish on Metal

Manila Gum	2½	lb.
Tung Oil	2½	pt.
Raw Linseed Oil	½	pt.
Zinc Sulphate	3	oz.
Lead-Manganese Drier	3	oz.
Turpentine	½	pt.
Varnolene	4	pt.

Melt gum to 625° F., cool to 575°. Heat again to 640° F., cool to 600°. Heat again to 650° F., cool to 600°. Heat again to 610° F.

Heat oils separately to 375° with the zinc sulphate, add to gum, then add drier; heat to 560° F., cool and add thinner at 375° F.

Bookbinder's Varnish

Venice Turpentine	5	kg.
Bleached Shellac	11	kg.
Alcohol	35	kg.

Anti-Rust Varnish

Coumarone China Wood Varnish	25	parts
White Spirit	15	parts
Lead Chromate	¼	part

Varnish, Anti-Skinning Agent for

The addition of 0.1% guaiacol diminishes "skinning."

Bottle Varnish

Rosin	65
Ceresin	5
Japan Wax	5

Melt and stir until uniform. While stirring and heating add slowly

Barytes (Powder)	25

Allow to cool to 90° C. and add slowly with stirring

Alcohol	2

taking care that it does not boil off. Other pigments may be used in place of barytes. This varnish is applied hot. It may also be used for bottle cork capping.

Varnish, Electrical Conducting

Varnish	54
Lithopone	37.8
Lampblack	8.2

The conductivity is increased by increasing lampblack and reducing lithopone.

Varnish, Emulsion

1. Proflex	5
2. Water	50
3. Varnish (4 hour)	40

Allow (1) and (2) to soak ½ hr. and warm and stir until all particles disappear. Put in a vessel fitted with a high-speed mixer and run (3) into it slowly, while stirring vigorously. Stir until uniform.

Varnish, Flat

Linseed or Chinawood Oil	15–30%
Calcium or Aluminum Stearate	15–30%
Kerosene 40° Bé.	33–40%
Naphtha	Balance

Hard Cold Made Varnish

Bleached Shellac	20	lb.
Sandarac	38	lb.
Pale Manila Gum	32	lb.
Rosin WW	10	lb.
Denatured Alcohol	16	gal.
Carbon Tetrachloride	4	gal.

Mix in tumbling barrel until dissolved.

Insulating Varnish

Coumarone Resin	30	parts
Ester Gum	16	parts
Wood Oil	114	parts
White Spirit	132	parts
Kerosene	57	parts
Linseed Oil	48	parts
Cobalt Acetate	0.05	part

Orange Shellac Varnish

T. N. Orange Shellac	200 lb.
Alcohol	40 gal.
Powd. Oxalic Acid	20 oz.

Tumble in barrel for 6–8 hrs. until dissolved; strain through cheese-cloth.

Quick Drying Floor or Interior Varnish

Beckacite Extra Hard	200	lb.
Chinawood Oil	30	gal.
Heavy-bodied Oil	7½	gal.
Mineral Spirits, depending on body desired	75–85	gal.

Directions: Heat gum and the Chinawood Oil to 565° F. This operation takes approximately 45 minutes. Remove kettle from fire and material will automatically rise in temperature to 575° F. Hold heat at 575° until liquid attains desired body. At this point chill with 7½ gal. of Heavy-bodied Oil. Allow material to cool to about 375° F. and thin with Mineral Spirits. When cold add about 4 gallons of lead manganese liquid drier. This formula makes approximately 145 gallons.

Quick Drying Spar Varnish

Beckacite Extra Hard	160	lb.
Chinawood Oil	50	gal.
Heavy-bodied Oil	10	gal.
Mineral Spirits, depending on body desired	75–85	gal.

Directions: Heat gum and Chinawood Oil to 565° F. This operation takes approximately 45 minutes. Remove kettle from fire and material will automatically rise to 575° F., at which time add the Heavy-bodied Oil. To chill back and prevent polymerization, cool material to about 375° F. and add thinner. Then add about 3 gallons of liquid drier. This formula makes approximately 165 gallons.

Straw Hat Varnish

Elemi	50 lb.
Rosin	45 lb.
Sandarac	30 lb.
Shellac	5 lb.
Castor Oil	12 lb.
Alcohol	860 lb.

Transfer Varnish

Gum Mastic	6 lb.
Rosin	12 lb.
Sandarac	25 lb.
Limed Rosin	1 lb.
Venice Turpentine	25 lb.
Alcohol	75 lb.

Violin Varnish

Gum Sandarac	78 lb.
Gum Elemi	31 lb.
Gum Mastic	98 lb.
Castor Oil	48 lb.
Alcohol	980 lb.
Venice Turpentine	20 lb.

Water Shellac Varnish

Borax	20 lb.
Shellac	60 lb.
Water	167 lb.

Warm with stirring until dissolved.

Concrete Silos, Varnish for Interior of

This simple coating is suggested as a wash coat for concrete silo interiors since it will resist the alkaline action of the concrete and the organic acids and other reactive liquids which, generated in the ensilage, have a destructive action on the concrete.

Cumar V-3	100 lb.
Xylol	5 gal.
V. M. and P. Naphtha	15 gal.

Dissolve the Cumar by agitation with the solvent mixture in a vessel provided with a mechanical mixer or in a tumbling barrel. The solution possesses a comparatively low viscosity.

Stir in about 300 pounds of Portland cement and apply with a heavy brush. It will be understood that if a glaze coat is required less cement will be used. If a flatter finish is desired a greater amount of cement can be added.

The mixture is applied with a heavy brush.

Alkali Resisting Varnish

Where a varnish of maximum alkali resistance is desired the following formula is suggested.

China Wood Oil	10–12 gal.
Cumar W	100 lb.
Cobalt Linoleate of 5% Metal Content (or equivalent)	8 oz.
Mineral Spirits	28 gal.

Quick Drying Rubbing Varnish

Beckacite Extra Hard	300 lb.
Chinawood Oil	22½ gal.
Thinner	75 gal.
Liquid Drier	2½ gal.

Directions: Heat gum and Chinawood Oil to 565° F. This operation takes approximately 45 minutes. Remove kettle from fire and material automatically rises in temperature to 575° F. Cool material to about 375° F. and add thinner. Then add about 3 gallons of liquid drier.

Rubber Shoe Varnish

Limed Rosin	10 lb.
Stearin Pitch	30 lb.
Asphalt	30 lb.
Coal Tar	10 lb.
Benzol	100 lb.
Light Naphtha	20 lb.

Allow to settle and decant before using.

Wood Preservative Finish

Creosote, Oil	4
Alcohol	1
Turpentine	2
White Lead	3
Paste Wood Filler	4

Waterproofing Varnish for Fishing Lines

	By Weight
Pyroxylin	100 parts
Castor Oil	250 parts
Amyl Acetate	400 parts
Magnesium Carbonate	2 parts
Methyl Alcohol	600 parts

Paper Varnish

Sandarac	5
Venice Turpentine	3
Denatured Alcohol	15

Wood Finishes

Furniture, caskets, toys, pianos, etc., that are made from raw wood are given certain types of stains and finishes that beautify them and protect them against wear.

The general procedure is first to apply stain, then a wood filler and finally one or more coats of lacquer or varnish.

Stains

Stains may be divided into four main groups, benzol stains, water stains, oil stains and finally lacquer or varnish stains.

Benzol stains are oil soluble dyes dissolved in benzol and are blended to give the desired shades, such as walnut, mahogany, oak, maple, etc. The concentration of dye will vary from perhaps one ounce up to say one pound per gallon of benzol depending largely upon the strength of the dye or dyes and upon the depth of the shade desired.

These solutions are applied by either brushing, spraying or dipping and are then permitted to dry. The wood is then filled with a pigment paste filler. A coat of sanding lacquer sealer is then applied over the filler and when it is dry, it is sanded to give a smooth surface. It is then followed by a coat of shellac to prevent the dyes from bleeding through subsequent applications of lacquer and varnish. Over the shellac one or more coats of either lacquer or varnish are applied in a gloss, semi-gloss or flat finish depending upon the effect desired. The final coat may be rubbed and polished if a hard rubbing finish is used.

Wood Filler

The wood filler is a paste that is first thinned with turpentine or substitute turpentine in the proportion of about ten to fifteen pounds of filler to one gallon of thinner. It is then brushed on the wood and permitted to set up for about twenty to thirty minutes and then the excess wiped off. This is done by wiping with a soft cloth against the grain of the wood so as not to remove the pigment particles from the wood pores.

Wood fillers are often colored with pigments to impart not only a filling but also a staining quality. A typical filler is as follows:

Japan Drier	3 gal.
Turpentine	5 gal.

Linseed Oil	10 gal.
Silica	250 lb.
Asbestine	150 lb.

The above may be colored by adding to or partially substituting umbers, siennas, carbon black and other colored pigments for the asbestine. This may be used to stain and fill at the same time.

Water Stains

Water stains are similar to benzol stains except that water soluble dyes are used instead of oil soluble dyes and water is the vehicle instead of benzol. Also, no shellac is necessary to seal these stains as these do not bleed as the benzol stains do. Water stains are also more resistant to fading than are the benzol types. One must consider all of these factors in choosing the type of stain for one's work.

Oil Stains

These stains are made by grinding colored pigments in oil and then diluting rather excessively with turpentine or substitute turpentine. The combination of pigments that are used depends of course upon the color desired. For example, one shade of walnut stain would be:

| Burnt Umber | 80 | lb. |
| Linseed Oil | 7½ | gal. |

(grind and add)

| Drier | 1 | gal. |
| Turpentine | 24 | gal. |

This is applied the same as the other stains, permitted to dry overnight and then finished with lacquer or varnish. No filler is used with this stain as the pigment serves to both fill and stain at the same time. Also no shellac is used as in the case of benzol stains since no bleeding can take place. This stain is employed only in cheaper work where it is desired to eliminate the extra operations of filling. The grain of the wood is not as clearly revealed as by the other stains.

Varnish and Lacquer Stains

For very cheap work, particularly for cheap toys, merely one application of a varnish or lacquer in which dyes are dissolved are used. This stains and finishes in one operation. Of course the result is not comparable to the finish obtained by the more complete methods, but it is satisfactory for the purpose for which it is used.

Lacquer and Varnish Finish

After staining and filling a sanding lacquer is applied. This is a lacquer that is designed to sand easily and to seal the imperfection in the wood. When dry it is sanded with very fine sandpaper to give a smooth surface. Over the sanding surface a coat of clear wood lacquer is sprayed. For the best type of finish a rubbing lacquer is used high in solid content with a high ratio of nitrocellulose. This is first water sanded with fine sandpaper then rubbed with a rubbing compound and then waxed. This gives the highest possible type of finish of glass-like hardness and smoothness. Where the labor of rubbing and polishing is to be eliminated a clear lacquer high in solid content and lower in nitrocellulose is used. This may be either in a high gloss, semi-gloss or flat finish.

Varnishes are also employed in finishes of varying degrees of gloss. With varnish it is possible to obtain a higher gloss than with lacquer. They cannot, however, be made as hard as lacquer and will not rub to as smooth a finish as lacquer. Also, all varnishes will impart a yellow tinge to the surfaces and cannot be made as clear and water white as lacquer. This discoloration is particularly true of the bakelite types which otherwise are the most suitable varnishes for this type of work.

These varnishes can be converted into flat finishes by grinding Aluminum or Zinc Stearate and Magnesium Carbonate into them. For example:

Aluminum Stearate	4	oz.
Magnesium Carbonate	4	oz.
Varnish	1	gal.
Turpentine	¼	gal.

Water Stains
Red Mahogany

Azo Rubine	4	oz.
Pylam Red	4	oz.
Pylam Black	½	oz.
Acid Orange	3½	oz.

Dissolve in 3 gal. hot water.

Brown Mahogany

Azo Rubine	4	oz.
Pylam Red	4	oz.
Nigrosine Powder	2½	oz.
Acid Orange	5½	oz.

Dissolve in 4 gal. hot water.

Dark Walnut

Pylam Black	5 oz.
Acid Orange	1 oz.
Pylam Yellow	1 oz.

Dissolve in 2. gal. hot water.

Light Walnut

Pylam Black	2 oz.
Acid Orange	2 oz.

Dissolve in 1 gal. hot water.

Oak

Pylam Black	1 oz.
Metanil Yellow	7 oz.

Dissolve in 4 gal. hot water.

Spirit Stains
Red Mahogany

Pylam Spirit Black	½ oz.
Bismarck Brown	3 oz.
Basic Fuchsine	½ oz.

Dissolve in 1 gal. denatured alcohol.

Brown Mahogany

Pylam Spirit Black	4½ oz.
Pylam Spirit Orange	3 oz.
Basic Fuchsine	½ oz.

Dissolve in 2 gal. denatured alcohol.

Walnut

Bismarck Brown	3 oz.
Pylam Spirit Black	1 oz.

Dissolve in 1 gal. denatured alcohol.

Oak (Dark)

Pylam Orange	10 gm.
Bismarck Brown	3½ gm.
Malachite Green	2 gm.

Dissolve in 1 pint denatured alcohol.

Oak (Golden)

Pylam Orange	1 oz.
Auramine	1 oz.

Dissolve in 1 gal. denatured alcohol.

The above are soluble in alcoholic shellacs and lacquers containing alcohol.

Coloring Wood
Water Stain

½ oz. of any Basic Color
1 quart of Water

This raises the grain. Gives best penetration.

Spirit Stain

½ oz. of any Basic Color
1 quart of Denatured Alcohol.

Good penetration. Raises the grain somewhat.

Oil Stain

½ oz. of Oil Soluble Color
1 quart of Benzol

Does not raise grain. Penetration—poor.

Varnish Stain

½ oz. of Oil Soluble Color
1 quart Varnish

Stir until thoroughly dispersed and allow to stand overnight.

Shellac Stain

Same as spirit stain. Substitute shellac solution for denatured alcohol.

Interior Wood Stains

Staining Interior Wood.—In staining new interior wood a coat of liquid composed of equal parts of raw linseed oil and turpentine, particularly if the wood is soft, should first be applied to make an even foundation for the stain. If this precaution is not taken, the stain will strike in here and there, appearing dark in some spots and light in others. When this coat is dry, the stain should be applied over it. After the stain has been on the surface for 5 or 10 minutes wipe off the surplus with a dry rag or waste.

Acid Proof Wood Stain
Solution A

Copper Sulfate	12½
Pot. Chlorate	12½
Water	100

Solution B

Anilin Oil (Light)	15
Hydrochloric Acid (Conc.)	18
Water	100

The wood surface must be freed thoroughly from paint, varnish, grease and dirt. Heat solution A to a boil and give wood two coats while hot, allowing first coat to dry before applying second.

Apply two coats of solution B in the same way. When surface is thoroughly dry wash well with soap and water. Dry and rub well with linseed oil.

◆

Wood Stains, Non-Grain Raising
Water or Spirit Soluble

Dye	4– 6 oz.
Ethylene Glycol	15–25 oz.

Heat on water bath until dissolved; cool and add

Methanol	1 gal.

◆

Black Walnut Stain

Gilsonite	2 lb.
Turpentine	2 lb.

◆

Ebony Stain

Nigrosine (water soluble)	16 lb.
Oxalic Acid	7 lb.
Water	640 lb.

◆

Clear Shingle Stain

Creosote Oil	1 gal.
Kerosene	1 gal.
	2 gal.

◆

Colored Shingle Stain (Red)

Red Oxide	45	lb.
Asbestine	15	lb.
Linseed Oil	3	gal.

Grind and add

Creosote Oil	12	gal.
Kerosene	12	gal.
	29½	gal.

Similarly other colored shingle stains can be made by changing the colored pigments.

Oil Soluble Stain
Red Mahogany

Sudan Red	2 oz.
Pylakrome Black No. 319	3 oz.
Azo Orange 30	1 oz.

Dissolve in two gallons benzol.

Brown Mahogany

Azo Oil Yellow 408	2 oz.
Pylakrome Oil Green 430	½ oz.
Sudan Red	1 oz.
Azo Orange	2½ oz.

Dissolve in two gallons benzol.

Walnut

Azo Oil Yellow 408	7 gm.
Sudan Red	½ gm.
Pylakrome Green 430	1 gm.
Azo Orange	4 gm.

Dissolve in one pint benzol.

Oak

Azo Yellow	15.5 gm.
Pylakrome Black 319	.5 gm.

Dissolve in two pints of benzol.

The above also soluble in waxes, acetone, turpentine and lacquers.

◆

Antique Finish for Wood

Chrome green (dark) (teaspoonfuls)	3
Van Dyke brown (teaspoonfuls)	2
Lampblack (teaspoonfuls)	2
Turpentine (pints)	1
Linseed oil (boiled) (pints)	1
Japan drier	a few drops

This should be coated over the entire moulding; when it starts to set, wipe off the high places with a clean rag. Go over the rest of the moulding and take off all excess glaze that is not needed.

◆

Wood Bleaches

As a wood bleach sodium perborate is probably superior to any of the others now used (including the old stand-by oxalic acid). It has the great advantage over the acid bleaches that it can be mixed directly with sodas and alkalies, since it is stable in alkaline solution. A soluble silicate should be present as a stabilizer. A good mixture is 90% sodium metasilicate and 10% sodium perborate. Some of the metasilicate may be replaced by trisodium phosphate. This is a combination paint and varnish remover and wood bleach. Use 1 lb. to 1 gallon of boiling water. Mop or brush on, and let stand 20 to 30 minutes. Then rub off and rinse well with water.

◆

Wood, Plastic

Nitrocellulose	15–20
Ester Gum	5– 9
Castor Oil	1– 5
Wood Flour	15–30
Lacquer Thinner	79–66

Wood Filler Powder

Silica Powder	200 lb.
China Clay	32 lb.
Linseed Oil	44 lb.
Turpentine	40 lb.
Liquid Drier	24 lb.

Filler for Cast Iron

This material is used to fill in the rough surfaces on cast-iron motor blocks, engines, machine-parts, etc., to obtain smooth surface, before enamel or lacquer is applied.

Japan Varnish	1½ gal.
Spar Varnish	½ gal.
Keystone Filler	4 lb.
Aluminum Silicate Flake	20 lb.

Filler for Automobile-Body Work

Rubbing Varnish	2 gal.
Blown Linseed Oil	¼ gal.
Japan Varnish	¾ gal.
Keystone Filler	4 lb.
Sublimed White Lead	4 lb.
Aluminum Silicate Flake	20 lb.

Fixative for Drawings

Mastic and celluloid are used in the preparation of a solution for "fixing" crayon and pastel drawings. The formula suggested is as follows:

Mastic	24 grains
Amyl acetate	3 oz.

Dissolve by shaking, and allow to stand for twenty-four hours before use.

Celluloid	7 grains
Amyl acetate	3 oz.

Dissolve, mix the two solutions when both are clear, and keep in a tightly corked bottle. Apply by means of a spray.

Butter Tubs, Coating for

To eliminate woody odor in butter, the inside of tubs is sprayed with

Casein	50
Caustic Soda	4
Water	170

followed by 4% formaldehyde.

Protecting Coating for Wax Finishes

Copal Varnish	6 lb.
Boiled Linseed Oil	6 lb.
Turpentine	10 lb.

Mix above together, and apply a thin coat to the wax finish. This will protect it from damp without dulling the finish.

Removing Plastic Paint

Mix one pound sal soda and two pounds hydrated lime and one-fourth of a pound of table salt. Add enough water to this mixture to produce a fairly heavy paste. Apply the paste with a fiber brush, and leave it on until the old material is softened, when it may be scraped off. If the paste material should become nearly dry before the old material is soft enough to be easily scraped off, apply the paste material again, but always be sure you do not get this caustic paste on the woodwork or floors, as it would injure them. When all the old material has been scraped off, wash the surface and rinse it until it is perfectly clean, and allow it to become dry before applying the first coat of paint.

Water Resistant Shellac

Add 2-3% of urea or thiourea to solution of shellac in alcohol.

Flat Indoor Shellac Lacquer

Copal } Mixed	13½	oz.
Alcohol	13½	oz.
Shellac T.N.	7	oz.
Alcohol } Mixed	18	oz.
Bone Oil	3	oz.

Flat Outdoor Shellac Lacquer

Shellac, Orange T.N.	50	oz.
Alcohol	200	oz.
Bone Oil	5	oz.
Oxalic Acid	½	oz.

Finishing Shellac Lacquer

Shellac, White Refined	100	oz.
Alcohol	125	oz.
Butyl Alcohol	4	oz.
Bone Oil	1	oz.

Floor Refreshener

5 lb. Shellac "Cut"	¼ gal.
Denatured Alcohol	¾ gal.

This mixture is applied with a mop. The alcohol cleans and at the same time there is left a thin film of shellac which adds lustre to the floor.

CHAPTER IV

COSMETICS AND DRUGS

ALTHOUGH the use of cosmetics goes back for thousands of years, it is only during the last decade that the use of cosmetics has become universal. The number of preparations now being marketed are myriad in number. Startling and unfounded claims are made for many cosmetics, but nevertheless many of them fill a legitimate need. Their cost of preparation is usually very low as compared to their selling prices. The greatest cost is that for distribution, advertising and selling. The trend toward individual production of cosmetics is on the increase.

Face Powders and Talcs

The most important consideration in the manufacture of face powders and the various talcum powders is in the selection of the raw materials. First, and most important, is the talc employed which should be judged on the basis of slip and smoothness, grit, color, mica content, fineness, acid soluble materials, and specific gravity both actual and apparent.

These properties should be carefully considered in the selection of the talc and more so after a selection has been made in checking of subsequent shipments from the raw material source. For the better products the Italian and Manchurian talcs are to be recommended. French talcs find their use in the medium grade products, and particularly in compacts. Californian talcs are also suitable for medium grade products, while the various other talcs are employed in low-priced products.

------◆------

Talcum Powders

Dusting Powder No. 1

Talc	95 lb.
Boric Acid	2 lb.
Magnesium Carbonate	3 lb.
Perfume	4–8 oz.

Dusting Powder No. 2

Talc	85 lb.
Magnesium Carbonate	10 lb.
Boric Acid	2 lb.
Zinc Stearate	3 lb.
Perfume	4–8 oz.

After-Bath Powder No. 1

Talc	80 lb.
Zinc Stearate	10 lb.
Boric Acid	3 lb.
Magnesium Carbonate	7 lb.

After-Bath Powder No. 2

Talc	85 lb.
Magnesium Carbonate	7 lb.
Zinc Stearate	7 lb.
Boric Acid	1 lb.

The zinc stearate is used for the adhesiveness and softness which it imparts to the powder. The boric acid is used for its antiseptic action. The magnesium carbonate is used for securing lightness and fluffiness. Substitutions, such as the use of magnesium stearate in place of zinc stearate, the use of a light precipitated chalk in place of magnesium carbonate can be made. The incorporation of other antiseptic bodies, such as methyl-para-hydroxy-benzoic-acid, tertiary-chlor-butanol, chlor-meta-Xylenol, is effected by melting these materials into the perfume oil (with the addition of a small amount of alcohol if desired). The perfume is incorporated in the usual manner.

The procedure in the manufacture of these talcum products is as follows:

Dry materials are mixed (usually in a horizontal type enclosed mixer) for a period of time. The perfume is added to a quantity of magnesium carbonate or of the mixed powder, equivalent to twenty (20) times the weight of perfume oil, mixed and then brushed through a forty (40) mesh wire screen, and then

49

through at least a ninety (90) mesh silk screen. The perfume mixture is then added to the full batch of dry materials and bolted through a silk screen of at least one-hundred (100) mesh. A mesh of two-hundred (200) should be used for the highest quality product. At times a pebble mill is used for a simultaneous mixing and grinding operation, preparatory to sifting. The powders may be tinted slightly by using the same color as used in a face powder to secure a rachel, peach, flesh or any other desired shade. The amount of color usually used is about 20% that used in the equivalent face powder shade. However, in talcums for men, the colors are about equivalent in intensity to regular face powder shades.

Face Powders
Formula No. 1—Heavy

Talc	40
Magnesium Carbonate	5
Zinc Oxide	10
Zinc Stearate	5
Rice Starch	10
Kaolin	30
Color—See Coloring	
Perfume	6–14 oz.

Formula No. 2—Medium

Talc	50
Zinc Oxide	15
Zinc Stearate	10
Kaolin	20
Precipitated Chalk	5
Color—See Coloring	
Perfume	6–14 oz.

Formula No. 3—Light

Talc	65
Zinc Oxide	10
Zinc Stearate	10
Kaolin	10
Precipitated Chalk	5
Color—See Coloring	
Perfume	6–14 oz.

Coloring for Face Powders and Talcs

The colors necessary to secure most shades of face powder are yellow ocher, geranium lake, persian orange lake, orange lake, burnt umber, burnt sienna, ultramarine blue, violet lake and green lake. These colors are diluted to make color bases as follows:

Rachel Color Base

Yellow Ocher	1
Talc	4

Flesh Color Base

Geranium Lake	1
Talc	9

Peach Color Base

Persian Orange Lake	1
Talc	3

Orange Color Base

Orange Lake	1
Talc	9

Grey Color Base

Burnt Umber	1
Talc	5

Tan Color Base

Burnt Sienna	1
Talc	3

Blue Color Base

Ultramarine Blue	1
Talc	9

Lavender Color Base

Violet Lake	1
Talc	10

Green Color Base

Green Lake	1
Talc	10

To secure the various shades of face powder, the following are used in conjunction with the above bases.

Light Cream

Base	100 lb.
Rachel Color Base	24 oz.

Rachel

Base	100 lb.
Rachel Color Base	48 oz.

Medium Rachel

Base	100 lb.
Rachel Color Base	64 oz.
Tan Color Base	8 oz.

Dark Rachel

Base	100 lb.
Rachel Color Base	48 oz.
Tan Color Base	16 oz.
Lavender Color Base	4 oz.
Grey Color Base	8 oz.

Various intermediate shades can be secured by combining two or more of the above formulae, or by increasing or decreasing the components of these formulae, or by the addition or deletion of items from these formulae. The procedure in the manufacture of these powders is: mix the dry materials, including the color base, in a horizontal type enclosed mixer, or else in a pebble mill, until all the materials are thoroughly distributed. The perfume is worked into about twenty (20) times its weight of either magnesium carbonate or powder base, and sifted through a wire screen and then a silk cloth as mentioned under talcum powders. The perfume mass is then added to the dry materials and the entire mixture is brushed through a sixty (60) mesh wire screen and then bolted through at least a one-hundred and twenty (120) mesh silk, with the finer mesh screens being more advisable. If it is so desired, the powder may be ground through a hammer mill or similar apparatus.

In the formulation of the powder, these factors are considered:

Talc is used for slip and its lubricating effect.

Kaolin—there are two (2) types available—the osmo or colloidal kaolin, and the bolted kaolin.

The *osmo kaolin* is much whiter and drier than the bolted, and has its main use in bodying the powder, giving it slip and coverage, and for its powers of absorption, especially in regard to perspiration and moisture.

The *bolted kaolin* is greener and much more moist than the osmo kaolin, and it is used to secure additional slip and a creaminess that can not be secured otherwise.

Rice Starch is used for the smoothness it imparts. Its absorption of moisture or perspiration with a subsequent swelling of the particles, resulting in enlarged pores, is a question of reasonable doubt, and can not be answered with entire satisfaction.

Magnesium Carbonate is used for securing lightness and bulkiness of the powder, and also as an absorbent of perfumes.

Zinc Oxide is used for its tinting covering powers.

Titanium Oxide is used for its covering powers which it possesses to a much greater degree than zinc oxide, but its tendency to pack and "ball" a powder in which it is used makes it a product not usually found in face powders; zinc oxide still being considered more favorable for securing cover. However, if care is taken in the formulation of a product, excellent powders can be made with titanium oxide as a component of the powder.

Zinc Stearate is used in securing adhesiveness of the powder, as well as softness and lightness.

Magnesium Stearate has properties similar to zinc stearate but it is heavier. It finds its main use to replace zinc stearate where the use of zinc stearate is prohibited.

Precipitated Chalk is used in securing the smoothness that rice starch imparts without running into the difficulties that might be encountered in the use of rice starch by reason of its property of swelling.

◆

Compact Powders

Compact powders are made in the same manner as are compact rouges, except that coloring is done with color bases, the same as are used in face powders. About 50% of the amount of color used in face powders is usually sufficient to give the same intensity of color.

Liquid Face Powder

Liquid face powders are suspension of chalk and zinc oxide in a water, alcohol and glycerin solution. The coloring is done with the color bases that are used in face powders.

◆

Liquid Powder

Zinc Oxide	3 lb.
Precipitated Chalk	3 lb.
Glycerine	1 pt.
Alcohol	4 pt.
Perfume	4 oz.
Water	4 gal.

Color
(See Face Powder)

Rachel—1 oz. Yellow Ocher Base
Tan—1 oz. Burnt Sienna Base
Flesh—1 oz. Geranium Base
Peach—½ oz. Persian Orange Base

◆

Further ideas for coloring may be had by referring to the various shades and the combinations necessary to secure them.

The zinc oxide may be replaced in whole or in part with titanium oxide.

Diethylene glycol may be used in place of glycerin.

Another type product is:

Liquid White (for Skin)

A lotion for hand and arms contains 2,500 parts witch hazel extract, 5,000 parts rose water, 1,000 parts alcohol, 1,800 parts glycerin, 100 parts tallow, 100 parts magnesium carbonate, 50 parts magnesium stearate and 1,000 parts antipyrine. First, the antipyrine is dissolved in the witch hazel extract and rose water. Then glycerin is added. The perfume used is absorbed by the magnesium carbonate, magnesium stearate and tallow. Then alcohol is added. This suspension is strongly shaken for two days. The milk is filtered through coarse filter paper. The two preparations are united with vigorous stirring and decanted. This preparation is applied with cotton. The skin is rubbed and the preparation is allowed to dry. The skin remains white the entire evening. The advantage of this preparation over ordinary liquid powder is that a dull white effect is obtained, lasting 4 to 6 hours.

Dry Rouges

Dry rouges were originally made by mixing talc and carmine with a mucilage of tragacanth, placing the plastic mass in metal trays or cups and allowing them to harden. These products were then sold in these trays or cups.

The manufacturing technique was then developed to the point wherein with a slight modification of the formula, the plastic mass was placed on discs, allowed to dry and then turned into shape by a cutting tool.

The present day manufacture of rouges is done in several different ways: The most simple is a mixture of the following:

Talc	40 lb.
Kaolin	35 lb.
Zinc Oxide	15 lb.
Precipitated Chalk	10 lb.

Mix the above with sufficient dry color to give the desired shade (see following), grind in a ball mill for a period of time sufficient to distribute all color particles throughout the entire mass. The material is then bolted through at least a 140 mesh silk, and is then moistened with a tragacanth solution of a strength of 1/4 oz. gum tragacanth and 1/4 oz. boric acid to 1 gallon water. This solution is added to the powder mass at the rate of 1 oz. to every pound of dry material. The perfume oil is also incorporated at this stage.

The slightly wetted powder is brushed through at least a 30 mesh wire screen and bolted through at least a 60 mesh silk screen. The rouge material is then ready for pressing.

The type press used for this particular formula is a foot press which has a fast downward and upward stroke. The metal disc or cup onto or into which the rouge is to be pressed is moistened with a tragacanth solution of a strength of 1 oz. of gum tragacanth to 1 gallon of water.

Colors for Rouges
To Be Added to a 100 Pound Batch

Orange

Scarlet Lake	7 lb.
Yellow Ocher	8 oz.
Indian Red	4 oz.

Light No. 1

Indian Red	3 lb.
Burnt Sienna	1 lb.
Brilliant Lake	2 lb.
Maroon Lake	8 oz.

Light No. 2

Scarlet Lake	4 lb.
Brilliant Lake	2 lb.
Yellow Ocher	8 oz.
Indian Red	8 oz.
Geranium Lake	4 oz.

Geranium No. 1

Geranium Lake	5 lb.
Scarlet Lake	7 lb.
Brilliant Lake	3 lb.

Geranium No. 2

Geranium Lake	12 lb.
Orange Toner	4 oz.
Crimson Lake	8 oz.

Medium No. 1

Brilliant Lake	18 lb.
Scarlet Lake	2 lb.
Maroon Lake	1 lb.

Medium No. 2

Brilliant Lake	12 lb.
Geranium Lake	4 lb.
Maroon Lake	3 lb.
Indian Red	3 lb.

Lip Rouge—Indelible
Formula No. 1

Castor Oil	7 lb.
Lanolin	1 lb.
Beeswax	1 lb.
Bromo Acid	12 oz.
(tetrabrom fluorescein)	
Lake Color	12 oz.
Perfume	as desired

Formula No. 2

Castor Oil	6 lb.
Cetyl Alcohol	1 lb.
Stearic Acid	4 oz.
Lanolin	1 lb.
Glycerol Mono Stearate	1 lb.
Bromo Acid	8 oz.
(tetrabrom fluorescein)	
Lake Color	4 oz.
Perfume	as desired

The lake color mixtures used in the previous formulae may be used to secure the various shades. The procedure is the same as in the above formulae.

Lip Pomade

Mineral Oil	6	lb.
Petrolatum	2	lb.
Paraffin	2	lb.
Ozokerite	¼	lb.
Beeswax	¾	lb.
Perfume	½	oz.

The materials are melted together and poured into a suitable mold. These sticks are intended for use in softening of lips and preventing chapping of the lips.

Materials such as: lanolin, absorption base, olive oil, cocoa butter, may also be introduced into the formula. From 2–4 oz. of zinc oxide ground into the product will give a whiter stick and will aid as a healing agent.

Materials such as: menthol, camphor, thymol and similar medicants, may be used in small quantities and most conveniently introduced into the product by molding them into the perfume and then adding the mixture to the melted oils and waxes.

Lipstick (Non-Indelible)

For Theatrical Use

Petrolatum	4	lb.
Paraffin	2	lb.
Mineral Oil	1	lb.
Carnauba Wax	6	oz.
Lanolin	8	oz.
Lake Color	1	lb.
Perfume	as desired	

The same lake color mixtures as are used in the greasy cream rouges are suggested to secure the various shades. The procedure is the same.

Lipstick—Indelible

Formula No. 1

Castor Oil	4	lb.
Cocoa Butter	2	lb.
Stearic Acid	1½	lb.
Paraffin	2	lb.
Beeswax	1¾	lb.
Carnauba Wax	2	oz.
Lanolin	8	oz.
Bromo Acid	12	oz.
(tetrabrom fluorescein)		

Lake Color	12	oz.
(see cream rouge)		
Propyl-para-hydroxy-benzoate	1	oz.
Perfume	2	oz.

Formula No. 2

Castor Oil	6	lb.
Glyceryl-mono-stearate	2	lb.
Stearic Acid	½	lb.
Cetyl Alcohol	½	lb.
Bromo Acid	10	oz.
(tetrabrom fluorescein)		
Erythrosine	¼	oz.
Oil Soluble Red	16	grains
Perfume	1½	oz.

The oils and waxes are heated together and the oil soluble color is dissolved therein. The bromo acid and erythrosine are then added and the entire mass is ground through an ointment mill. For this particular type lipstick, various shades may be secured by the use of various of the oil soluble colors, such as: yellow and orange in combination with the red, and the use of other water soluble dyestuffs in place of the erythrosine, such as: tartrazin, ponceau, etc.

Formula No. 3—Changeable Orange

Castor Oil	5	lb.
Cocoa Butter	3	lb.
Ceresin	3	lb.
Beeswax	2	lb.
Bromo Acid	2	oz.
(tetrabrom fluorescein)		
Propyl-para-hydroxy-benzoate	2	oz.
Perfume	2	oz.

The oils and waxes are melted. The bromo acid and the benzoate are then added and the entire mixture is filtered hot.

Mascara—Soapless Type—Poured

Triethanolamine	14	lb.
Stearic Acid	20	lb.
Oleic Acid	5	lb.
Ricinoleic Acid	5	lb.
Carnauba Wax	30	lb.
Ozokerite	15	lb.
Petrolatum	6	lb.
Perfume	1	lb.

These materials are all melted together.

The following colors are ground into the molten mass to secure the various shades:

Black

Charcoal Black	5 lb.

Brown

Burnt Umber	10 lb.
Burnt Sienna	1 lb.
Indian Red	3 lb.

Blue

Ultramarine Blue	12 lb.
Titanium Oxide	2 lb.

Creams

Cold creams are the most basic and still the most important creams that are sold. Cold creams are usually formulated using mineral oil as a softening and cleansing agent, and emulsifying with water by the action of borax on beeswax.

A rather soft but exceptionally smooth cream is made as follows:

Mineral Oil	1 gal.
Beeswax	1¾ lb.

Heat the above to 160° F. Dissolve 1½ oz. of borax in 5 pints of water, heat to 160° F. and add this solution to the oil and wax with rapid stirring. When the temperature drops to 140°, add 1 oz. of perfume oil and pour the cream at about 120°.

These basic formulae may be modified by replacing up to half of the beeswax with paraffin, ceresin, ozokerite or spermaceti.

The oil may be replaced in part by petrolatum or by the vegetable oils. If vegetable oils are used, a preservative should be employed.

Materials such as lanolin and absorption base may be introduced in small quantities.

Examples of modified formulae are as follows:

Cold Cream (Inexpensive)

Spermaceti	125
White Wax	120
Liquid Petrolatum	560
Borax	5
Distilled Water	190
Oil of Rose, Synthetic	q.s.

Melt the wax and spermaceti on the water bath and add the liquid petrolatum. Heat the distilled water and in it dissolve the borax. Add this warm solution to the melted mixture while both are warm and at about the same temperature. Beat rapidly; as soon as it begins to congeal add the oil of rose and beat until congealed. Dispense preferably in pure tin tubes.

Cold Cream

Beeswax	540 grams
Spermaceti	300 grams
Mineral Oil	1730 grams
Stearin	430 grams
Water	720 cc.
Borax	100 grams
Sodium Benzoate	10 grams
Perfume.	

The fat bases should be melted with mineral oil. The borax and benzoate of soda dissolved in water and brought to the boil and stirred while still hot into the molten fats. Allow to cool with slow agitation. Add perfume.

Cold Cream

Mineral Oil	54 %
White Wax	18 %
Absorption Base	5.5%
Borax	1 %
Water	21 %
Perfume	.5%
	100.0%

Melt the white wax, add the mineral oil. Dissolve borax in part of water with heat. Add to melted fats. Heat rest of water, stir in absorption base until smooth and mix with fats. Agitate thoroughly and when just above solidifying point, add perfume.

Cleansing Cream

A second type of cold cream is based on the action of triethanolamine on stearic acid.

The following are examples of this procedure:

Cleansing Cream

1.	Mineral Oil	78 lb.
	White Wax	5 lb.
	Spermaceti	28 lb.
	Trihydroxyethylamine Stearate	20 lb.
2.	Perfume	1 lb.
3.	Glycerin	4 lb.
	Water	92 lb.

Heat Nos. 1 and 3 separately to 200° F.; then add Nos. 1 to 2 slowly, stirring thoroughly. When the cream begins to set, the perfume is added and stirred in. Allow to stand over night. Stir thoroughly the next morning and package. This cream will not sweat oil during hot weather and will maintain its consistency.

A third type of cream is that in which the emulsifying agent is either glyceryl monostearate or glycosterin.

These creams are emulsions of oil in water and for that reason evaporate quickly, and produces a cooling effect. They are much more water soluble than the beeswax type creams. These creams should be packed in air tight jars as there is a tendency for a small amount of water to separate from them.

The following are examples of this type product:

Cold Cream (Non-Greasy)

1.	Glycosterin	22 lb.
2.	Petrolatum	16 lb.
3.	Paraffin Wax	12 lb.
4.	Mineral Oil	30 lb.
5.	Water	100 lb.

Heat first four ingredients to 170° F. and stir together. Then slowly with stirring pour in the water which has been heated to the same temperature. Stir thoroughly and then allow to stand (hot) until air bubbles are gone. Add perfume and stir and pour at 110–130° F. Cover jars as soon as possible.

Neutral Cleansing Cream

1.	Mineral Oil	80 lb.
2.	Spermaceti	30 lb.
3.	Glyceryl monostearate	24 lb.
4.	Water	90 lb.
5.	Glycerin	10 lb.
6.	Perfume to suit.	

Heat 1, 2 and 3 to 140° F. and stir into it slowly 4 and 5 heated to same temperature. Add perfume, at 105° F., stir slowly until cold; after allowing to stand for 5 minutes stir until smooth and pack.

A four purpose cream that cleans, nourishes, stimulates and acts as a powder base is made as follows:

Mineral Oil	3 pt.
Petrolatum (white)	½ lb.
(heat to 140° F.)	
Water	4½ pt.
Glycerin	5 oz.
Preservative	½ oz.

Heat to 140° F. and add slowly with stirring to oil mixture. As the temperature falls, a gelatinous mass forms at 120° F. 1 oz. perfume oil is added while stirring and the gelatinous mass changes to a white cream. Slow stirring is continued until cold. This cream may be packed either in tubes or jars.

This cream can be modified by various coloring agents and perfume as under cold cream to obtain specialty creams. Since it is neutral there may be incorporated in it viosterol, or gland or hormone extracts.

Liquefying Cleansing Creams

This type of cream is composed of approximately 50% mineral oil together with petrolatum to give sufficient viscosity so that when the cream liquefies on the skin, it suspends the dirt which is removed from the pores.

The following formulae give excellent results:

Formula No. 1—Soft Translucent

Mineral Oil (light or medium)	56
Paraffin	25
Petrolatum (white)	19

Formula No. 2—Medium Translucent

Mineral Oil (light or medium)	50
Paraffin	18
Petrolatum (white)	23
Spermaceti	9

Formula No. 3—Medium Opaque

Mineral Oil (light or medium)	50
Paraffin	30
Petrolatum (white)	20

Formula No. 4—Hard Opaque

Mineral Oil (light or medium)	45
Paraffin	25
Petrolatum (white)	20
Spermaceti	10

The ingredients are melted and stirred together on a water-bath and 4 oz. of perfume is added per 100 lb. These creams are poured at the lowest possible temperature and allowed to stand undisturbed until solid.

Nourishing Cream

1.	Beeswax	15	parts
	Mineral Oil	45	parts
	Lanolin		
	(Anhydrous)	12	parts
	Ceresin	15	parts
2.	Water	25	parts
	Borax	1¼	parts
	Benzoate of Soda	½	part
3.	Perfume	½	part

Heat Nos. 1 and 2 separately to 200° F., then add 1 to 2 slowly with stirring in an emulsifier or beater. When the cream begins to set add the perfume. Allow to stand over-night; stir the next morning and package.

This cream possesses exceptional penetrating powers and is absorbed very readily by the skin.

Vanishing Creams

Vanishing creams are greaseless creams, essentially stearic acid soaps with excess suspended stearic acid dispersed in water. Pearliness is the effect produced by the crystalline stearic acid in suspension.

For a soft type vanishing cream, triethanolamine is used as a saponifying agent.

Stearic Acid (heated to 170° F.)	24 lb.

Heat the following to 170° F.:

Triethanolamine	1 lb.
Glycerin	8 lb.
Water	8 gal.

and add to the melted stearic acid slowly, while stirring rapidly for a few minutes until emulsification is completed.

When the temperature falls to 135° F. add 4 oz. perfume oil and stir intermittently at slow speed until cold.

Allow to stand for a few days stirring slowly at least once a day for a few minutes.

For harder products, potassium carbonate or hydrate is used as a saponification agent. For example:

Stearic Acid (heat to 170° F.)	24 lb.

Then heat, also to 170° F., the following:

Potassium Carbonate	5 oz.
Glycerin	8 lb.
Water	12 gal.

and add to the melted stearic acid. The procedure is the same as above. Use 5 oz. of perfume.

Massage Creams

The following examples are the more popular types of massage creams:

Massage Cream

Glycerin	1 oz.
Borax	2 drachms
Boracic Acid	1 drachm
Oil Rose Geranium	30 drops
Oil Anise	15 drops
Oil of Bitter Almonds	15 drops
Milk	1 gal.

Heat the milk until it curdles and allow it to stand 12 hours. Strain it through cheese-cloth and allow it to stand again for 12 hours. Mix in the salts and the glycerin, and triturate in a mortar, finally adding the odors and the coloring. The curdled milk must be as free from water as possible in ordei to avoid separation.

Rolling Massage Cream

1.	Stearic Acid (Triple Pressed)	6.75 lb.
	Cocoa Butter	13.50 oz.
	Mineral Oil	2.25 lb.
2.	Corn Starch	12.00 lb.
	Boric Acid	2.40 lb.
	Water	5.60 gal.
	Preservative	1.50 gm.
3.	Glycerin	45 fl. oz.
	Ammonia 26 Baumé	12 fl. oz.
	Perfume (Rose)	4 oz.
	Color (Rose)	1 oz.

Mix the corn starch with cold water until smooth (no lumps). Add the boric acid. Heat until it forms a thick translucent paste, stirring continually, taking care to avoid overheating and burning the bottom of the pan. Take off the heat and add No. 3. Stir. Then add No. 1, which has previously been melted together at 200° F. Stir rapidly foi about 1½ to 2 hours. Add color and perfume, and 2 oz. sodium benzoate dissolved in 4 oz. water. Pack cold.

Witch Hazel Jelly

Boric Acid	1 oz.
Tragacanth	2 oz.
Witch Hazel	1 gal.

Allow to stand over night; stir and pack when smooth.

Bleach Cream

White Wax	1½ oz.
White Petrolatum	12½ oz.
Ammoniated Mercury	1¼ oz.
Bismuth Subnitrate	¾ oz.
Oil of Red Rose	40 drops

Melt the white wax in a double boiler. Add the petrolatum and stir until melted. Cool. Mix the ammoniated mercury and bismuth subnitrate. Add ¼ pound cold petrolatum mixture and mix in a paint mill. When smooth, add the balance of the petrolatum mixture and perfume.

Lotions (Astringent)

Astringent lotions are usually based on alcohol as the active ingredient or on an alcohol containing product, such as: witch hazel. For example:

Astringent Lotion—Mild

Menthol	4 oz.
Zinc phenol sulphonate	5 lb.
Camphor	4 oz.
Perfume	8 oz.
Alcohol	10 gal.

All of the above are dissolved together and 90 gallons of witch hazel is added to it. The product may be colored slightly by the use of a water soluble color.

Astringent Lotion—Medium Strength

Alcohol	35 gal.
Borax	2 oz.
Zinc phenol sulphonate	4 lb.
Perfume	2 lb.
Gum Camphor	8 oz.
Glycerin	3 gal.

All of the above are dissolved together and water sufficient to bring the volume to 100 gallons is added.

Astringent Lotion—Strong

Alcohol	50 gal.
Ethyl amino benzoic-acid	8 oz.
Parachlor Meta Xylenol	8 oz.
Menthol	8 oz.
Thymol	4 oz.
Lavender Oil	3 lb.
Glycerin	5 gal.
Vanillin	8 oz.

All of the above are dissolved together. Water is added to make 100 gallons.

Other materials, such as: resin, benzoin, peru or styrax may be used in small quantities.

Propylene glycol, or diethylene glycol may be used in place of the glycerin.

LOTIONS

Face Lotions

Skin milks and milky face lotions are made with trihydroxyethylamine stearate or else with mucilages of the various gums, as for example:

Liquid Cleansing Cream

The following milky cream is stable and an effective cleanser. It will even remove indelible lipstick and rouge from the skin in addition to the usual grime and dirt. It leaves the skin clean, fresh and stimulated and serves as a perfect powder base without any harmful effects.

Stearic Acid	3 lb.
Mineral Oil (Heat to 170° F.)	2 gal.
Water	3 gal.
Triethanolamine	1 lb.

Diethyleneglycol	3 lb.
Diethyleneglycol Ethyl Ether	2 lb.

Heat to 170° F. and add slowly with rapid stirring to the melted stearic acid and mineral oil. Continue stirring until temperature falls to 150° F. when 2 oz. of perfume is added. Stir until cool. A thicker cream is made by replacing part of the mineral oil by petrolatum.

Pearly Finishing Astringent Lotion

Gum Tragacanth	½ oz.
Water (Warm)	5 pt.

Allow to stand for a day and stir into it

Alcohol	3 pt.

To 12 pounds of soft vanishing cream (see Vanishing Cream formulae) which has stood long enough to develop a pearly sheen, there is added slowly a gallon or more, if desired, of the above gum solution. Stirring must be slow but thorough. Filter through cheesecloth and bottle.

This lotion may be colored if desired. This lotion is shimmering and pearly. It is quick drying and leaves the skin in a fresh, soft condition ready for the application of powder.

Skin Milks

Milky preparations for use on skin can be made with lanolin, cucumber milk and almond milk. The following are some examples:

Formula No. 1

Lanolin	50
Pure Castile Soap	3
Glycerin	20
Rose Water	300
Tincture of Benzoin	5
Perfume Bouquet	10
Water	612

Formula No. 2

Lanolin (Melt on water bath)	30

Then add the following mixture, gradually:

Warm Rose Water	200
Potash Soap (pure)	10
Glycerin (in solution)	20

Then add the following:

Perfume Composition	10
Tincture of Benzoin	30

Remove the entire mixture from water bath and mix with

Cucumber Juice	700

Freshly percolated, warmed

The mixture is then agitated until it cools off.

Formula No. 3

Shelled almonds, 70 parts, are crushed with addition of sufficient rose water to give stiff paste. Then the following mixture is added:

Tincture of Benzoin	20 parts
Benzaldehyde	2 parts
Rose Oil	1 part
Borax	7 parts

and 50 parts glycerin in sufficient rose water to give total of 1,000 parts.

Mixture is allowed to stand several days and then filtered through hair sieve.

Liquid Cleansing Cream (Non-Greasy)

1.	Beeswax	1.5
2.	Spermaceti	6.5
3.	Cherry Kernel Oil	6.0
4.	Glycosterin	4.0
5.	Water	122.0
6.	Alcohol	3.0
7.	Galagum	1.0
8.	Borax	3.0
9.	Perfume	3.0
10.	Glycerin	4.0

Melt together 1, 2 and 3. Heat while stirring 4, 5, 7 and 8 together until uniform. Mix these two solutions stirring until uniform. Stir in 6, 9 and 10 and mix until uniform.

Almond Cream Liquid

Oil Sweet Almonds	1 lb.
Spermaceti	2 lb.
Beeswax	2 lb.
Castile Soap Powdered	3 lb.
Borax	2 lb.
Quince Jelly	1 lb.
Alcohol	1 pt.
Water	4 pt.

Melt the spermaceti and wax together. Dissolve the soap and borax in hot water. Mix these together and add balance of ingredients. Stir and filter through cloth.

Lemon Juice Lotion

Pectin	2.5
Lemon Juice	9.5
Water	88
Preservative	0.15

Skin Smoothener

Boric Acid	3 drams
Tragacanth	8 grams

Glycerin	3 drams
Distilled Water	16 oz.

Boil—stir until a clear jelly is obtained.

Hand Lotions

Hand Cleanser and Conditioner

1.	Mineral Oil	70 lb.
2.	Olive Oil	8 lb.
3.	Trihydroxyethylamine Stearate	14 lb.
4.	Water	70 lb.
5.	Perfume	2 lb.

Heat Nos. 1, 2 and 3 together to 140° F. and stir until homogeneous. Add No. 4 slowly while stirring and then stir in the perfume. Continue stirring until cool. By varying the amount of water a thicker or thinner preparation will be formed. The thicker preparations are put up in tubes and are now carried by men and women, especially motorists, who, when water is not available, merely put a little of this cleaner on their hands, rub it in and then wipe off with it the grease, oil, paint or dirt present. Not only is this an excellent detergent but it leaves the skin smooth, and produces a cooling sensation and prevents chapping during cold weather.

Hand Lotion

Macerate 3 oz. of Quince Seed in 2 quarts of cold water for 24 hours. Strain through linen cloth with force and add 1 quart of water to the strained mucilage. Mix: Bay Rum, 16 oz.; Glycerin, 8 oz.; Orange Flower Water, 12 oz.; Alcohol, 26 oz. and add to the mucilage, followed by sufficient water to make 1 gal. of finished product.

Hand Lotion

Boric Acid	1 dram
Glycerin	6 drams

Dissolve by heat and mix with

Lanolin	6 drams
Petrolatum	1 oz.

The borated glycerin should be cooled before mixing. Add any perfume desired.

Hair "Restorers"

These products are sold and used on faith. Some are helped by the accompanying massage and later washing. There is nothing known which actually will grow hair where it is wanted.

Preparations for Baldness

Ointment

Pilocarpine Hydrochloride	20 oz.
Precipitated Sulfur	120 oz.
Parachol	60 oz.
Balsam of Peru	60 oz.
Resorcinol Monoacetate	30 oz.
Petrolatum	900 oz.
Water	60 oz.
Perfume to suit.	

Procedure: Dissolve the pilocarpine in water and mix with absorption base. Mill the sulphur and the monoacetate with part of the petrolatum. Melt the rest and stir in the absorption base and add finally the sulphur mass. Mix thoroughly.

———◆———

Hair Lotion

Mercuric Chloride	1 oz.
Salicylic Acid	5 oz.
Chloral Hydrate	5 oz.
Glycerin	25 oz.
Acetone	10 oz.
Alcohol	200 oz.
Water	825 oz.
Perfume to suit.	

Procedure: Mix together all ingredients but the water. Stir for a while; then add the water and stir until dissolved.

———◆———

BRILLIANTINES AND POMADES

Dressing for "Kinky" Hair

Beefsuet	16 oz.
Yellow Beeswax	2 oz.
Castor Oil	2 oz.
Benzoic Acid	10 gr.
Perfume	sufficient

Melt the suet and wax, add the castor oil and acid, allow to cool, and add perfume.

———◆———

Liquid Brilliantine

Mineral Oil	100
Chlorophyl (Oil Soluble)	to Suit
Perfume	to Suit

———◆———

Solid Brilliantine

Petrolatum	100 lb.
Chlorophyl (Oil Soluble)	2 oz.
Perfume Oil	8 oz.

HAIR PREPARATIONS

Hair Tonic

Tannic Acid U.S.P.	0.5
Salicylic Acid U.S.P.	1.0
Castor Oil U.S.P.	24.5
Resorcinol Monoacetate	5.0
Alcohol	69.0
Perfume	sufficient

———◆———

Eau de Quinine Hair Tonic

Tincture of Cantharidin	6 oz.
Quinine Hydrochloride	1 oz.
Tincture of Capsicum	2 oz.
Glycerin	3 oz.
Bay Rum	6 gal.
Tincture of Cudbear	sufficient to color

———◆———

SHAMPOOS

Milky Hair Wash (Kerosene)

1. Trihydroxyethylamine Stearate	10 lb.
2. Kerosene	150 lb.
3. Pine Oil	6 lb.
4. Water	250 lb.

Heat Nos. 1 and 2 to 140° F. and stir until dissolved; then stir in No. 3. Now allow No. 4 to run in slowly while stirring. If the pine oil is objectionable, however, any other oil may be substituted for it. It may be colored beautifully by means of any water-soluble dye free from salt.

———◆———

Soapless Shampoo

Sulfonated Olive Oil, concentrated	40 parts
Sulfonated Castor Oil, concentrated	10 parts
White Mineral Oil	15 parts
Water	35 parts
25% Solution of Caustic Soda to Clear	

Mix all the ingredients with the exception of the caustic soda, warm to 45–50° C. and add enough of the caustic soda solution (1 or 2%) until the mixture turns bright. Perfume as desired.

———◆———

Soapless Shampoo
A.

1. Turkey Red Oil	10 lb.
2. Mineral Oil	10 lb.
3. White Oleic Acid	10 lb.
4. Alcohol	2–10 lb.
5. Perfume	4 oz.

Mix the above materials in the order given. If desired, the cost can be reduced by adding an additional amount of water. The water should be added carefully with stirring. The addition of water should be stopped just before a cloudiness appears.

These shampoos are used by pouring a little into the hand and rubbing to a creamy consistency with water and then applying to the hair which must be wet.

B.

Sapinone	10
Water	900
Alcohol	100
Perfume	15

Shampoo

Oleic Acid	55 lb.
Cocoanut Fatty Acids	40 lb.
Triethanolamine	50 lb.
Diethylene glycol monoethyl ether	55 lb.
Perfume	1 lb.

The product prepared in this way is a liquid soap of a clear red color, which can be diluted with water to any desired consistency or concentration. Glycerin and/or alcohol may replace in whole or in part the diethylene glycol ethyl ether.

Olive Oil Shampoo

Olive Oil	4 lb.
Oleic Acid	8 lb.
Cocoanut Oil	8 lb.
Caustic Potash	5 lb.
Alcohol	3 pt.
Water to make	10 gal.

Dissolve the caustic potash in water. Mix and heat the oils to 120° F. Pour in the alkali solution and stir until saponified. Add two pints of the alcohol and heat to 160° F. Meanwhile prepare the following mixture and add the foregoing:

Glycerin	16 oz.
Borax	16 oz.
Potassium Carbonate	8 oz.
Oleic Acid	1 oz.

Dissolve the oleic acid in one pint of alcohol. Dissolve borax and potassium carbonate in glycerin with heat, mix thoroughly and add oleic solution. Add this mixture to soap base while still quite hot. Transfer to a refrigerating tank the day after soap has been finished, refrigerate to 40° F., filter and fill at once.

Dry Shampoo Powder

Cocoanut Oil Soap Powder	30%
Sodium Carbonate Monohydrated	45%
Borax	25%
Henna Leaves Powder	trace
Aniline Yellow	trace
Perfume	to suit

Mix together and sift. Keep in closed containers.

Lemon Rinse

1. Lemon Oil	3	oz.
2. Alcohol	14	lb.
3. Citric Acid	3½	lb.
4. Tartaric Acid	4½	lb.
5. Water	16	lb.

Dissolve 1 in 2 and add to it slowly with stirring 3 and 4 which have been dissolved in 5.

Hair Fixative

Water	20 gal.
Gum Tragacanth	1 lb.
Boric Acid	1 lb.
Preservative	sufficient

Allow to stand overnight and stir until uniform: then stir in

Perfume Oil	4 oz.
Color	to suit

Hair Setting Preparation

Psyllium Seed	1 oz.
Distilled Water	5 gal.
Water soluble Perfume	

Prepared by boiling for five minutes, straining and mixing with an equal bulk of alcohol.

Permanent Wave Solution

Borax	3.75
Sodium Bicarbonate	3.50
Linseed Oil	0.17
Starch	0.40
Water	99.00

Finger Wave Lotion

Borax	600 gm.
Acacia	80 gm.
Boiling Water	18 liters
When cold add:	
Spirit of Camphor	75 cc.
Perfume	as desired

SUN PREPARATIONS
Sun Tan Oil

Mineral or Olive Oil	95–98
Quinine Ricinoleate	5– 2
Oil Soluble Red or Orange	to suit

Artificial Sunburn Liquids

A.
Powdered Cudbear	20 lb.
Powdered Henna	4 lb.
Peanut or Almond Oil	32 lb.

Macerate at 120° F. for 3 hours and filter.

B.
Quinine Sulfate	2 lb.
Witch Hazel	5 lb.
Lanolin	10 lb.
Peanut Oil	92 lb.

C.
Peanut Oil	60 lb.
Olive Oil	35 lb.
Bergamot Oil	1 lb.
Laurel Berry Oil	3 lb.
Chlorophyl	1 lb.

Formulae B and C above require exposure of skin to sun.

Sunburn Liniment
Formula:

Water-White Steam-distilled Pine Oil	75%
Medicinal Olive Oil	25%

The finished product is an effective treatment for sunburn. The product is applied by rubbing directly on the sunburned surface of the skin.

DEODORANTS AND ANTI-PERSPIRATION PRODUCTS
Deodorant Pencil

Zinc Phenolsulfonate	5	10
Zinc Oleate	10	10
Aluminum Palmitate	7.50	7.5
Absorption Base	20.00	30
Ceresin	40.00	30
Titanium Dioxide	——	15

Rub first three ingredients to fine powder and add to liquefied wax the Absorption Base mixture. Stir until just before solidification and pour into molds.

Anti-Perspiration Cream

1. Lanolin Hydrous	1
2. Benzoinated Lard	90
3. Zinc Oxide	6.5
4. Salicylic Acid	1.2
5. Benzoic Acid	0.9
6. Perfume Oil	0.4

Dissolve (4) and (5) in small amount of alcohol; mix into (1) and then work into (2). Grind in (3) until smooth and then work in (6).

Anti-Perspiration Powder

Oxyquinoline Sulfate	1
Talc	10

Anti-Perspiration Liquid

Oxyquinoline Sulfate	1
Rose Water	500

Perspiration Deodorants
A. Liquid Type

Salicylic Acid	2 gm.
Aluminum Chloride	4 gm.
Cologne Spirit	30 mil.
Rose Water	54 mil.
Glycerin	10 mil.
Rose Color	a trace

Dissolve the salicylic acid in the Cologne spirit, and the aluminum chloride in the rose water. Mix and add the glycerin. A more delicate perfume may be used.

B. Paste Type

Salicylic Acid	10 gm.
Levigated Zinc Oxide	60 gm.
Greaseless Cold Cream	480 gm.
Perfume to Suit.	

Freckle "Removers"

Two grams of zinc sulphophenylate, 30 grams of distilled water, 2 grams of ichthyol, 30 grams each of anhydrous lanolin and petroleum jelly and 2 grams of lemon oil or other suitable perfume, will give good results.

Preparations with a bleaching action are made containing 1500 grams of wool grease, 530 grams of almond oil, 110 grams of beeswax, 150 grams of borax, 150 grams of hydrogen peroxide (100% by volume) and 10 grams of yellow petrolatum.

BEAUTY MASKS AND CLAYS
Face Clay

Clay	100 lb.
Water (Cold)	20 gal.
Tincture of Benzoin	3 pt.
Perfume	3 oz.

Add the water to the clay and grind till smooth. Evaporate until 150 lb. remain. Run through mill to smooth clumped particles; cool and mix in the benzoin and perfume. Fill in collapsible pure tin tubes.

Beauty Pack

Tragacanth	25
Alcohol	40
Calamine	80
Zinc Oxide	30
Zinc Stearate	50
Glycerin	60
Lime Water	1000

Mix the tragacanth in the alcohol. Then add to the lime water. Rub up zinc stearate, zinc oxide and calamine with glycerin. Add tragacanth, alcohol, lime water mixture to calamine, zinc oxide, zinc stearate and glycerin mixture.

Leg and Arm Blemish Covering

Stearic Acid	4 lb.
Diethylene Glycol	16 lb.

Heat to 180° F. and to this add while stirring the following solution heated to 140° F.

Caustic Potash	4 oz.
Water	16 pt.

When uniform work in following:

Zinc Oxide	15 lb.
Yellow Lake	12 oz.
Persian Lake	4 oz.
Perfumed Oil	4 oz.

The colors may be varied to give more suitable shades.

Mole and Blotch Covering

Collodion	1 gal.
Zinc Oxide	1 lb.
Geranium Lake	½ oz.
Yellow Ocher Lake	1½ oz.

MOSQUITO PREPARATIONS

The following application is suggested as a means of preventing insect bites:

Cedar Oil	2 dr.
Citronella Oil	4 dr.
Spirits of Camphor	ad 1 oz.

This should be smeared on the skin of the exposed parts as often as is necessary. Cod-liver oil used in the same way has been highly recommended, and in combination with quinine it makes an effective

"sunburn and midge cream," a formula being as follows:

Quinine Acid Hydrochloride	5 parts
Cod-liver Oil	20 parts
Anhydrous Wool Fat	75 parts
Oil of Lavender (or geranium)	a sufficiency

The irritation of a mosquito or fly bite may be allayed by gently rubbing the puncture with a moist cake of soap, or by applying a 1 per cent alcoholic solution of menthol, or 1–20 aqueous carbolic lotion. Hydrogen peroxide or weak ammonia solution dabbed on is also useful. If the bite shows signs of sepsis, constantly renewed hot boric fomentations should be applied, or if a limb is implicated, hot saline arm or leg baths.

Mosquito Cream

Good results can be secured from a composition containing 5 parts powdered wheat starch, 10 parts water, 45 parts glycerin 28° Bé., 30 parts lanolin and 5 to 10 parts oil of clove. Starch is rubbed into smooth paste with water; glycerin is mixed in and mass converted into jelly-like consistency by heating and agitating; it is then allowed to cool.

BATH PREPARATIONS
Liquid Toilet Ammonia
(For Bath)

Ammonium Stearate (Paste)	8 oz.
Ammonia 28°	6 oz.
Water	50 oz.
Glycerin	2 oz.

Perfume to suit, avoiding the use of aldehydes and unstable esters.

Borated Bathing Solution

Boric Acid	10.0 gm.
Alum powdered	2.5 gm.
Camphor	1.5 gm.
Alcohol	120.0 cc.
Water enough to make	500.0 cc.

Perfume to suit, dissolved in Alcohol.

BATH SALTS AND WATER SOFTENERS

The most widely sold bath salts are products that are based on sodium-sesquicarbonate. Sodium Bicarbonate and Sodium Chloride are also used.

Formula No. 1

To 100 lbs. of sodium-sesqui-carbonate is added a mixture of

Dye Stuff   1 oz.
  (to give desired shade)
Perfume Oil   12 oz.
Alcohol   1 pt.

The entire mass is mixed until the color and perfume are thoroughly dispersed.

Other bases are as follows:

2. Sodium Bicarbonate   100 lb.
* 3. Sodium Bicarbonate   50 lb.
   or Sodium-Sesqui-Carbonate 50 lb.
4. Sodium Chloride   100 lb.

All of the above bases are colored and perfumed as shown previously. Care should be taken in the selection of the crystals as regards crystal size, appearance and uniformity. In securing dye stuffs, it is necessary that the type base for which they are intended be specified.

Pine Needle Concentrate

(For Bath)

Many pine-needle oil preparations now marketed, do not take into account that when they are put into water the oil floats on top and only makes contact with a very small portion of the body. By using the following formula the oil is emulsified and spreads uniformly through the bath, giving the entire body the benefit of the pine needle oil.

1. Pine Needle Oil   10 lb.
2. Sodium Sulforicinoleate   10 lb.
3. Water   5 lb.
4. Fluorescein   To Suit

Mix 1 and 2 until dissolved. Add 3 slowly with stirring. Add 4 and stir until dissolved.

The above formula when thrown into water disperses uniformly to give a milky green solution. Other oils may be substituted for Pine Needle Oil. If a lower cost is desired, part of the pine oil may be replaced by mineral, olive or cottonseed oil and a larger amount of water may be added.

Pine Needle Milk

The first step in the process is to prepare a 5% solution of 80% soda soap in 95% alcohol.

Then, triturate the following items:

Pulverized White Gum Tragacanth (Finest)   5 parts
Soap Solution   100 parts

Then, add the following mixture to the paste:

Pine Needle Oil   45 parts
Juniper Oil   5 parts
Alcohol 95%   125 parts

Add the following next:

Water at 30° C.   550 parts

and the entire mixture is agitated for a long time. A thick emulsion is formed, resembling a cod liver oil emulsion. It is ready for use and can be added directly to the bath.

Astringent substances, such as oak bark extract, may be added to the emulsion, but this must be done during the manufacturing process.

REMOVING OF TATTOO MARKS

I.

Pepsin and papain have been proposed as applications to remove the epidermis. A glycerol solution of either is tattooed into the skin over the disfigured part; and it is said that the operation has proved successful. Papain, 5; water, 25; glycerol, 75; diluted hydrochloric acid, 1. Rub the papain with the water and hydrochloric acid, allow the mixture to stand for an hour, add the glycerin, let it stand for three hours and filter.

II.

Apply a highly concentrated tannin solution to the tattooed places and treat them with a tattooing needle as the tattooer does. Next vigorously rub the places with a lunar caustic stick and allow the silver nitrate to act for some time until the tattooed portions have turned entirely black. Then take off by dabbing. At first a silver tannate forms on the upper layers of the skin, which dyes the tattooing black; with slight symptoms of inflammation a scurf ensues, which comes off after fourteen or sixteen days leaving behind a reddish scar. The latter assumes the natural color of the skin after some time. The process is said to have good results.

Obviously such treatments are heroic and carry along with them the risk of permanent scarring. It is therefore a job for a trained dermatologist rather than for a layman.

PERFUME BASES

Unless the amount of perfume base used in manufacturing is considerable, it is not advisable to make them on a small scale. It will probably be cheaper to buy small amounts of these bases from reputable manufacturers than to make them. To make them it is necessary to make a considerable investment in numerous materials. The slightest error in compounding will result in expensive spoilage. The beginner cannot attempt to try out new compounding without excessive waste of time and materials.

Eau De Cologne

Italian Lemon Oil	20 grm.
Bergamot	20 grm.
Neroli or Neroli Synthetic	35 grm.
Italian Sweet Orange Oil	10 grm.
Lavender 40–42% Ester	10 grm.
Orris Root Tincture	2 grm.
Ambreine	3 grm.

Use 100 grams to 1 gallon 70% alcohol. Allow to stand for one week. Chill and filter while cold.

Perfumes

The compounding of perfumes and perfume oils is rather complex. These products are made from mixtures of natural oils together with synthetic aromatic chemicals and natural isolates, as well as certain animal derivatives.

Certain of the aromatic chemicals are necessary to secure reproductions of certain of the natural flower odors, and they, when blended properly with the natural flower oils, give products of the desired character.

In the preparation of extracts, an oil is added to alcohol at anywhere from 8 to 16 ounces per gallon of alcohol, although in certain cases up to 20 ounces are used.

For pre-fixing alcohol, small amounts of the natural resins or gums, or small amounts of the animal derivatives, such as: ambergris, civet or castorium, are allowed to stand in alcohol for at least a month before it is used. The addition of small amounts of water to an alcoholic extract will reduce the tendency towards the alcoholic sharpness.

Toilet Waters

Toilet waters are made in a similar fashion to the perfume extracts, except-ing that a 60–70% alcoholic concentration is used, and from 3–6 ounces of oil are used per gallon of 60–70% alcohol.

Shaving Creams

Shaving creams are special types of soap.

A shaving cream must

1. Lather freely and rapidly.
2. Lather in hot or cold water.
3. Be dense and firm.
4. Be capable of being worked into a dense and voluminous lather.
5. Must not form too soluble a lather which would wash off with excess water.
6. Lather must not dry rapidly but should remain moist for some time.
7. Must be a powerful emulsifying agent, cut surface tension and have good degreasing properties.
8. Must be stable in tube or jar and not dry out or turn hard and gummy and maintain the same consistency for all reasonable temperatures.

The problem of the shaving soap is a problem of balance, so as to obtain a combination which most nearly gives the desired result.

The addition of a sufficient amount of glycerin will help keep the lather moist. The amount generally used is about 10% of the finished cream.

Analysis of the average shaving cream will generally show as follows:

Actual soap content	40%
Water	50%
Glycerin	10%

For the rapid lather a very "soluble" soap is required. If the cream consists entirely of rapid lathering soap, it will be too soluble and will wash away in hot water or on vigorous rubbing, therefore, a large quantity of the "less soluble" soap is required. The more soluble soaps are made from the more soluble oils. These are represented by coconut oil and palm kernel oil. Because of their solubility, they will give a rapid lather, will lather up in cold water or in hard water, but will wash away in hot water or on vigorous rubbing. Because both coconut and palm kernel contain lower molecular weight acids, they will irritate the face if used in too high concentrations. They are generally limited to about 15% or less of the total fat content. While both are satisfactory, coconut is the more widely used, since the odor of palm kernel is more likely to occur in the finished soap. However, a type of deodorized palm kernel has recently been made available.

The soap required to give a more lasting lather, which will retain its body in hot water, must contain a soap such as tallow, stearic acid or palm oil. If a very dense, persistent lather is required, fats containing large amounts of behenic acid may be used.

The consistency desired is obtained not only by a balancing of soaps according to the fatty acids contained, but also by the proper balancing of sodium and potassium soaps. Too much sodium soap cannot be used because of its hardness.

The proper blends of soaps, glycerin and water, is all that a shaving cream consists of. Some contain borax and other fillers. A typical shaving cream formula would be as follows:

Coconut Oil	9
Tallow	3
Stearic Acid	28
Sodium Hydroxide	1.0
Potassium Hydroxide	7.0
Glycerin	10
Water	45

Sodium Hydroxide is prepared as a 20° Bé. solution, using part of the water, in the formula.

Potassium Hydroxide is prepared as a 35° Bé. solution.

Glycerin, coconut oil and tallow are melted in the tank. The sodium hydroxide is run in slowly making sure that saponification is complete.

The excess fat is now saponified with potash, ½ the potash is added to the tank and the mass agitated until saponification appears to be complete. The stearic acid is melted and added and finally the remainder of the potash solution. The mass is stirred until neutralization is complete, and then adjusted to the amount of free stearic acid desired. Three per cent excess stearic acid is commonly used.

This soap when made will be very thick while hot, but will soften on cooling. It is possible to keep the soap thin while hot as by finishing with a large excess of stearic acid which may be later neutralized by adding the appropriate amount of potash solution to the cold soap with suitable agitation.

Shaving Stick

Stearic Acid	40
Coconut-oil	10

Caustic Potash 38° Bé.	23
Caustic Soda 38° Bé.	6
Glycosterin	4

Fats must be saponified at 70° Celsius. The reaction is rather strong, therefore the lye must be added more quickly than usual; to the saponified mass add Glycosterin and leave to the self-induced heating process for three hours, but stir through hourly. Put into forms or pass through a drying machine. A soap put into forms takes very long to harden. Good drying is necessary. The freshly machined sticks are too soft for cutting and must be left to harden several hours. After cutting wrap in tinfoil for preserving their soft and pliant quality.

Latherless, or Brushless Shaving Creams

1

Stearic Acid	50	lb.
Lanolin (anhydrous)	9	lb.
Glycerin	3	lb.
Triethanolamine	1.5	lb.
Borax	1.7	lb.
Water	135	lb.

2

Stearic Acid	40	lb.
Lanolin (anhydrous)	7	lb.
Mineral Oil (white)	18	lb.
Glycerin	3	lb.
Triethanolamine	3.3	lb.
Borax	3.7	lb.
Water	125	lb.

Preparation

Melt the stearic acid, which should be the purest grade obtainable, either alone or with the mineral oil depending upon which formula is followed. Add the lanolin and bring the temperature to about 70° C. Heat the water, Triethanolamine, glycerin, and borax in a separate container and when at the boiling point, add the acid solution. Stir vigorously until a smooth emulsion is obtained and then add the perfume. During the further cooling of the cream, stir gently but continuously taking care to avoid rapid stirring, as this tends to aerate the cream.

Properties

Cream No. 1 is a white, pearly product somewhat like a vanishing cream and is preferable for oily skins. Cream No. 2 is a smooth white cream of greater body than the other, and is preferred for use on dry skins. Both creams are redaily applied to give a smooth coating on the face, have a soothing after-effect and are readily washable. The consistency of these creams can be varied by altering the proportion of water, and other changes can be made along the lines indicated by the difference in the two formulae. A cream of good consistency can be made by combining the two formulae as above.

3

Stearic Acid	50 gm.
Cocoa Butter	9 gm.
Sodium Carbonate Monohydrated	10 gm.
Borax	20 gm.
Glycerin	40 cc.
Alcohol	32 cc.
Water	400 cc.
Perfume q.s.	

Procedure.—Dissolve the sodium carbonate, borax, and glycerin in hot water. Melt the fats and waxes and add the alkali solution. Stir briskly until effervescence ceases and a smooth white soap is formed. Stir slowly until cold; then add the perfume mixed with alcohol.

Liquid Creams

4

Stearic Acid	200 g.
Triethanolamine	10 g.
Water	800 g.

Thicker Creams

5

Stearic Acid	200 g.
Triethanolamine	10 g.
Anhydrous Sodium Carbonate	10 g.
Water	800 g.

AFTER SHAVING PREPARATIONS
Almond Cream for After Shaving

1. Potassium Carbonate 1 oz. 130 gr.
    Distilled Water   15 oz.

Dissolve *Potassium Carbonate* in water, filter

2. | | |
|---|---|
| Gum Tragacanth | 175 gr. |
| Glycerin | 10 gr. |
| Borax | 1 oz. |
| Distilled Water | 64 oz. |

In 20 oz. hot water dissolve Borax then add Gum Tragacanth and Glycerin. Allow to stand 12 hours, stirring frequently. When gum has formed mucilage add the remaining 44 oz. of water while stirring and strain through muslin.

3. | | |
|---|---|
| Stearic Acid triple pressed | 5 oz. 260 gr. |
| Oil Sweet Almond | 3 oz. |
| Ethyl Amino Benzoate | ½ oz. |

Melt acid and oil together and add Ethyl Amino Benzoate. Stir until dissolved and adjust temperature to 70° C.

After Shave Lotion

Menthol	1	dr.
Boric Acid	2½	oz.
Glycerin	5	oz.
Alcohol	5	qt.
Water, to make	5	gal.
Perfume		

Dissolve menthol in alcohol. Add Boric Acid, perfume, and glycerin. Stir thoroughly until everything is dissolved. Add water. Filter. This preparation may be colored by adding enough color to give shade desired.

STYPTICS
Styptic Pencils

The following are the methods adopted for the manufacture of alum pencils: White: Liquefy 100 gm. of potassium alum crystals by the aid of heat. Remove any scum and avoid overheating, particularly of the sides of the vessel in which liquefaction is being carried out. The molten liquid should be perfectly clear. Triturate a mixture of French chalk in fine powder, 5 gm., glycerin 5 gm. to a paste, incorporate with the liquefied alum and pour into suitable molds. A white appearance can be imparted to the resulting pencils by the addition of more French chalk. Clear: Carefully liquefy potassium alum crystals so as to avoid loss of water of crystallization, adding a small amount of glycerin and water (about 5 per cent) until a clear liquid is obtained. This is poured, whilst hot, into suitable molds, previously smeared with fat. The solidified pencils are rendered smooth by rubbing them with a moistened piece of cloth.

Styptic Powder

An excellent styptic powder results from the mixture of 50% powdered talc and 50% phthalyl peroxide. The latter often contains up to 40% of its weight as phthalic acid; this is beneficial and acts as a stabilizer. The mixture is antiseptic.

Nail Polish

The formulation of a suitable nail polish presents problems peculiar in itself.

The properties desired in the finished product are:

1. Ease of application
2. Drying time
3. Appearance of dry film
4. Permanency

Ease of application is essential. If the polish is too thin, it will tend to flow too readily when applied to the nail, and will give difficulty in securing a smooth even coat. If the polish is too thick, a lumpy, streaky finish will result. In other words, the viscosity of the polish should be such that it will allow an even film to be brushed upon the nail. The drying time should be such that when the nails of the second hand are finished, the coat on those of the first hand should be sufficiently dry to permit the second application. Naturally, this applies only to the so-called "2 coat polishes."

The dry film should present an even appearance, any ridges, streaks, or even pinholes being absent. Finally, a good nail polish should remain on the finger nails for at least 5–7 days with little diminution of its original brilliance, and should show no signs of cracking and peeling.

True solvents, such as acetone, butyl acetate, amyl acetate, etc., give free flowing solutions whose viscosity can be influenced by increased concentration of low viscosity cotton, or by the addition of non-solvents, such as toluene, xylene, etc. Commercial nitrocellulose is manufactured in various viscosities, ½ second, 4 seconds, 15–20 seconds, 40 seconds, etc. However, ½ second regular soluble nitrocellulose generally furnishes the basis of nail polishes. This permits the incorporation of a sufficient solid content, whereas the higher viscosity cottons, even in small quantities, give a much too viscous product.

"Regular Soluble Cotton" is nitrocellulose soluble in acetone, amyl acetate, etc., but not in ethyl alcohol. There is another type of nitrocellulose produced, the alcohol soluble type. This type of cotton is sometimes used in formulating nail polishes where a high alcohol content is desired. However, the film of this type of cotton is not as strong as that of an equivalent amount of Regular Soluble Cotton. Where the incorporation of a large percentage of low boiling solvent is desired, the use of R. S. Cotton and ethyl acetate is preferable.

The solvents most commonly used in nail polishes are: ethyl acetate, absolute denatured ethyl alcohol, butyl acetate, normal butyl alcohol, amyl acetate, glycol ethers (cellosolve, cellosolve acetate, butyl and methyl cellosolve) and acetone oil. The non-solvents are toluene, benzol and VMP Naphtha. Most polishes contain little or none of these non-solvents as they have a disagreeable odor which is objectionable in the finished product. The evaporation rate of solvents is related, in most cases, to their boiling points.

In formulating, it is necessary to make sure that at all times there is sufficient true solvent for the cotton present. Otherwise, although the polish may be clear, the resultant film deposited may be cloudy due to the "throwing out" of solution of the nitrocellulose. The presence of resins further complicates this problem, as the solvents must also be balanced to insure sufficient solvents being present to prevent the resins from being thrown out.

The solvents boiling below 100° C. generally constitute 50% more of the total solvents of a nail polish. This insures a sufficiently rapid evaporation rate. If the solvents are very volatile and the air humid, the rapid evaporation cools the air about the film to below the dew point and the condensing moisture whitens and makes the film opaque. This is commonly termed "blushing." A film that has "blushed" quickly peels. This condition is alleviated by incorporation of small amounts of high boiling solvents that have the property of absorbing the condensed moisture, preventing the precipitation of cotton and resins, and causing the water to evaporate with the constituents of the polish. These compounds are the glycol ethers and acetone oil.

The manner in which nitrocellulose is deposited from solution depends upon the solvent used. In many cases, the resultant film is ridged and rippled. Certain solvents have the ability to "flat" the film and markedly alleviate the above condition. Such solvents are: ethyl alcohol, butanol, cellosolve and methyl cellosolve.

Proxylin solutions have, to a pronounced degree, the property of contracting upon drying, and this causes it to buckle away from the surface to which it has been applied. To prevent this, substances called "plasticizers" are incorporated. These plasticizers are high boiling organic solvents which very slowly evaporate from the film. But small amounts of these substances are used, as their too liberal use would retard the setting of the film. The commonly used ones are castor oil, tricresyl phosphate, dibutyl pthalate, butyl stearate and camphor. The one that gives more plasticizing value, ounce for ounce, than any of the others is tricresyl phosphate. It tends to discolor and blacken with age, changing the color of the polish.

Castor Oil is widely used, but in slight excess it softens the film. Camphor is objectionable because of its odor and the fact that the luster of polishes containing camphor rapidly diminishes and in some cases the surface of the film soon presents a dull pitted appearance. The best of the lot is dibutyl pthalate, as it gives a good plasticizing effect, is stable and relatively odorless.

A number of resins can be used in pyroxylin lacquers and it is here that the formulator has his evident choice as well as one of his greatest troubles. The resins are two types, natural and synthetic.

Many of the natural resins and gums must be treated before incorporation in pyroxylin lacquers, as they contain waxes and other constituents which are incompatible with nitrocellulose. Each has its own treatment, to remove the insoluble matter.

The synthetic resins need no previous treatment before incorporation in lacquers, but in the main they have the drawback that they are colored compounds, yielding lacquers that are suited only for dark colored polishes.

All resins should be used as stock solutions in appropriate solvents, and the solutions assayed for resin strength from time to time thus insuring the proper percentage of this ingredient in the finished product.

All components should be combined on a weight basis as this alleviates any errors due to expansion or contraction. The total solids (cotton, resin and plasticizers) constitute from 10–30% of the polish, depending upon the desired thickness of final film. Plasticizers are added in the ratios of 20–30% of the weight of dry cotton, or 5% of the resin content. Actual mixing is done in glass or tin lined containers. All motors or shafting should be grounded and adequate ventilation provided.

1.
½ Sec. R.S. Wet Cotton	24 oz.
Ethyl Acetate	25 oz.
Butanol	5 oz.
Toluene	48 oz.
Dammar Solution	19 oz.
Cellosolve Acetate	4 oz.
Dibutyl Phthalate	2 oz.
Tricresyl Phosphate	2 oz.
Butyl Acetate	25 oz.

2.
Dry Alcohol Soluble Cotton	12 oz.
Shellac	1 oz.
Castor Oil	1 oz.
Ethyl Alcohol	50 oz.
Ethyl Acetate	20 oz.
Butanol	5 oz.
Amyl Alcohol	6 oz.
Acetone Oil	5 oz.

In coloring polishes, it is best to secure dye stuffs for this particular purpose, from dye stuff houses.

Nose and Throat Spray

(1)
White Mineral Oil	4 ounces
Menthol	5 grains
Camphor	10 grains
Eucalyptol	5 grains

(2)
White Mineral Oil	99%
Ephedrine	1%

Sore Throat Relief

Mix tincture of ferric chloride, 30 gm., alcohol, 30 gm., and potassium chlorate, 60 gm., with sufficient water to make 250 cc.

Antiseptic Inhalant

Eucalyptol	20.0 c.c.
Menthol	7.5 gr.
Oil of Rosemary	10.0 c.c.
Oil of Pine Needles	10.0 c.c.
Oil of Lavender	3.0 c.c.
Oil of Jack Rose Comp.	2.0 c.c.
Brilliant Green	trace
Ethyl Alcohol (S.D.) q.s.	100.0 c.c.

Dissolve the menthol in the oils. Make a strong solution of brilliant green in alcohol. Use enough to give finished product a green tint. Add the remaining alcohol to make 100 c.c.

Antiseptic for Telephone Mouthpiece

1.
Stearic Acid	6.00
Alcohol	20.00

2.
Sodium Hydroxide	1.35
Alcohol	10.00
Water	5.00

Glycerin	5.00
Alcohol	10.00
Fluorescein	0.01
Menthol	1.00
Camphor	1.00
Oil Eucalyptus	5.00
Oil Lavender	5.00

Mix 1 and 2 at 60° C. Then add the remainder and before it cools pour into molds.

----◆----

Antiseptic Toothache Drops

Beechwood Creosote	15 oz.
Oil Clove	30 oz.
Cinnamic Aldehyde or Oil Cassia	20 oz.
Chloroform	30 oz.
Ethyl Amino Benzoate	5 oz.

Mix Creosote with oils and Chloroform then add Ethyl Amino Benzoate and stir until dissolved.

----◆----

Toothache Gum

Yellow Beeswax	60 oz.
Venice Turpentine	10 oz.
Gum Mastic Powder	10 oz.
Ethyl Amino Benzoate	5 oz.
Dragon Blood Powder	10 oz.
Oil Clove	5 oz.

Melt Beeswax and Venice Turpentine together and add Gum Mastic. Stir until dissolved. Then add Ethyl Amino Benzoate and, when dissolved, Dragon Blood. Stir until cooled to about 50° C. then add Oil Clove and mold into sticks.

----◆----

Dental Plate Adhesive
I.

Vanillin	0.5
Boric Acid Powd.	5.0
Powdered Acacia	
Powdered Tragacanth	
of each enough to make 100.0	

II.

Powdered Acacia	
Powdered Agar-Agar	
of each	0.05
Powdered Tragacanth to make 10.00	

In making these preparations, it is essential that all of the ingredients be in the form of a very fine powder.

----◆----

Tooth Paste

Glycerin	41.0	parts
Pure Water	37.0	parts
Calcium Chloride	1.5	parts
Powdered Gum Tragacanth	2.0	parts

Powdered Neutral White Soap	15.0	parts
Calcium Sulfate	82.0	parts
Soluble Saccharin	0.2	parts
Oil Peppermint	2.0	parts

Procedure:

Mix the glycerin and gum tragacanth. Dissolve the calcium chloride in the water and add to the glycerin-gum tragacanth mixture, stir and let stand until the gum is thoroughly hydrated (approximately one hour).

Now mix all the powdered ingredients and sieve thru 40 or 60 mesh and add these and the essential oils to the elixir and mix until the paste is smooth.

The consistency can be changed as desired by adding more or less of the Calcium Sulfate but this should never be changed greatly.

A smoother, creamier paste will be produced if ground thru a paint or ointment mill before tubing.

----◆----

(Acid) Tooth Paste

Glycerin	200.0	parts
Flavor	9.6	parts
* Acid Solution	64.0	parts
Benzoic Acid	0.8	parts
Calcium Chloride	2.4	parts
Cerelose	40.0	parts
Powdered Gum Tragacanth	6.4	parts
Powdered Gum Karaya	7.2	parts
Calcium Sulfate	304.0	parts
Tricalcium Phosphate	90.4	parts
	724.8	parts

* The acid solution is made as follows.

5 parts each of citric, boric, and tartaric acids dissolved in 100 parts cold water.

Procedure:

(a) Mix the glycerin, flavor, acid solution benzoic acid, calcium chloride, and cerelose. Mix for 15 minutes.

(b) Mix the powdered gums, Calcium Sulfate and the Tricalcium Phosphate.

(c) Add (b) to (a) and mix at least two hours.

Mill through a paint or ointment mill before filling tubes.

Flavor is composed of 8.0 parts Oil Peppermint, 1/.1 parts Oil Spearmint, 0.3 parts Menthol and 0.4 parts Oil Cassia.

----◆----

Tooth Paste (Soapless)

Glycerin	30	parts
Powdered Karaya Gum	0.3	parts

Powdered Tragacanth	0.3	parts
Glycosterin	3.0	parts
Crysalba (Swann Calcium Sulfate)	40	parts
Tricalcium Phosphate (Swann)	5	parts
Water	27	parts
Saccharin	.05	parts
Benzoic Acid	1	parts
*Flavor	.5	parts

*Flavor has the following composition:

Oil Peppermint	10	parts
Oil Spearmint	2	parts
Oil Cassia	.2	parts

Melt the Glycosterin. Mix the powdered gums with the Glycerin. Add Benzoic Acid and Saccharin. Finally, the water. Mix for 5 minutes. Heat above melting point of Glycosterin and add to the latter with constant stirring. After mixing for about 5 minutes add the mixed Crysalba and Phosphate with stirring, until a smooth paste is produced. When the temperature is about 30° C. add the flavor mixture with stirring, and pour into tubes.

There seems no special difficulty in the preparation of this paste, and a smoother product will be obtained if the abrasives are mixed into the paste at a temperature sufficiently high to be above the melting point of the Glycosterin. And after all has been added it is passed thru an ointment mill. This paste does not seem to harden in the tube nor become friable after exposure for 24 hours. The flavor can of course be modified to suit individual taste.

----◆----

Tooth Powders

Titanium Dioxide	115 gr.
Calcium Carbonate Heavy	600 gr.
Pulverized Neutral White Soap	100 gr.
Sodium Carbonate Monohydrated	140 gr.
Flavor (Oil of Wintergreen)	18 c.c.

Procedure: Rub up the oil with part of the calcium carbonate until finely dispersed. Add the other ingredients and mix thoroughly. Sift.

Calcium Carbonate	500 gr.
Tricalcium Phosphate	150 gr.
Calcium Chloride	20 gr.
Bicarbonate of Soda	50 gr.
Pulv. Neut. Soap	55 gr.
Confectioner's XXX Sugar	100 gr.
Flavor to Suit	8 gr.

Procedure: Mix the flavoring with the sugar thoroughly. Add the soap and mix again. Add the bicarbonate and the calcium chloride. Mix. Add the tricalcium and the chalk and mix thoroughly and sift.

----◆----

Magnesia Tooth Powder

Magnesium Carbonate	150 oz.
Magnesium Hydrate	600 oz.
Precipitated Chalk	250 oz.
Wintergreen Oil	8 oz.
Eucalyptus Oil	3 oz.
Saccharin	1 oz.

----◆----

Pyorrhea Astringent

Potassium Iodide	15 parts
Iodine Crystals	20 parts
Glycerin	25 parts
Zinc Phenolsulphonate	15 parts
Distilled Water, a sufficient quantity to make	100 parts

----◆----

Mouth Wash

Benzoic Acid	1 lb.
Boric Acid	2 lb.
Borax	1 lb.
Alcohol	1½ gal.
Eucalyptus	3 fl. oz.
Oil of Thyme	1 fl. oz.
Oil of Wintergreen	2 fl. oz.
Water	15 gal.
Caramel Coloring	1¼ fl. oz.

The boric acid and borax are added to part of the water and dissolved by boiling. The solution is cooled by the addition of the rest of the water and left to become quite cold. The benzoic acid is dissolved in half the alcohol, and the essential oils in the remaining half, and the two mixed and added to the water solution. The caramel color is added while stirring, and thorough mixing is continued for four hours.

----◆----

Mouth Wash

Benzoic Acid	12 parts
Tincture or Rhatany	60 parts
Alcohol	400 parts
Oil of Peppermint	3 parts

A teaspoonful in a small wine-glassful of water.

----◆----

Alkaline Mouth Wash

This is made as follows:

Potassium Bicarbonate	21.0 gm.
Sodium Borate	20.0 gm.

Sassafras Oil	1.0 c.c.
Thymol	0.5 c.c.
Eucalyptol	1.0 c.c.
Methyl Salicylate	0.5 c.c.
Cudbear	2.0 gm.
Alcohol	50.0 c.c.
Glycerin	90.0 c.c.
Magnesium Carbonate	10.0 gm.
Water	to 1,000 c.c.

Mix the potassium bicarbonate and sodium borate with 100 c.c. of water. When the effervescence ceases, add this solution to 500 c.c. of water. This is then added to the alcohol in which the essential oils have been previously dissolved. The tincture of cudbear and the rest of the water are next added with the magnesium carbonate. The whole is mixed thoroughly for 2 hours and allowed to stand for 48 hours, chilled, and filtered. Purified talc may be used in place of the magnesium carbonate.

Athlete's Foot Relief

Dissolve one-tenth of a gram of basic fuchsine in 10 grams, and add to it 225 c.c. water, phenol, 5 gm., boric acid, 1 gm., acetone, 5 gm., and resorcinol, 10 gm. Paint the affected parts with this solution, letting it dry in before putting on stockings or shoes.

"Athlete's Foot" Treatment

An anesthetic ointment is liberally applied to all exposed areas. Considerable patience is necessary in the selection of the proper ointment for each case. The following, alone or in various combinations, have proved satisfactory: Ethylaminobenzoate ointment, 10 per cent; nupercaine ointment, 1 per cent; camphorphenol ointment, 1 per cent of each. Following the application of the ointment, a protective gauze dressing is applied. Dressings may be changed once or twice a day, depending on the severity of the case.

Foot Powder

Zinc Stearate	60 gm.
Aluminum Acetate	10 gm.
Menthol	½ gm.

Foot Powder

The ordinary old-time foot powder is composed principally of some such base as talc and starch, together with a little boric or salicylic acid. A modification of this old formula is as follows:

Salicylic Acid	6 dr.
Boric Acid	3 oz.
Powdered Elm Bark	1 oz.
Powdered Orris	1 oz.
Talc	36 oz.

Oxygen-liberating liquids and powders seem to be in favor for cleansing wounds and feet. A typical formula for such a powder is:

Sodium Perborate	3 oz.
Zinc Peroxide	2 oz.
Talc	15 oz.

Solutions for Perspiring Feet

Formic Acid	1 dr.
Choral Hydrate	1 dr.
Alcohol, to make	3 oz.

Apply by means of absorbent cotton.

Boric Acid	15 gr.
Sodium Borate	6 dr.
Salicylic Acid	6 dr.
Glycerin	1½ oz.
Alcohol, to make	3 oz.

For local application.

Hay Fever Ointment

Lanolin Anhydrous	50 oz.
Yellow Petrolatum	25 oz.
Ethyl Amino Benzoate	5 oz.
Menthol	½ oz.
Epinephrin Solution 1-1000	2 oz.
Distilled Water	23 oz.

(1) Triturate Ethyl Amino Benzoate and Menthol with a portion of the Yellow Petrolatum until smooth. Gradually add the remainder of the Petrolatum and the Lanolin.

(2) Mix Epinephrin Solution with Distilled Water and add this aqueous solution slowly under trituration to No. 1 and mix until homogeneous.

Stainless Iodine Ointment

Iodine, in moderately coarse powder	5 parts
Paraffin	5 parts
Oleic Acid	20 parts
Petrolatum	70 parts

Liniment

Camphor Oil	74 oz.
Oil Laurel, Expressed	10 oz.
Oleoresin Capsicum USP (VIII)	5 oz.

Ethyl Amino Benzoate	2	oz.
Camphor Powder	2	oz.
Oil Rosemary	2	oz.
Chloroform	5	oz.
Oil Mustard, USP	½	oz.

Athletic Liniment

Oil of Camphor	25	gm.
Emulsone B	3.5	gm.

Rub together in mayonnaise type mixer and add

Glycerin	7.5	gm.
Water	46.5	c.c.

Allow to soak for 1 hour and while beating add

Glycerin	7.5	gm.
Water	46.5	

Beat intermittently for 1 hour.
This produces a heavy fluid emulsion which is very stable.

White Liniment

Ammonium Linoleate	18	lb.
Water	15	gal.

Stir until dissolved and add

Ammonium Hydroxide	4	gal.
Water	30	gal.

Stir mechanically and add slowly

Turpentine	12	gal.
Oil of Camphor	12	lb.
Cottonseed Oil	8	lb.

Stir 10–15 minutes and add

Ammonium Carbonate	20	lb.

Stir until uniform.

Sore Muscle Liniment

Olive Oil	60	c.c.
Methyl Salicylate	30	c.c.

Mix the two ingredients thoroughly and keep in a well-stoppered bottle until desired. Apply externally.

"Chest Rub" Salve

Vaseline (brown)	1	lb.
Paraffin Wax	1	oz.
Oil Eucalyptus	2	oz.
Menthol Crystals	½	oz.
Oil Cassia	⅛	oz.
Turpentine	½	oz.
Carbolic Acid	⅛	oz.

Melt the vaseline and paraffin wax together, then add the menthol crystals and stir till dissolved. Remove from fire, and while cooling add the oils, turpentine and acid. Pour into one-ounce tin boxes when it begins to thicken. Uses: similar to Vapo-Rub.

Burn Ointment

Picric Acid	80
Ethylaminobenzoate	120
Olive Oil	400
Lime Water	400
Lanolin Anhydrous	2000
Petrolatum to	4000

Rub the picric acid and ethylaminobenzoate together with the olive oil and add the lime water slowly. Work to a smooth emulsion and incorporate this mixture into the anhydrous lanolin and then enough petrolatum to make the required quantity.

Burn Treatment

Gum Tragacanth	30
Gentian Violet (1% sol.)	1000

Allow to swell; warm and stir. Applied to burns this leaves a thin moist, cooling, protective layer and rapid healing results.

Lotion for Hives or Prickly Heat

Menthol	2 gm.
Alcohol	3 oz.
Sodium Bicarbonate	10 gm.
Witch Hazel	3 oz.
Water	to make to one pint

Dissolve menthol in alcohol, add sodium bicarbonate and witch hazel. When dissolved add the water stirring vigorously. Should not be used near the eyes, delicate skin, cuts, etc.

Healing Plaster

Take two parts of beeswax, four parts of pine tar, and four parts of rosin. Melt these materials together, when almost cool mold into sticks.

In its use, it is found best to heat the wax and apply a thin coat on a piece of muslin and place over the injury.

Antiseptic Cure for Poison Ivy

Wash infected parts well with strong soap and water to remove poisonous oils. Also use ether and chloroform or gasoline.

Then apply 5% solution ferric chloride mixed with 50–50 alcohol and water.

Pat generously on infected part.

Calamine Lotion

Calamine	8.00
Zinc Oxide	16.00
Glycerin	15.00
Lime Water	60.00
Rose Water q.s.ad.	120.00

Haemorrhoid Ointment
(Pile Ointment)

Yellow Petrolatum	53 oz.
Lanolin Anhydrous	30 oz.
Yellow Beeswax	5 oz.
Ethylaminobenzoate	5 oz.
Bismuth Subgallate	5 oz.
Thymol Iodide	2 oz.

Melt Yellow Petrolatum, Lanolin and Beeswax together and allow to cool. Mix the three powders and triturate with a portion of the ointment base until smooth. Then add gradually the remainder of the base and mix until ointment is homogeneous. Note: This ointment must not come in contact with iron as discoloration will result so only porcelain or wooden utensils should be used.

Corn Cures: are solutions of Pyroxylin, generally in mixtures of esters and alcohols to avoid the unpleasant hydrocarbon action on the body. An 8 oz. Pyroxylin solution in a mixture of 25% Butyl Acetate, 20% Butanol, 15% Ethyl Acetate and 40% Denatured Alcohol characterizes them. The corn cures contain a small amount of Salicylic Acid and occasionally a trace of Hemp.

Corn Remedy

Acetone	168 oz.
Castor Oil	3 oz.
Venice Turpentine	6 oz.
Celluloid	10 oz.
Salicylic Acid	40 oz.
Ethylaminobenzoate	10 oz.

Dissolve the Salicylic Acid and Ethyl Amino Benzoate in the Acetone. Then add the Castor Oil and Venice Turpentine and finally the Celluloid. Allow this mixture to stand, stirring it now and then until the Celluloid is completely dissolved. Then add sufficient Oil Soluble Chlorophyl to color it dark green.

Emulsion of Liquid Petrolatum with Agar

Heavy Liquid Petrolatum	500.0 c.c.
Agar	5.5 gm.
Sugar	120.0 c.c.

Acacia (fine powder)	30.0 gm.
Tragacanth (fine powder)	4.0 gm.
Tincture of Vanilla	8.0 c.c.
Tincture of Lemon	2.0 c.c.
Oil of Cassia	0.5 c.c.
Water, to make	1000.0 c.c.

Mix the agar and the sugar with 300 c.c. of boiling water and when they are dissolved strain the resulting solution and set it aside to cool. Triturate the powdered gums with the liquid petrolatum, then add the agar solution and whip the mixture with an egg beater. Finally add the tinctures and the oil and lastly enough water to make 1000 c.c.

Mineral Water Crystals

Epsom Salts	29	lb.
Sodium Sulfate	2	lb.
Ferrous Sulfate	½	lb.
Sodium Bicarbonate	30	lb.
Potassium Bicarbonate	10	lb.
Calcium Sulfate	¼	lb.
Ammonium Chloride	2	lb.

All of the above should be powdered; mix and sift twice through a 20 mesh screen. All containers should be dry and airtight.

Medicinal Mineral Salts

A tartar-dissolving mineral salt is Karlsbad Sprudel salt. This is, chemically:

Lithium Carbonate	0.2%
Sodium Bicarbonate	36.1%
Potassium Sulfate	3.1%
Sodium Sulfate	42.4%
Sodium Chloride	18.2%

The Spanish "La Toja" mineral salt, famed for its re-mineralizing action, contains the following as its principal ingredients:

Sodium Chloride	78.3%
Potassium Chloride	7.1%
Calcium Chloride	8.1%
Magnesium Chloride	1.6%
Calcium Sulfate	1.6%
Calcium Bicarbonate	1.9%

In addition, it contains traces of lithium chloride, ammonium chloride, strontium sulfate, iron bicarbonate, magnesium bicarbonate, sodium bromide, and sodium arsenate.

Antiseptic Solution, Double Strength
(Listerine type)

Boric Acid	25	g.
Thymol	1	g.

Eucalyptol	5 c.c.
Methyl Salicylate	1.2 c.c.
Oil Thyme	.3 c.c.
Menthol	1.0 g.
Sodium Salicylate	1.2 g.
Sodium Benzoate	6 g.
Alcohol	300 c.c.
Water	to make 1 liter

"Lysol"-Type Disinfectant
(Phenol Coefficient about 2.5)

Straw Colored Cresylic Acid 50 parts
 (Phenol Coefficient about 5.0)
Sulfonated Castor Oil, Con. 25 parts
25% Caustic Potash
 Solution   15 parts

Add the caustic potash while stirring to a mixture of the other two, and adjust either with alkali or red oil (oleic acid) until a sample dissolved in alcohol is neutral to phenolphthalein.

Feminine Hygiene Jelly
1.

Water	76.85 c.c.
Sodium Chloride	3.00 gm.
Lactic Acid	2.00 gm.
Glycerin C.P.	15.00 gm.
Parachlormetaxylenol	0.10 gm.
Oxyquinoline Sulfate	0.10 gm.
Tragacanth Gum	2.75 gm.

Dissolve the lactic acid and sodium chloride in the water. Add the parachlormetaxylenol and oxyquinoline sulfate to the Glycerin. Warm till thoroughly dissolved, then add the tragacanth and stir till thoroughly mixed. To this, add the salt, and lactic acid solution slowly with hand stirring till cold. Allow to stand overnight, and stir the following day.

If a heavier jelly is required, reduce the amount of Glycerin.

2.

Gum Tragacanth	80 gm.
Boric Acid	55 gm.
Water	1200 c.c.
Glycerin C.P.	60 c.c.
Lactic Acid 85%	13 c.c.

Dissolve Boric Acid in 500 c.c. boiling water. Take up Tragacanth in balance of water; add other ingredients and stir until a gel results.

Antiseptic Vaginal Jelly

| 1. Gum Tragacanth | 6 |
| 2. Glycerin | 10 |

| 3. Water | 100 |
| 4. Boric Acid | 5 |

Mix 1 and 2 and add 3 and 4 slowly with stirring; let stand overnight.

Smelling Salts

Phenol	1
Menthol	1
Camphor	2
Weak Solution of Iodine (2.5%)	1
Oil of Pumilio Pine	1
Oil of Eucalyptus	1
Strong Solution of Ammonia	3
Ammonium Carbonate	90

The ammonium carbonate should be packed into the bottle, the strong solution of ammonia added, then the other ingredients, previously mixed. Sodium sesquicarbonate is sometimes substituted for ammonium carbonate.

Migraine Salve

Ten parts beeswax and 46 parts anhydrous lanolin are melted and 180 parts distilled water added. Mass is well mixed and then mixture of 15 parts menthol, 16 parts methyl salicylate and 2 parts rosemary oil are worked in and uniform salve obtained. In another preparation 5 parts menthol are dissolved in 6.5 parts acetic ester, 4.2 parts absolute alcohol, 1.85 parts triple strength ammonia liquor and solution is worked up into salve with 45 parts anhydrous lanolin, 36.5 parts white petrolatum and perfumed with 0.5 part lavender oil and 1 part essence of eau de cologne.

Migraine Pencil

| Stearic Acid | 70 |
| Menthol | 30 |

Melt together on water bath and cast in molds.

Menthol Pencil or Crayon

Menthol	100
Benzoic Acid	10
Eucalyptol	3

Melt together and cast in forms.

Aspirin Tablets

Aside from other properties acetylsalicylic acid tablets must have good appearance and must dissolve rapidly in the stomach. Such tablets are made with base of 240 parts pulverized arrow-root starch and 240 parts heavy magnesium oxide. Base is well mixed and screened. Then it is moistened with solution of coconut oil, 10 parts in about 400 parts ether, and moistened mass screened again. Powder is spread on paper and ether evaporates. Acetylsalicylic acid, 2000 parts, are added and mixture carefully mixed to perfect homogeneity. Then it is mixed with acetone as required, about 30 parts to 250 parts powder. After drying and heating for 2 hours at 50° C., 2530 parts of the granulated mass are mixed with 30 parts pulverized agar-agar, 60 parts arrow-root starch and 80 parts pulverized talc. When unit of weight used is gram, 4000 tablets can be prepared from final mixture, each tablet weighing 0.7 gram and containing 0.5 gram of acetylsalicylic acid. To prevent powder from tablets from penetrating into lower die on tablet-making machine, latter is covered with cotton threads impregnated with paraffin oil.

Artificial Petrolatum

| Ceresin or Paraffin | 15–20 |
| White Mineral Oil | 85–80 |

Wart Remover

| A. Salicylic Acid | 2 |
| Glacial Acetic Acid | 20 |

| B. Trichloracetic Acid | 90 |
| Water | 10 |

Aseptic and Analgesic Dusting Powder for Wounds

Urea Crystals	80 oz.
Ethyl Amino Benzoate	5 oz.
Thymol Iodide	5 oz.
Boric Acid Powder	5 oz.
Bismuth Subgallate	5 oz.

Mix and grind in a ball or pebble mill and sift through a No. 120 mesh sieve. Fill into cans with sprinkler top.

Cesspool Deodorant and Disinfectant

1.

Calcium Oxide (QuickTime) Powder	10 lb.
Chlorinated Lime (Bleaching Powder)	2 lb.
Potassium Carbonate	1¼ lb.

Directions: Mix and sift thoroughly the three ingredients. Put up in airtight metal or glass containers.

Sprinkle liberally over the excrement which will liquefy and deodorize. The colloidal matter will be completely destroyed.

Note: Be careful not to get the compound in contact with the skin.

2.

Calcium Hydroxide	9 lb.
Ammonium Chloride	6 lb.
Water	1 gal.

Directions: Mix the Calcium Hydroxide with the water. This is best done by placing a pail, containing the gallon water in cold water bath before the calcium hydroxide is added as a great amount of heat is evolved when the calcium is added to the water. When cold, add the Ammonium Chloride and stir until completely dissolved. Pour off the clear solution after settling and keep in tightly closed containers.

Pour a liberal amount over the excrement which will completely liquefy and deodorize.

Note: Be careful not to get the compound in contact with the skin.

Pine Oil Disinfectants

Pine Oil Disinfectants are commonly made according to the Hygienic Laboratory Formula:

Parts by Weight

Pine Oil	1000
"I" Wood Rosin (Acid Number—165)	400
Sodium Hydroxide (25% Solution)	200
	1600

It is prepared in the following manner: The Pine Oil and "I" Wood Rosin are heated together at a temperature of 80° C. in a jacketed steam kettle, the degree of heat is maintained until the rosin is thoroughly dissolved in the Pine Oil. The temperature is then dropped to 60° C. at which point the Sodium Hydroxide (25% solution) is added by stirring in very slowly. Saponification should be complete in thirty (30) minutes. This product has a predicted phenol coefficient of 3.5 to 4 determined by the Food and Drug Act Method against B-Typhosus.

The following formula was developed using a vegetable oil soap base:

| Vegetable Oil Soap Base | 20% |
| Pine Oil | 80% |

Pine Oil is added to the vegetable oil soap and stirred in slowly. No heating is required for this blend. This product has a predicted phenol coefficient of 5.2 determined by the Food and Drug Act Method against B-Typhosus.

The following label has been approved for disinfectants by the Government:

Pine Oil Disinfectants

Active Ingredients

Pine Oil	1
Soap or Base	2

Inert Ingredients

Moisture	3

(Moisture not to exceed 10% of total.)

Food and Drug Act Test—Phenol Coefficient (4). (Fill in blanks (1)—(2)—(3)—(4) to correspond with the disinfectant manufactured.)

Directions

In the bathroom.—To wash the bathtub, basin and toilet, apply in a 1 to 40 dilution in water.

In public places. — Schools, Hotels, Theatres, Stores, Office Buildings, Colleges, etc. Spray freely one part to forty parts of water.

In garbage receptacles.—To check the development of putrefactive action and breeding of flies. Spray the receptacle with a 1 to 40 dilution in water.

In the stable.—To help promote sanitation and destroy stable odors. Spray a 1 to 40 dilution in water.

In kennels, chicken houses, etc.—To kill lice, spray a 1 to 40 dilution with water on roots and dropboards; to kill fleas, wash dogs in a 1 to 40 dilution in soapy water.

The Government has strict regulations to prevent labeling a product as a disinfectant if an adulterant is present.

Manufacturers should have a representative sample of their disinfectant tested for determination of phenol coefficiency.

The above procedures, if followed, insure the manufacturer of having a disinfectant labeled within the Government regulations.

A steam-distilled Pine Oil Disinfectant made according to the prescribed rules and regulations insures the following:

1. Has a clear sparkling amber color.
2. Produces a snowy white emulsion in water.
3. Does not burn body tissues.
4. Is non-corrosive and non-toxic to humans.
5. Does not stain when in diluted form.
6. Leaves a clean piney odor wherever applied.

7. Kills typhoid, scarlet fever, diphtheria and cholera germs, etc.
8. Is free from suspended matter. This denotes uniformity.
9. May be used as an antiseptic for minor cuts and bruises as a wet dressing.

Deodorant Spray
For theatres, lavatories, etc.

Pine-needle Oil	
Formalin	of each 2 oz.
Acetone	6 oz.
Isopropyl Alcohol	to 20 oz.

For use as a spray 1 oz. is mixed with a pint of water.

Fumigating Pastilles

Charcoal	62
Cascarilla Bark	17
Gum Benzoin, Siam	15
Saltpetre	6

The above should be powdered finely and thoroughly mixed and then sprayed oils such as sandalwood, patchouli, vetiver, cassia or other suitable perfuming materials. Usually 10–15% of perfume is used to this powder. As a binder an acacia mucilage is used before molding Drying is best done slowly.

Fly Catching Mixture

Rosin	56
Ester Gum	1
Heavy Mineral Oil	40

Melt together and stir until dissolved. Remove from heat and stir in

Glycerin	2½
Honey	1½

Fly Paper

Rosin	32
Rezinel No. 2	20
Castor Oil	8

Heat above and stir until uniform. Apply hot to suitable paper.

Increasing rosin content gives a heavier faster drying coating. Decreasing rosin gives a thinner stickier coating which remains sticky for longer periods.

Fly Paper

Water	21
Glucose	16
Sodium Silicate	11
Glycerin	½

First soak coated paper in a weak

alum solution; dry and then coat with above.

Fly Paper

Rosin	32 gm.
Flexoresin El	20 gm.
Castor Oil	8 gm.

Melt together, and dip paper into warm mixture.

Fly Spray

1.

This is made by macerating 500 gms. of pyrethrum with 4 liters of kerosene (followed by expression) after 24 hours. Perfume by adding 90 c.c. of methyl salicylate to each 4 liters of solution.

Pyrethrum	240 gm.
Kerosene	2000.0 c.c.
Gasoline	2000.0 c.c.
Naphthalene	30.0 gm.

Macerate the pyrethrum in the petroleum liquids for 48 hours, then strain, express and then add the naphthalene.

2.

Deodorized Kerosene	89
Methyl Salicylate	1
Pyrethrum Powd.	10

Percolate a few times and filter.

Moth Killer

1.

Ethylene Dichloride	74 parts
Carbon Tetrachloride	25 parts
Paradichlorbenzene	1 part
and Diglycol Oleate	1 part

2.

Camphor 10, naphthalene 40, capsicum 100, oil of cloves 10, turpentine 100 and alc. 900 parts are macerated for 48 hours and strained.

3.

(Non-Staining)

Sodium Aluminum Silico-fluoride	0.52
Water	98.48

4.

Sodium Fluoride	0.5
Sodium Taurocholate	0.2
Carbon Dioxide	
	to saturation point of water
Water	100

5.

Paranitro Chlorbenzol	10–20
Paradichlorbenzol	80–90

Codling Moth Bands

Bands are treated with a solution obtained by heating

Beta Naphthol	1 lb.
Red Engine Oil (300 sec.)	1½ pt.
Aluminum Stearate	½ oz.

Rodent Poison

Strychnine	0.55
Saccharin	0.15
Flour	98.30

Strychnine	0.35
Anise Oil	0.15
Sugar	20.50
Flour	79.00

Non-Poisonous Rat Destroyer

Gypsum	100
Rye Flour	300

Dry thoroughly in oven and add 0.1 oil of anise. Keep in air-tight containers.

Mouse Exterminator

Barium Carbonate	100
Oatmeal	300
Saccharin	1
Water	enough

Make a stiff dough, force through a coarse sieve, and dry in an oven.

"Silverfish," Poison for

White Arsenic	30 gm.
Flour	500 c.c.
Water	to make paste

Insecticide (Bed Bugs)

Cresol	3 fl. oz.
Dichlorbenzene	13 fl. oz.

Use one pint of this mixture to five pints kerosene.

Insecticide, Bed Bug

Kerosene	90
Clovel	5
Cresol	1
Pine Oil	4

Bed Bug Exterminator

Insect Powder	150
Colocynth	50
Phenol	50
Oil of Turpentine	100
Alcohol	1000

Macerate the crude drugs in the alcohol for eight days, express, and filter, then add the phenol and oil.

Bed Bug Killer

Kerosene	96–98
Phenol	4– 2

Use as spray in cracks and on springs.

Bed Bug Spray

Lysol	1 oz.
Carbon Tetrachloride	75 parts
Refined Kerosene	25 parts

Mix.

Insecticide for Mexican Bean Beetle

Spray with

Barium Silicofluoride	5 lb.
Water	50 gal.

Cockroaches, Exterminant for

(1)

	Parts by Weight
Powdered Borax	4
Flour	2
Chocolate Powder	1

(2)

	Parts by Weight
Powdered Borax	10
Insect Powder	1
Starch	1

(3)

	Parts by Weight
Kieselguhr	22
Sodium Fluoride	40
Sodium Chloride	10

The ingredients in the finest powder are thoroughly mixed and the powder sprinkled about runs of the insects.

(4) Freshly burnt plaster of Paris and fine oatmeal (dry) in equal parts are thoroughly mixed and the powder is dusted around places infested by roaches.

Insect Powder (Cockroach)

Powdered Borax	8 lb. 10 oz.
White Hellebore	8 oz.
Dalmatian Powder	8 oz.
Ground Cloves	4 oz.
Cayenne Pepper	2 oz.

Hay Fever Remedies

Formula No. 1

Ephedrine (Dried)	0.1 g.
Petrolatum, Liquid	10 cc.

Use as nasal spray.

No. 2

Ephedrine Sulphate	1 g.
Calcium Lactate	4 g.

Place in No. XXX capsules; use one or 4 times daily.

Sea-Sickness Remedy

Antipyrin	4 g.
Sodium Bromide	8 g.
Sugar	2 g.

Use once every three hours.

Appetite Stimulant

Tincture of Capsicum	2 cc.
Tincture of Nux Vomica	16 cc.
Tincture of Gentian Compound	72 cc.

Dose: Three teaspoonfuls daily.

Bronchitis Inhalant

Menthol	½ g.
Chloroform	4 cc.
Tincture of Benzoin	120 cc.

Inhale twice daily, using one teaspoonful to pint of boiling water.

Menthol Inhalator

Eucalyptus Oil	4 cc.
Menthol	2 g.
Paraffin Oil	94 cc.

Disinfectant for Telephones

Solution 1

Oil of Wintergreen	0.5 g.
Oil of Eucalyptus	0.25 g.
Denatured Alcohol	15 g.

Solution 2

Formaldehyde	25 cc.
Water	225 cc.

Add solution 1 to solution 2 and dilute with water to 1000 cc.

Laryngitis Spray

Thymol	0.15 g.
Menthol	1.2 g.
Eucalyptus Oil	3 g.
Petrolatum, Liquid	300 cc.

Tonsilitis Gargle

Potassium Chlorate	8 g.
Tincture Ferric Chloride	12 cc.
Glycerin	60 cc.
Water	240 cc.

CHAPTER V

EMULSIONS

ORDINARY emulsions may often be made by shaking (in a bottle) the proper proportions of oil, emulsifying agent and water. Thick emulsions are made by mixing with a fork or an egg-beater. When using the latter it must not be "run" too quickly or much air will be beaten in which will give a false body which will sink later. For best results on emulsions made with emulsifiers other than gums, an electric malted milk mixer or a propeller attached to a shaft on a small motor should be used. For emulsions using gums a paddle mixer (of the mayonnaise or ice-cream freezer type) is best employed. With the latter type long and intermittent beating is desirable.

Emulsions should not be kept at temperatures above the boiling point of water or below freezing as the former will cause evaporation of the water and the latter will freeze the water. Both of these conditions will cause "breaking" of the emulsion with a resultant separation of the ingredients.

Theory

Since the theory and practice of emulsions is still in a highly disorganized state the theoretical side will be touched on but lightly.

An emulsion may be considered as a homogeneous suspension of tiny droplets of oil in water or water in oil. The oil in water type may be represented by the usual furniture polish (milky) and the water in oil type by butter. The term "oil" includes oils (mineral, vegetable, animal or essential), fats, greases, waxes, hydrocarbons (benzol, naphtha, turpentine, etc.), synthetics (ethylene dichloride, nitrobenzol, etc.)—that is, something which does not mix with water.

Emulsification formulae and methods have been evolved chiefly through practice—by actually making innumerable emulsions. Because of the vagaries and eccentricities of emulsions practical workers have made greater technical advances in this field than the pure research chemists. Too often the trained chemist does not achieve as good emulsions as the lay worker—because the former rebels instinctively against empirical formulae and does not follow instructions as implicitly as the man "who knows he doesn't know." Moreover each new emulsion represents a new problem having numerous variable factors. These should not be underestimated if a good stable emulsion is desired. The technique and preparation of any particular formula should first be mastered before any variations are attempted.

Methods

Just as one man's food may be another's poison—so one method, which will give a perfect emulsion in one case, may produce a perfect failure in another. Thus no one method or emulsifying agent will serve universally. Specific technique will be given later in the case of the different emulsifying agents recommended.

When an emulsion of a solid melting above 100° C. is desired, it should first be melted with sufficient solvent or oil to reduce the combined melting point below 100° C. For example naphthalene with naphtha or other hydrocarbons; synthetic resins with hydrocarbons or vegetable oils.

Uses

Technical emulsions are used in numerous ways in many fields. The following are but a few of a large number of uses. Polishes, beauty creams, lotions, water-proofing, agricultural sprays, mayonnaise, cleaning compounds, lubricants, etc. Many new specialty emulsions are likewise being created.

Summary

It must be borne in mind, however, that perfect results cannot be gotten until a few experimental emulsions are made in order to become familiar with working conditions. That is why experience shows that one of the given formulae should be mastered before attempting any variations.

Variations in raw materials, procedure, errors in proportions, etc., produce poor results. The formulae given have been repeated many times and will work if they are strictly adhered to.

Of course these formulae cannot fill every individual requirement. Variations are therefore necessary. In order to work out successful formulae, patience is essential. That which is worth while getting is worth while striving for. It is suggested that only one ingredient or proportion be varied at a time. This enables one to know exactly what produces the change in the finished product.

Emulsifying Agent
Ammonium Linoleate Paste

A cream-colored paste; ammoniacal odor.

This is an excellent agent for emulsifying vegetable and fish oils, waxes, fat, resins, hydrocarbons and many other water insoluble products. When emulsifying a water insoluble product having a melting point of over 100° C., the latter should be first dissolved in naphtha, ethylene dichloride, turpentine or similar solvent. Alcohol as a rule should not be used as it breaks down most emulsions. Similarly acids, esters and salts must be avoided.

Procedure

Using proportions given in the following table, first dissolve the indicated amount of water in the Ammonium Linoleate Paste. This is done by covering the Ammonium Linoleate with the required amount of water and allowing it to soak over night. Work in slowly the next day until dissolved completely. Do not attempt to dissolve in any other way or lumps will result. To this add slowly with vigorous agitation the indicated amount of oil and continue stirring until homogeneous.

When a wax is to be emulsified the wax is melted and considered as an oil. In this case the water must be heated above the melting point of the wax. Most trouble is encountered in making wax emulsions because the solution of Ammonium Linoleate in water and the melted wax are not heated sufficiently. To play safe keep each of these solutions between 95 and 100° C., not allowing the temperature to drop below the melting point of the wax while adding one to the other. These formulae have been repeated numerous times with uniformly good results. If your emulsion is grainy or forms a film of wax on the surface, then the fault is in manipulation and not in the Emulsifier. Good wax emulsions cannot be made by hand or with a slow moving paddle. The vigorous agitation of a fast electric stirrer is essential.

Emulsions of the various inflammable hydrocarbons produce products of high cleansing powers and of a much higher flash-point.

In many synthetic reactions where better contact is desired between an aqueous and a water insoluble liquid recourse is had to emulsions. Similarly a water soluble solid may be dissolved in water and then emulsified with the water insoluble liquid.

Formulae

(All parts by Weight)

No.	Material Emulsified	Parts	Parts of Water	Parts Ammonium Linoleate Paste
1.	Kerosene..........	90	90	8
2.	Naphtha..........	90	100	7
3.	Benzol...........	90	100	7
4.	Gasoline..........	90	100	7
5.	Pine Oil..........	90	90	10
6.	Carnauba Wax.....	90	620	12
7.	Beeswax..........	90	500	12
8.	Ozokerite.........	90	400	14
9.	Turpentine........	90	100	8
10.	Nitrobenzol.......	90	100	8
11.	Orthodichlorbenzol..	90	100	8
12.	Methyl Salicylate...	90	100	8

The above formulae can be lessened in cost by reducing the amount of emulsifier used. The minimum can be determined by experiment. Increasing the amount of water will give thin emulsions. Certain oil emulsions are improved by the addition of 1% or so of ammonia dissolved in water when making the emulsion.

Emulsifying Agent
Trihydroxyethylamine Stearate
(T. S. for short)

A light brown wax. Faint fatty odor. In the formulae given below proceed as follows:

Melt the T. S. with the oil and add this to the water (some prefer to use warm water) slowly while stirring vigorously with an electric mixer. Warm water and very rapid stirring produce uniformly stable emulsions.

Formulae

	Material Emulsified	Parts	Parts Water	Trihydroxy-ethylamine Stearate
A.	Mineral Oil.......	75	185	15
B.	Pine Oil..........	75	85	14
C.	Turpentine........	75	85	14
D.	Paraffin Wax......	85	200	10
E.	Eucalyptus Oil....	75	85	14
F.	Balsam Copaiba...	75	85	14
G.	Gasoline..........	75	85	14

Emulsifying Agent
Di-glycol Stearate

A light colored wax. Practically odorless (m.p. 55–56° C.). This is non-alkaline.

One part of Di-glycol Stearate when melted in 10–30 parts of boiling water produces, on stirring, while cooling, a uniform milky dispersion of the wax in water which is very stable. The consistency varies with the amount of water used. They may be also used as lubricants to be squirted between spring-leaves or other inaccessible places. On evaporation of the water a film of non-flowing wax remains behind as a lubricant. These make excellent suspending media for titanium dioxide, carbon black, graphite, silica and other abrasives.

Formulae

A		10	Pine Oil.....40	Water	40
B	Di-glycol	10	Mineral Oil..50	Water	500
C	Stearate	10	Paraffin Wax.40	Water	250
D		10	....	Water	50
E		10	....	Water	300

Procedure

The oil or wax is melted with the Di-glycol Stearate. The water is heated to a temperature above the melting point of the wax and added slowly while stirring vigorously. Continue stirring until cool. By varying the amounts of water, emulsions of varying consistency are obtained. They are very white in color and stable. Other oils and waxes may be emulsified in a similar way.

Formulae (A), (B), (C), all useful as polishes.

Formula (A) serves as a liniment, disinfectant or deodorant. The pine oil may be replaced by turpentine, citronella oil or perfume compounds.

Formula (B) with a little perfume dissolved in the oil makes an excellent lotion or liquid cleansing cream.

Formula (E) with a little perfume is used as a lotion or powder base.

Formula (D) serves as a greaseless ointment in paste rouge base.

Fuel or Lubricating Oil Emulsion

Fuel or Lubricating Oil	88
Triethanolamine Oleate	6
Oleic Acid	5½
Water	90

Lard Oil Emulsion

Lard Oil	88
Triethanolamine Oleate	9
Oleic Acid	4
Water	76

Soya Bean Oil Emulsion

Soya Bean Oil	86
Triethanolamine Oleate	6
Oleic Acid	6
Water	85

Sperm Oil Emulsion

Sperm Oil	82
Triethanolamine Oleate	6
Oleic Acid	6
Water	82

Black Wax Emulsions

To color any non-edible wax emulsion black, stir into it, with a high-speed mixer about 10 parts colloidal carbon black per every 100 parts of wax present in the emulsion.

Rosin Emulsions

Rosin	700 gr.
Water	2100 cc.
Glue	150 gr.

Melt glue in water and while boiling hot, slowly add melted rosin, agitating violently. Continue agitation until perfectly smooth.

Rosin	700 gr.
Water	2100 cc.
Gelatine	150 gr.

Melt Gelatine in water and while boiling hot, add melted rosin slowly, agitating violently. Continue agitation until perfectly smooth.

Rosin, Turpentine Emulsion

Rosin	11.0 gm.
Turpentine	2.5 gm.
Ammonium Linoleate	2.0 gm.
Water	50.0 cc.
Ammonia	15.0 cc.

The ammonium linoleate and water are taken up in the usual way. Heat and mechanically agitate (high-speed mixer). The rosin and turpentine are heated together and added to the ammonium linoleate dispersion in water to which has previously been added the 15 cc. of ammonia. Stirring is continued until cool. This gives a paste emulsion.

Soluble Oil Emulsions

The soluble oil method is particularly applicable for medium viscosity mineral oils and is not successfully applied to other oils or solvents. With such mineral oils, however, the method yields excellent emulsions which are quite stable. These oils usually require from 3.5 to 4.0 per cent Triethanolamine, depending upon the stability desired in the emulsion. The amount of oleic acid lies between 8 and 11 per cent, the amount varying especially with the type of oil. The more refined oils are the most difficult to emulsify as will be seen from the following table:

Soluble Oils

Type of Oil	Color	Oil	Oleic Acid	Triethanolamine
Cutting Oil...	Yellow	88 lb.	8.0 lb.	3.7 lb.
Textile Oil....	Bloom	87 lb.	8.8 lb.	3.5 lb.
Medicinal Oil.	White	86 lb.	10.0 lb.	4.0 lb.
Rayon Oil....	White	85 lb.	10.4 lb.	4.0 lb.

Formulation by this method requires great exactness, and it is always necessary to derive formulae for the specific oil to be emulsified because of the great variation in commercial petroleum products. Given an unknown oil, take 88 grams, add 8.0 grams of oleic acid and stir to a clear solution. Now measure carefully 4.0 grams of Triethanolamine into this solution and stir thoroughly. On holding this mixture up to the light, it will usually be cloudy or show minute suspended droplets. Now add oleic acid drop by drop, stirring thoroughly after each addition until the mixture becomes

clear. It will now emulsify in water, but a few drops further of acid will give a slightly superior soluble oil. The total oleic acid can now be calculated and the whole formula reduced to the basis of 100 pounds.

Olive Oil Emulsions

Olive Oil	88 lb.
Oleic Acid	10 lb.
Triethanolamine	2 lb.
Water	80 lb.

Preparation

Working at ordinary temperatures add the Triethanolamine, oleic acid and 30 lb. of the olive oil to the agitator. As soon as these three ingredients have been added, but not before, stir vigorously until the mixture is fairly homogeneous. Then slowly add with constant stirring 33 lb. of water, obtaining a thick smooth emulsion.

Continuing with the same stirring rate, first add the remainder of the oil in small portions, and finally the remaining water in a similar manner. Emulsification is complete when the oil and water are evenly distributed.

Pine Oil Emulsion

Pine Oil	91 lb.
Oleic Acid	6 lb.
Triethanolamine	3 lb.
Water	100 lb.

Preparation

Add the oleic acid, Triethanolamine and 30 lb. of the pine oil to the mixer and stir until the product is clear. Then add very slowly an equal volume of water, stirring vigorously meanwhile.

When this mixture has become a smooth uniform emulsion, the remainder of the oil is gradually added with constant agitation. The rest of the water is next similarly added until emulsification is complete.

Light Mineral Oil Emulsion

Mineral Oil	88 lb.
Oleic Acid	8.0 lb.
Triethanolamine	3.7 lb.
Water	

Formulation

The above formula was derived for a particular low viscosity lubricating oil and is typical of the formulation for a cutting oil.

Preparation

Weigh out the oleic acid and 8 lb. of the mineral oil and stir together to obtain a uniform solution. Then add the exact amount of Triethanolamine and stir until the solution is clear. Some warming will occur during the reaction of the acid and amine.

This soluble oil base is dilutable with the remainder of the oil at any time. Simply stir the remaining 80 lb. of the oil into the base, or four parts by weight of the oil to one part of the base.

------◆------

Linseed Oil Emulsion

Linseed Oil	88 lb.
Oleic Acid	10 lb.
Triethanolamine	2 lb.
Water	80 lb.

Preparation

Working at ordinary temperatures, thoroughly mix the oleic acid, Triethanolamine and 30 lb. of the linseed oil. Add 33 lb. of water to this mixture slowly with constant, vigorous stirring. This procedure yields a thick, smooth emulsion.

The remainder of the oil is then added in small portions, maintaining the same stirring rate, and the rest of the water is added similarly. Stirring is discontinued as soon as the last of the water has been evenly dispersed.

------◆------

Kerosene Emulsion

Kerosene	89 lb.
Oleic Acid	8 lb.
Triethanolamine	3 lb.
Water	100 lb.

Preparation

The preparation of this emulsion is typical of the procedure used for any liquid. In one container weigh out the above quantities of kerosene and oleic acid and mix these two liquids thoroughly. In a separate container stir together the water and Triethanolamine until a homogeneous solution is obtained.

The oil solution is now poured into the water solution, and the resulting mixture is stirred or agitated vigorously. After the emulsion is well formed, it should be stirred occasionally, a few minutes at a time.

Carnauba Wax Emulsion

Carnauba Wax	87 lb.
Stearic Acid	9 lb.
Triethanolamine	4 lb.
Water	400 lb.

Preparation

Weigh out the stearic acid, water and Triethanolamine, and heat the mixture in a kettle to 100° C. After the acid has melted completely and the solution is boiling gently, stir carefully until the acid has been dissolved and a smooth soap solution is obtained.

In a separate steam-heated container melt the carnauba wax until a temperature of 85–90° C. is reached. Do not allow the temperature to rise above 95° C., or the wax will be darkened in color. Now add the molten wax to the boiling soap solution and stir vigorously until an even dispersion of the wax results. Stir gently, but continuously, until the emulsion has cooled to room temperature.

------◆------

Carnauba Wax, Kerosene Emulsion

Carnauba Wax	16.0 gr.
Kerosene	20.0 cc.
Ammonium Linoleate	2.4 gr.
Water	200.0 cc.

The ammonium linoleate was placed in a vessel and covered with the water (cold) and allowed to stand overnight. The following day it was warmed and stirred until completely dispersed in the water, taking care that no lumps were left. This was taken to 90° C. and stirred by means of a high speed mixer. The wax was melted, taken to 100° C., and the kerosene added and stirred until the wax was dissolved in it. This was then added to the hot ammonium linoleate dispersion and the agitation continued until the emulsion was cool. This gave a fluid emulsion.

------◆------

Carnauba Wax, Mineral Oil Emulsion

Mineral Oil (Spindle)	19 cc.
Carnauba Wax	18 gr.
Ammonium Linoleate	2.4 gr.
Water	102 cc.

The ammonium linoleate and water were allowed to stand overnight as above. Then heated to 90° C. and stirred by means of high-speed mixer. The wax and oil were heated together until the wax dissolved in the oil, and taken to 100° C. This solution was then added to the ammonium linoleate dispersion in water, and stirred rapidly. This gave a paste emulsion.

Paraffin Wax Emulsion

Paraffin Wax	88 lb.
Stearic Acid	9 lb.
Triethanolamine	3 lb.
Water	300 lb.

Preparation

Mix the water, Triethanolamine and stearic acid and heat to 100° C., allowing the mixture to boil gently. Then stir carefully so that a smooth soap solution is obtained with a minimum of foam. In a separate container melt the paraffin wax and bring its temperature to 90° C. Add the hot wax immediately to the boiling soap solution and stir vigorously until the wax is evenly dispersed. Continue to stir the emulsion slowly while cooling.

Emulsifying Agent
(Potassium Oleo-Abietate)

A viscous paste; resinous odor. Alkaline reaction.

Used in place of Turkey Red or Sulfonated oils where an acid product is undesirable. For making ''soluble'' oils.

The following formulae give clear solutions without heating. When these solutions are thrown into water they diffuse rapidly to give milky emulsions.

A. Pine Oil	6 lb.
Potassium Oleo-Abietate	1 lb.
B. Pine Oil	5 lb.
Kerosene	1 lb.
Miscibol	1 lb.
Water	1 lb.

Asphalt Emulsions

Asphalt	500 gr.
Water	500 cc.
Bentonite	30 gr.
Quebracho	30 gr.
Soda Ash	10 gr.

Combine bentonite, Quebracho, soda ash and water and heat to 200° F. While stirring, add asphalt which has been heated to approximately 200° F. Continue stirring until asphalt is dispersed.

Asphalt	2800 gr.
Water	2800 gr.
Rosin Soap (50%)	118 gr.
Pine Oil	40 cc.

Add rosin soap to water and heat to 200° F. Heat asphalt to 200° and add pine oil. While agitating, slowly pour asphalt into water and continue agitating until a smooth emulsion is formed.

Lanolin Emulsion (Fluid)

Diglycol Oleate (Light)	10 gm.
Lanolin (Anhydrous)	30 gm.

Warmed till dissolved. Added to the above with rapid agitation

Water	60 cc.

made slightly alkaline with Caustic soda (¼%). Stir five to ten minutes.

Paradichlorbenzene Emulsion

Paradichlorbenzene	12 gm.
Glycol Stearate	3 gm.
Water	150 cc.

Melt the glycol stearate in the water (about 90° C.). Stir rapidly (high-speed mixer). Melt the paradichlorbenzene, preferably on water bath and add slowly to the stearate dispersion in water. Continue stirring until cool.

Raw Tallow Emulsion (50%)

Raw Beef Tallow (Good Quality)	80–100 lb.
Trihydroxyethylamine Stearate	9 lb.
Water	90–100 lb.

(6–8 ounces of Trisodium phosphate added to water may prove advantageous if water used is of a high degree of hardness.)

This is a substitute on an equal basis for commercial 50% Sulfonated Tallow in sizing preparations.

Cresylic Acid Emulsion

Potassium Abietate	10
Cresylic Acid	20

Warm together and dissolve by stirring. Add water as desired while stirring.

CHAPTER VI

FOOD PRODUCTS, BEVERAGES AND FLAVORS

D URING the last decade the kitchen has moved from the home to the factory. Many food products formerly made at home are now purchased by the housewife. Most of the products originated in small kitchen "factories" and to-day our food factories are merely enlarged kitchens run most efficiently, hygienically and economically. The products which they produce are legion—everything from "soup to nuts."

Included in the beverage section are over six hundred formulae for making divers alcoholic liquors. Non-alcoholic beverage formulae are likewise given.

Mayonnaise

1.

Whole Eggs	4	
Egg Yolks	16	
Liquid Pectin	2½	oz.
Mustard Powder (yellow)	¼	oz.
Sugar	1½	oz.
Salt	1	oz.
Vegetable Oil	1	gal.
Mayonnaise Flavor	2	cc.
Tincture Capsicum (optional)	4	cc.
Lactic Acid	4	cc.
Vinegar	6½	oz.
Water	6½	oz.

2.

Cottonseed Salad Oil	70.25
Egg Yolk	10.00
Vinegar (50 grain)	10.00
Water	3.90
Salt	1.45
Sugar	3.50
Mustard	0.80
White Pepper	0.10

This formula gives good resistance to freezing, keeps well and has good flavor and appearance.

3.

Egg Yolk	8	oz.
Vinegar	8	oz.
Sugar	1¾	oz.
Oil	96	oz.
Salt	1½	oz.
Mustard	½	oz.
Water	10	oz.

Build up and run on colloid mill.

Vanilla Bean Flavoring Powder

Ground Vanilla Bean	25 parts
Confectioners Powdered Sugar	74 parts
Oil of Bitter Almond	1 part

Mix the above ingredients very thoroughly. Place in sifter top cans and use as powdered flavor over ice cream, cereals and baking.

Vanilla Sauce Powder

Corn Flour	100
Vanillin	0.5
Yellow Food Color	0.05

Butter Substitute

1. Water	120
2. Carob Bean—Best Quality Powder	1
3. Cottonseed Oil	40
4. Caustic Soda	0.02
5. Butter Flavor	to suit

Dissolve 4 in 1 and strew 2 on surface; bring to a boil while stirring; run 3 and 5 into it slowly with high speed intermittent stirring.

Coffee Substitute

Coffee Bean Powdered	33
Sugar Powdered	5
Roasted Peanuts Powdered	62

Butter Coloring

A solution of annatto in oil constitutes probably the best and most satisfactory butter coloring. The following formula may be used:

| Annatto, Powder | 3 oz. |
| Cottonseed Oil | 16 oz. |

Mix, heat to 212° F. for some time, set aside for 24 hours, strain and filter.

Purified annatto, or annattoin, yields a still finer preparation, somewhat less being required in proportion, although "strength" is a matter for individual judgment.

The following compound annatto powder is also largely used:

Annattoin	5 oz.
Turmeric, Powder	6 oz.
Saffron, True	1 oz.
Lard Oil, Odorless	16 oz.
Alcohol	4 oz.

Rub the annattoin and turmeric with the oil, which may be deodorized by filtration through charcoal and macerate for several days. Prepare a tincture with the alcohol and saffron. After sufficient maceration separate the solids from the oil by filtration, adding more oil through the filter to keep up the quantity, and mix with it the tincture of saffron, afterwards driving off the alcohol with gentle heat.

Household Baking Powders

1.

Sodium Bicarbonate	28 parts
Mono Calcium Phosphate	35 parts
Corn Starch	27 parts

Mix the above powders thoroughly and store in airtight containers.

2.

Sodium Bicarbonate	28 parts
Calcium Acid Phosphate	29 parts
Sodium Aluminum Sulfate	19 parts
Starch Corn	24 parts

3.

Sodium Bicarbonate	28 parts
Mono Calcium Phosphate	22 parts
Sodium Aluminum Sulfate	21½ parts
Starch Corn	38½ parts

4.

Sodium Bicarbonate	28 parts
Sodium Aluminum Sulfate	28 parts
Corn Starch	44 parts

Bakers' Baking Powder

5.

| Sodium Bicarbonate | 35 parts |
| Mono Calcium Phosphate | 9 parts |

| Sodium Aluminum Sulfate | 29 parts |
| Corn Starch | 27 parts |

6.

Sodium Bicarbonate	35 parts
Sodium Aluminum Sulfate	35 parts
Corn Starch	30 parts

7.

Sodium Bicarbonate	35 parts
Calcium Acid Phosphate	36 parts
Sodium Aluminum Sulfate	24 parts
Starch Corn	5 parts

8.

Sodium Bicarbonate	28 parts
Sodium Acid Pyrophosphate	20 parts
Mono Calcium Phosphate	22 parts
Corn Starch	30 parts

9.

Sodium Bicarbonate	27 parts
Cream of Tartar	60 parts
Corn Starch	13 parts

10.

Sodium Bicarbonate	27 parts
Cream of Tartar	45 parts
Tartaric Acid	6 parts
Corn Starch	22 parts

In all these formulas the powders are of course mixed thoroughly.

Ice Cream Roll Cake

Eggs	3½ lb.
Sugar	4½ lb.
Salt	1½ oz.
Invert Sugar	¾ lb.
Cocoa	1 lb.
Soda	¾ oz.
Milk	3½ lb.
Baking Powder	2¼ oz.
Flour	5¼ lb.

Bring the cocoa, milk and invert sugar to a boil and let cool. Dissolve soda. Then whip the sugar, eggs and salt until fairly stiff. Blend the flour with baking powder and add alternately with the cocoa solution. Bake lightly at about 425° F.

The Manufacture of Buttermilk from Skimmed Milk

The finest quality of buttermilk is probably that produced by churning clean-flavored cream which has been properly ripened with the aid of a pure culture of lactic acid. Surplus skimmed-milk, may, however, in many cases, be profitably converted into an artificial buttermilk of practically the same composition and quality as the natural buttermilk.

In making artificial buttermilk the skimmed-milk may or may not be pasteurized. In either case about 10% of clean flavored lactic acid culture should be added to the skimmed-milk which is maintained at a temperature of 70° F. until coagulation takes place. If the time required to produce coagulation is too long the process should be hastened by increasing the percentage of culture used, rather than by raising the temperature. Raising the temperature above 70° F. will usually result in a product of inferior flavor.

As soon as coagulation has taken place the curdled milk is transferred to the churn which is revolved for thirty to forty minutes as in churning cream. If the skimmed-milk is allowed to stand long after coagulation takes place before being churned, the whey and curdy matter of the finished product will show a greater tendency to separate. The churning breaks the curd into fine particles producing a smooth velvety buttermilk which is difficult to distinguish from a good natural product. As soon as the artificial buttermilk is drawn from the churn it should be strained to remove any particles of curd which may not have been broken up in the churning process. The temperature of the product should at once be reduced to at least 50° F. to retard the development of acidity and of undesirable flavors.

Artificial buttermilk may also be satisfactorily produced in a small way in the home. A clean fruit jar of suitable size may be partially filled with clean fresh skimmed-milk which is allowed to sour naturally at a temperature of 70° F. to 75° F. When coagulated, the milk should be vigorously shaken for a few minutes in the closed jar. It may now be strained to remove any lumps of curd not finely broken up by the agitation after which it should be kept in a cool place. If a clean pleasant flavor is obtained by such natural souring and the artificial buttermilk is to be made frequently, it is advisable to add a few ounces of the first artificial buttermilk to the next quantity of skimmed-milk to be soured. Thus the desirable flavor may be reproduced from time to time in the same manner as yeast is propagated.

The composition of such artificial buttermilk is practically the same as that of natural buttermilk, the only difference being that the latter usually contains slightly more milk fat. The percentage of milk fat in the artificial buttermilk may be increased to approximately that of natural buttermilk by adding to each one hundred pounds of skimmed-milk before souring, two quarts of whole milk.

Yogurt or Bulgarian Buttermilk

Propagate a small culture of the Bacillus Bulgaricus from day to day as indicated for the lactic culture for buttermilk. This culture may be obtained from various commercial laboratories. To prevent contamination by yeasts or gas-forming bacteria, it is necessary to carry this culture at a temperature of about 110° F. A small egg incubator may be used for this purpose.

Carry in a similar way a culture of the ordinary sour-milk organism, which may be obtained from many of the commercial laboratories.

Thoroughly pasteurize the milk to be fermented. If a small quantity—5 to 10 gallons, for instance—is to be made, it may be done by holding a can of milk in a tub or vat of water heated by a steam hose. If a larger quantity is made, one of the starter cans used in creameries will be found convenient These are essentially cylindrical vats with mechanical stirrers and a jacket which can be filled with steam for heating or water for cooling. The milk should be held at a temperature of at least 180° F. for not less than 30 minutes.

Cool the milk to about 100° F. Draw off one-half and inoculate it with the culture obtained in the second operation. Inoculate the remaining half with Bulgaricus culture obtained in the first operation. The amount to be added will depend on the quantity of milk to be fermented, the time at which it is desired to have it curdled, and the temperature maintained during the fermentation. This can best be determined by experience. One pint should be sufficient for any amount between 10 and 20 gallons.

Kefir or Koumiss

Use buttermilk or freshly curdled sour milk. This should be thoroughly agitated to break the curd into fine particles. Buttermilk containing Bacillus Bulgaricus will give a flavor too acid for most tastes.

Add 1% cane sugar (1½ oz. to the gallon). Add a small amount of yeast cake—one-fourth of a cake will be sufficient for 1 gallon of buttermilk. The yeast cake should be ground up in water so that it will be well distributed.

Bottle this preparation, leaving sufficient space to permit a thorough shaking of the contents. Strong round bottles of the type used for carbonated drinks should be used, as considerable pressure is developed by the fermentation. If the bottle is not provided with a sealing device the corks must be securely tied or wired in place.

Hold for 4 or 5 days at a temperature of 65 to 70° F., shaking every day to keep the curd well broken up. At the end of this time there should be considerable gas but not enough to blow the milk out of the bottle. It should have a pleasant acid taste with a slight bitterness. The fresh milk sometimes has a yeasty taste but this gradually disappears. If the milk is kept on ice it will remain in good condition for two weeks or more.

Churned Buttermilk

Catering to the ideas of certain individuals who believe that the products and practices of our childhood are better than those of today, many dealers have placed on the market within recent years a type of fermented milk termed churned buttermilk. This product has been made in numerous ways, but in general there are three methods.

Probably the more common method is to ripen thoroughly and pasteurize a 2% milk to about .75% acidity. The ripened milk is then churned at a sufficiently high temperature to produce butter granules in the usual length of time or even shorter. The churning is stopped when the granules are about the size of small rice grains. The buttermilk is then pumped over a cooler and bottled. If butter coloring is added to the milk before churning a more distinct granule will be obtained.

The second method is to churn a good grade of highly colored cream until butter granules of desirable size are secured. The granules are then chilled in a 40° F. room until firm and are then added to starter that has been cooled to at least 50° F., in sufficient quantities to be visible in the bottle. A small quantity of cream added to the starter will improve the flavor. The main objection to this method is the fact that the finished product lacks the buttermilk flavor. Its

main advantage is in the reduced volume of cream that must be churned.

Another method is to ripen 8–10% pasteurized sweet cream to an acidity of about .35%. Butter color is added and the mixture churned until granules of the proper size are secured. Enough cooled starter is then added to bring the fat content down to about 1%. This gives a product of good flavor and fairly light body. The advantage of this method over the first is the greater ease of churning and the reduced volume of cream that must be handled in the churn.

Buttermilk Lemonade

A refreshing and nutritious drink may be made by the addition of lemon juice and sugar to buttermilk, following the same procedure as in making ordinary lemonade. It will usually be found necessary to use more sugar and more lemon juice than in making lemonade with water. Buttermilk lemonade should be served very cold.

Malted Milk Powders

Powdered Malt Extract	50 parts
Powdered Skimmed Milk	20 parts
Cane Sugar	30 parts

Mix well. One teaspoonful when added to 8 ounces of a mixture of chocolate syrup, milk and ice cream and then mixed with the malted milk machine will make a delicious malted milk drink.

Carbonated Milk

The best results are secured when newly pasteurized milk or cleanly drawn fresh milk is treated with carbon dioxide in a tank, such as is used in bottling establishments in preparing carbonated drinks, and then placed in siphon bottles. When charged under pressures of from 70 to 175 pounds and kept at temperatures ranging from 35° to 60°, bottles of clean fresh milk or pasteurized milk kept from four to five months without perceptible increase in acidity.

Milk carbonated under a pressure of 70 pounds comes from the bottle as a foamy mass, more or less like kumiss that is two or three days old. It has a slightly acid, pleasant flavor, due to the carbon dioxide, and has a somewhat more salty taste than ordinary milk. In the case of carbonated milk pasteurized at 185° F., there is, of course, something of a "cooked" taste. Though the cream separates in the bottle, it is thoroughly remixed by a little shaking as the milk comes from the bottle and there is no appearance of separate particles of cream. All who have had occasion to test the quality of carbonated milk as a beverage agree in regarding it as a pleasant drink. In the case of milk bottled under a pressure of 150 pounds of carbon dioxide, the milk delivered from the siphon is about the consistency of whipped cream, but, on standing a short time, it changes into a readily drinkable condition. From the experience had, it would seem that carbonated milk might easily be made a fairly popular beverage.

Infants Milk

To make cow's milk more easily digestible by bottle-fed babies—one level tablespoon gelatine for each quart of milk is used. The gelatine is soaked for 10 minutes in ½ cup of cold milk taken from formula, then placed in boiling water and stirred until dissolution. Then add remainder of the milk.

Jelly Powders: In the manufacture of flavored gelatine, 10 parts gelatine is mixed with 85 parts sugar to which flavor, color and tartaric acid 2 parts are used to sharpen the flavor.

Gelatin in Ice Cream and other Food Products: ½ of 1% gelatine in ice cream prevents the formation of ice crystals by acting as an emulsifying agent and improves the texture and body of the finished product.

Cultured Milk

Three different organisms are commonly used in the manufacture of cultured milk drinks in this country. The most common product is that made by souring milk under control conditions with pure cultures of S. lacticus. Some manufacturers prefer a heavy body and a sharper flavor which they secure by adding a small proportion of L. bulgaricus starter to that made with S. lacticus. For the acidophilus drink a third organism is used called L. acidophilus. All three of there starters can be secured from any commercial culture laboratory.

In some cases no butterfat is added, but a much more palatable product can be secured by the addition of sufficient cream to make a total fat content of 1–2 per cent.

Essential to Have Good Starters

Probably the most essential requirement for the successful manufacture of cultured milk is that the starter be kept pure. This means that proper facilities must be available for growing the cultures, and a competent person must be in charge. Even with the best of care, starters occasionally "go off" and need to be replaced with new stock.

Mother cultures should be grown in the laboratory. From these mother cultures the bulk cultures can be set. In no case should the attempt be made to carry starters by transferring from one vat or can to another. The transfer should be carefully made, using only sterile equipment, from the mother culture to what is to be the next mother. Since the preparation of the three starers varies somewhat each one will be considered separately.

Preparation of S. Lacticus Starter

A. Mother culture.

1. Use only high quality skim milk.
2. Place milk in glass container such as fruit jar and heat to 190° F. for 30 minutes.
3. Cool slowly to 72° F.
4. Using sterile spoon or pipette transfer about 10 cc. of the last mother culture to each quart of the sterilized milk. Cover bottle immediately.
5. Incubate at about 72° F. for about 18 hours or until curd is well set up.
6. Place in 40° F. room until used.

B. Bulk starter.

1. Use only high quality skim milk.
2. Heat to 180° F. for 30 minutes.
3. Cool to 72° F.
4. Add 1½–2 per cent of the mother culture and mix well.
5. Incubate at about 72° F. for 18 hours or until acidity of about .75 per cent is reached.
6. Break curd and cool immediately to at least 50° F. by pumping over surface cooler.

Preparation of L. Bulgaricus Starter

A. Mother culture.

1. Use only high quality skim milk.
2. Place milk in glass container such as fruit jar and heat to 190–200° F. for 30 minutes.
3. Cool slowly to 100° F.
4. Using sterile spoon or pipette transfer about 10 cc. of the last mother cul-

ture to each quart of the sterilized milk. Cover bottle immediately.

5. Incubate at 100° F. for about 18 hours or until firm curd is formed.

6. Place in 40° F. room until used.

B. Bulk starter.

In case only small quantities of bulgarlac are to be made it will not be necessary to prepare any bulk starter of the bulgaricus culture, as a sufficient amount of the mother culture can be prepared to supply the quantity needed to mix with the lactic starter. Otherwise proceed as follows:

1. Use only high quality skim milk.

2. Heat to 190° F. for 30 minutes.

3. Cool to 100–105° F.

4. Add 1½–2 per cent of mother culture.

5. Hold at 100° F. for 18 hours or until acidity of about 1.00 per cent is obtained.

6. Break curd and cool immediately to at least 50° F. by pumping over surface cooler.

Occasionally bulgaricus starter is sold for a cultured milk drink, but its flavor is so sharp and its body so viscous that it is better to mix it with the lactic culture. A desirable drink can be prepared by adding one part of the bulgaricus to nine parts of the lactic culture together with the amount of cream necessary to supply 2 per cent fat in the finished product.

This product has the advantage of a distinct acid flavor, a smooth and fairly heavy body with little tendency to whey off.

Preparation of L. Acidophilus Starter

The preparation of acidophilus cultures requires considerable care as slight contamination will ruin the culture.

A. Mother culture.

1. Sterilize selected milk in autoclave by heating to 240° F. for 15 minutes.

2. Cool to 100° F.

3. Add about 10 cc. of mother culture using sterile pipette. Cotton plug should be flamed before returning to flask.

4. Incubate at 100° F. for 18 hours.

5. Use immediately if possible; otherwise store at about 50° F.

Acidophilus cultures should be examined microscopically occasionally to make sure the culture is pure.

B. Bulk starter.

1. Use selected milk.

2. Heat to boiling or slightly higher for 30 minutes.

3. Cool to about 100° F., hold 30 minutes and again heat to boiling for 10 minutes.

4. Cool to 105° F.

5. Add 1–1½ per cent mother culture.

6. Incubate at 105° F. for 18–20 hours or until an acidity of about .70 per cent is reached.

7. Cool as rapidly as possible to 50° F. Care must be taken to keep the temperature up to at least 100° F. during the incubation period. All possible sources of contamination should also be controlled as the culture must remain pure. These factors are so important that specially constructed vats are necessary for the successful manufacture of acidophilus milk on a commercial basis.

L. acidophilus cultures may be stored at 40° F. or lower for several days without affecting the number of living organisms.

Sour Cream

Commercial sour cream sometimes called Jewish cream, is the heavy bodied, smooth textured product of high acid flavor secured by processing and ripening sweet cream under control conditions. It is used as a spread for bread, as a dressing for vegetables, and in the making of sauces of various kinds.

There are several successful methods for preparing sour cream. Variations in plant equipment and plant conditions make it impossible to suggest a method applicable to all plants. Three general procedures will therefore be given.

A. Method for making sour cream without the use of a viscolizer or homogenizer.

   1. Using enzyme

      a. Pasteurize the cream (18–20 per cent fat) by heating to 175° F. for 30 minutes.

      b. Cool to 85° F. and add 3 per cent starter and .5 cc. of rennet (diluted with 30 volumes of water) to each 100 pounds of milk.

      c. Pour cream into shotgun cans.

      d. Incubate at 85° F. until a firm curd is formed.

      e. Cool rapidly without stirring by placing can in ice water or 40° F. room.

This is a fast method for making a sour cream of good body.

2. Using cheese curd
   a. Pasteurize 32 per cent cream by heating to 145° F. for 30 minutes and cool to 72° F.
   b. Add 3 per cent starter and incubate at 72° F. for 18 hours.
   c. Mix 4 parts of soured cream with 1 part of cottage cheese curd and 1.5 parts of good starter which have been previously mixed and strained to' remove curd particles.

3. Using skim milk powder
   a. Add 3 per cent skim milk powder to 20 per cent cream.
   b. Raise temperature gradually to 145° F. with constant stirring. Hold 30 minutes at 145° F.
   c. Cool to 72° F. and add 3 per cent starter and ⅓ cc. rennet (diluted with 3 volumes of water) to each 100 pounds of milk.
   d. Place in shotgun cans.
   e. Incubate 15 hours at 72° F.
   f. Cool without stirring by placing can in ice water or 40° F. room.

B. Method of making sour cream, using viscolizer or homogenizer.
   1. Pasteurize 18–20 per cent cream at 180° F. for 30 minutes.
   2. Homogenize at 180° F. using 3,000 pounds pressure on one valve. (Be sure homogenizer is thoroughly washed and sterilized previous to use.)
   3. Cool to 72° F. and add 3 per cent starter.
   4. When acidity of .6–7 per cent has been reached package and store at 40° F.

A slightly heavier body can be secured by adding 2 or 3 per cent of milk powder to the cream; or enough concentrated skim milk to increase the serum solids 2–3 per cent; or .25 per cent of high grade gelatin.

A better body can also be secured by ripening the cream in the final container if such a procedure can be made practical.

◆

Milk and Cream, Increasing Viscosity of

To increase the viscosity and improve the consistency of milk or cream, the material is heated to 40–42° C. in 20–30 minutes, cooled to 2–3° in 20–30 minutes and held at 2–3° for 1–2 days.

Brick Cheese

Perfectly sweet milk is set in a vat at 86° F. with sufficient rennet to coagulate it in 20 or 30 minutes. The curd is cut with Cheddar curd knives, is then heated to 110° or 120° F., and is stirred constantly. The cooking is continued until the curd has become so firm that a handful squeezed together will fall apart when released. The curd is then dipped into the mold, which is a heavy rectangular box with a bottom and with slits sawed in the sides to allow drainage. The mold is set on the draining table, a follower is put on the curd, and one or two bricks are used on each cheese for pressure. The cheeses are allowed to remain in the molds for 24 hours, when they are removed, the entire surface rubbed with salt, and the cheeses piled three deep. The salting is done each day for three days, after which the cheese is taken to the ripening cellar, which should be comparatively moist and have a temperature of from 60° to 65° F. Ripening requires two months.

◆

Brie Cheese

This is a soft, rennet cheese made from cows' milk. The cheese varies in size and also in quality, depending on whether whole or partly skimmed milk is used. The method of manufacture resembles closely that of Camembert.

The milk used is usually perfectly fresh. It is not uncommon, however, to mix the evening's milk, when kept cool overnight, with the morning's milk. Some artificial coloring matter is added to the milk, which is then set with rennet at a temperature of 80° or 85° F. After standing undisturbed for about two hours, the curd is dipped into forms or hoops, of which there are three sizes in common use. The largest size is about 15 inches in diameter, the medium size about 12 inches in diameter, and the smallest size about 6 inches in diameter, all varying in height from 2 to 3 inches. After drainage for 24 hours without pressure being applied, the hoops are removed, and the surface of the cheese is sprinkled with salt. Charcoal is sometimes mixed with the salt used. The cheese is then transferred to the first curing room, which is kept dry and well ventilated. After remaining in this room for about eight days the cheese becomes covered with mold. It is then transferred to the second curing room or cellar, which is usually very dark, imperfectly ventilated and has a temperature of about 55° F. The cheese remains there for from two to four weeks, or until the consistency and odor indicate that it is sufficiently ripened. The red coloration which the surface of the cheese finally acquires has been attributed to an organism designated *Bacillus firmaticus*. The ripening is due to one or more species of molds which occur on the surface and produce enzymes, which in turn cause a gradual and progressive breaking down of the casein from the exterior toward the center. The interior of a ripened cheese varies in consistency from waxy to semiliquid and has a very pronounced odor and a sharp characteristic taste.

Brinza Cheese

This cheese from sheep's milk, or a mixture of sheep's and goats' milk.

The cheese is made in small lots, from 2 to 4 gallons of fresh milk being used at one time. This is put into a kettle and when the temperature of the milk is from 75° to 85° F. sufficient rennet is added to obtain coagulation in 15 minutes. The curd is broken up and the whey dipped, and the curd is placed in a linen sack and allowed to drain for 24 hours. It is then cut into pieces and placed on a board, where with frequent turnings it is allowed to remain until it commences to get smeary, which requires about eight days. The pieces are then laid one on top of another in a vessel holding from 40 to 60 pounds, where they remain for 24 hours, after which they are removed, the rind cut away, and the curd or partially cured cheese broken up in another vessel. After 10 hours salt is stirred in and the curd run through a mill, which cuts it very fine, when it is packed in a tub with beech shavings.

Camembert Cheese

This is a soft, rennet cheese made from cow's milk. A typical cheese is about 4¼ inches in diameter, three-quarters of an inch or 1 inch thick, and in the market in this country is usually found wrapped in paper and inclosed in a wooden box of the same shape. The cheese usually has a rind about one-eighth of an inch in thickness, which is composed of molds and dried cheese. The interior is yellowish in color and waxy, creamy, or almost fluid in consistency, depending largely upon the degree of ripeness.

Camembert cheese is made from whole milk or from milk slightly skimmed. It is not advisable to skim the milk unless it tests more than 3.5 per cent butterfat. The temperature of setting is from 78° to 87° F., and the quantity of rennet added for this purpose is sufficient to get the desired degree of firmness in from two to five hours. The curd is then transferred, usually with as little breaking up as possible, to perforated tin forms or hoops about 4¼ inches in diameter and the same in height. These rest upon rush mats, which permit it free drainage. The filling of the forms may be done at two or three different times, short intervals being necessary for the curd to settle. Each form holds the equivalent in curd of about 2 quarts of milk. After draining for about 18 hours, preferably in a room having a uniform temperature of 65° or 70° F., the cheese is turned. This is repeated frequently for about two days, when it is removed from the forms and salted on the outside. After 24 hours the cheese is carried to the curing rooms, which are maintained at temperatures of from 53° to 59° F. and with a high relative humidity. Curing the cheese is the most difficult part of the manufacturing process, for not only must there be a uniform and progressive development of the ripening agents, but the curd must be gradually desiccated at the same time. Proper conditions of humidity and temperature must be maintained and subject to regulation in order to favor the development of the needful mold, *Penicillium camemberti*, the bacteria, and yeasts. Although the growth of the mold is necessary in order to bring about a gradual breaking down of the casein, this growth should not be too vigorous and luxuriant; otherwise the product will be rendered unfit for commercial purposes. Following the growth of the mold, other organisms develop, giving the resultant cheese a reddish appearance instead of a white and blue, as is the case in the initial mold fermentation. From 15 to 20 days are required to bring about the proper balance between the various forms of life. At the end of that time the cheese is allowed to complete its ripening at the lower limits of the indicated temperatures and with a minimum of ventilation.

Cheddar Cheese

The milk, morning's and evening's mixed, is set at 85° F. with sufficient rennet to coagulate to the proper point in from 25 to 40 minutes. At the time of setting the milk should have an acidity of about 0.18 or 0.20 per cent. Color may or may not be used. The curd is cut when it breaks evenly before the finger. The cutting is done with curd knives made up of blades set about one-third of an inch apart in frames. In one frame the knives are set perpendicularly and in the other horizontally. When well cut the curd is in uniform cubes of about one-third of an inch.

After being cut, the curd is heated slowly and with continued stirring until it reaches a temperature of from 96° to 108° F. With the use of mechanical agitators, as is the common practice, the curd should be heated about 4° higher than when stirring is done by hand. After heating, the stirring is continued intermittently until the curd is sufficiently firm. This is determined by squeezing a handful, which should fall apart immediately on being released. The whey is then drawn. At the same time the acid should have reached about 0.20 per cent, or one-fourth of an inch, the latter of which is determined by measuring the length of strings when the curd is touched to a hot iron. The curd is then matted about 4 inches deep, sometimes in the bottom of the vat, sometimes on racks covered with a coarse linen cloth. After it has remained there long enough to stick together it is cut into rectangular pieces easy to handle, which are turned frequently and finally piled two to four deep; in the meanwhile the temperature of the curd is kept at about 90° F. When the curd is broken down until it has the smooth feeling of velvet, which requires from one to three hours, it is milled by means of a machine, which cuts it into pieces the size of a finger. It is then stirred on the bottom of the vat until whey ceases to run, which requires from one-half to one and one-half hours, when it is salted at the rate of 2 or 2½ pounds of salt to 100 pounds of milk. It is then ready to be put into the press. The curd is put into tinned-iron hoops of the proper size, which are lined with cheesecloth bandages. The hoops are put into presses and great pressure is applied by means of screws. The next morning the cheese is removed from the hoops and put on shelves in a curing room. Formerly it was kept in a curing room as long as six months, but at the present time it is covered with a coat of paraffin and put into cold storage when from 3 to 12 days old. There is a growing demand on the part of consumers for mild cheese, and consequently ripening must be carried on at a temperature below 50° F.

An important point in the process of manufacturing Cheddar cheese is the development of the desired quantity of acid, which is responsible for the proper breaking down of the curd before milling and salting. The maximum quantity of acid that can be developed in the whey without injuring the texture of the cheese should, therefore, be aimed at. It is very probable that too much weight has been placed on the desirability of a maximum development of acid, and that practically as good cheese can be produced without the high acid.

Some of the details in the manufacture of Cheddar cheese are varied to some extent, and other names may be used to designate the cheese so made. A stirred-curd cheese is one in which the curd particles are not allowed to mat together after the whey is drawn. The curd is stirred occasionally to prevent this matting process, but it differs from the sweet-curd cheese, as acid is allowed to develop before salting and pressing. Formerly a comparatively large quantity of stirred-curd cheese was made, but very little, if any, is made at the present time.

A washed-curd cheese varies from the regular Cheddar process in having the milled curd subjected to cold water for a short period. This process is evidently practiced to force the curd to take up a small percentage of the water and increase the yield. It results in a cheese which apparently breaks down or ripens much more rapidly than cheese made in the ordinary way. This ripening is very likely not due to the excess of moisture but to some other unexplained reason. Some States have prohibited the use of the State brand on washed-curd cheese.

----◆----

Cheshire Cheese

This cheese is one of the oldest and most popular of the English varieties. It is a rennet cheese made from whole milk of cows, and is named for Chester County, England, where it is largely produced. It is made in cylindrical shape, from 14 to 16 inches in diameter, and weighs from 50 to 70 pounds. In making this cheese sufficient annatto is used to give the product a very high color. The process of manufacture varies in detail in different sections. Perfectly sweet milk, night's and morning's mixed, is set at a temperature of from 75° to 90° F. In one hour, the curd is cut usually with an instrument in which knives are set in a frame to cut cubes 1 or 1¼ inches square. This is pushed down through the curd and finally worked back and forth at an angle. This is continued for about an hour, or until the particles of curd are the size of peas. The curd is then allowed to settle and mat on the bottom of the vat for about an hour, when it is rolled up to one end, weighted down, and the whey drawn, after the desired degree of acidity has been obtained. The curd is cut in pieces of the right size to handle and is piled on racks. It is then run through a curd mill, salted at the rate of 3 pounds to 1,000 pounds of milk, and put into a hoop having a number of holes in the side, through which skewers can be thrust into the cheese to promote drainage. The cheese in the hoop is put into a heated wooden box called an oven, and sometimes light pressure is applied, the pressure increasing gradually until it reaches about 1 ton. The curing cellar or room is about 60° to 65° F. The time required for thorough ripening is from 8 to 10 months.

Cottage Cheese

Cottage cheese is sometimes made with a small amount of rennet, and the curd is heated to from 118° to 125° F. It may be made on a small or a factory scale. With this method the skim milk is pasteurized, cooled to 70° or 80°, and 1 to 5 per cent of a starter added. Rennet is then added at the rate of 1 c.c. per 1,000 pounds of milk. The curd is allowed to develop an acidity of about 0.55 in from 6 to 10 hours. The coagulum is then cut into ½-inch cubes. Water at a temperature of 115° is run over the curd in about an hour and the temperature of the wash water then gradually raised to 120°. The curd is then stirred until it will stand without breaking. It is then gradually cooked to a temperature of 118° to 126° in the course of one and one-half to three hours. When the curd may be squeezed in the hand and still retain its shape, the whey is withdrawn and the curd is washed

two or three times in cold water. After the washing the water is withdrawn, and the curd ditched along the side of the vat or kettle, and drained for one hour. It is then placed in a cooler for 12 hours. To each 100 pounds of curd, 70 pounds of a mixture of milk and cream containing 10 per cent cream is added. The curd is then stirred for a few minutes. After creaming the cheese is placed in a cooler at 30° to 40° until ready to use or ship.

When the cheese is made on a factory scale a drier product is desired in order that it may be marketed successfully. For this reason the curd is generally cooked at a higher temperature than when made on a smale scale. The main equipment necessary for making cottage cheese on a factory scale is a pasteurizing outfit and a channel-bottom Cheddar vat. Ordinarily from 5 to 10 per cent of a good lactic starter is added to skim milk, after which the milk is allowed to ripen at a temperature of 70° to 80° F. until curdled. The curd is then cut into cubes and gradually heated to from 115° to 125° in 30 to 45 minutes. When the whey has been removed, the curd is washed with cold water, drained, and piled along the sides of the vat. Ordinarily the cheese is salted at the rate of 3 to 4 ounces per 100 pounds of milk. Often the cheese is mixed with cream and then marketed in small, single service, paraffined paper containers, or in butter tubs.

With milk of a good quality a yield of 15 to 18 pounds of cheese per 100 pounds of skimmilk is obtained. Cottage cheese should always be kept in a refrigerator or in a cooler until disposed of.

Manufacturing Cream Cheese (Hot Process)

The new method of manufacturing cream cheese involves a new principle; namely, the aggregation of the fat globules into large clusters by proper homogenization. This is accompanied by a partial coagulation of the casein in these fat clusters so that the entire mass sets to a permanent condition which is not materially affected by temperature.

Sweet cream of good flavor containing 40 to 42 per cent of milk fat is the basis for this cheese. From 3 to 5 per cent of soluble dry skimmilk is stirred into the cream. Then 0.5 to 0.7 per cent of finely ground agar free from objectionable flavor or odor should be added to this mixture while it is being constantly stirred.

The mixture should then be heated to 180° to 185° F. and held for 5 to 10 minutes for the agar to dissolve. It should then be cooled to 110° F. Add 0.75 per cent of common salt and 0.5 to 1 per cent of good commercial starter depending upon the rate at which acidity is desired in the cheese. The mixture should then be passed thru a coarse strainer and homogenized at 3,000 to 4,000 pounds pressure per square inch. The mixture should leave the homogenizer at the consistency of soft butter and slightly firmer than ice cream as it leaves the freezer.

The mixture should be placed immediately into the final molds before the temperature lowers to 100° F. or less because the finest body and texture is secured if the cheese is not mixed after the agar has set. The cheese can be chilled in the refrigerator to 70° and then placed in a 70° room for 10 or 15 hours for the acid flavor to develop.

The quantity of acid developed in the cheese can be varied not only by the percentage of starter but by the quantity of dry skimmilk. The more dry skimmilk the higher the acidity will be. Acid develops somewhat slowly in this cheese so that it may be necessary to increase the percentage of starter under special conditions.

When relish, olives, etc., are mixed with the cheese it is generally not necessary to use starter since the relish gives plenty of tartness and flavor to the cheese. The quantities used vary from 10 to 30 per cent. The cream can be homogenized at 120° thus making it possible to pack a much warmer cheese with less danger of the agar congealing before packing. It is desirable in such cheese to use fully 5 per cent of dry skimmilk to help prevent any whey drainage. If there is much juice from the relish it may be desirable to add it to the warm cheese before homogenization but such a procedure increases the acidity in the cream thereby causing excessive fat clumping. This may be offset by the use of lower homogenization.

Cream Cheese

Genuine cream cheese is made from a rich cream thickened by souring or from sweet cream thickened with rennet. The cream for this cheese should always be pasteurized. This thickened cream is put into a cloth and allowed to drain, the cloth being changed several times during the draining, which requires about four days. It is then placed on a board covered with a cloth, sprinkled with salt, and turned occasionally. It is ready for consumption in from 5 to 10 days.

Another variety of cream cheese is made from cream with a low content of butterfat (6 or 8 per cent). A small quantity of a lactic-acid starter is added to the cream, and after the mixture is warmed to from 70° to 76° F. and thoroughly stirred, rennet is added at the rate of from 1 to 1½ ounces of commercial liquid rennet to 1,000 pounds of cream. Usually the cream is placed in shotgun cans holding about 30 pounds each. After setting for about 18 hours, the curd is poured, with as little breaking as possible, upon draining racks covered with cloths. After a few hours' draining the cloths are drawn together, tied, placed upon cracked ice, and allowed to remain overnight. The curd is then pressed, salted, and worked to a paste by means of special machinery or by suitable substitutes. The cheese is then molded into pieces weighing from 3 to 4 ounces, wrapped in tin foil and, without curing, placed upon the market. The standard package of cream cheese is 3 inches by 2 inches by 1 inch. It is a mild rich cheese which is relished most when eaten a few days after it is made. Cream cheese is now quite extensively made in the larger factories of the United States, where the ever-increasing demand for it makes it one of the most popular varieties of soft cheese.

Edam Cheese

The perfectly fresh milk is set at 82° to 84° F.; color is added and sufficient rennet is used to coagulate the milk in 30 minutes. The curdled milk is divided evenly with a knife. After 20 minutes the whey is partly removed. The curd is further divided; after 10 minutes another portion of the whey is removed and stirring is resumed for 10 minutes. Then the temperature of the mixture is increased to 92°. The curd is now allowed to settle and the whey removed; then the layer of curd is cut into pieces, each part having the size of a cheese. These are left to settle in the molds, and they are then turned a few times; after being wrapped in cloth they are pressed two or three hours. After this they are salted, either by rubbing in salt and putting them in molds without lids, or by immersion in brine for three days. They are then stored for ripening and turned at intervals, which is the cause of their flattened shape. When they are a few weeks old they are marketed and the ripening process continues in the warehouses of the cheese merchants.

Emmenthaler (Domestic Swiss) Cheese

This is a hard, rennet cheese made from cows' milk, and has a mild, somewhat sweetish flavor. It is characterized by holes or eyes which develop to about the size of a cent in typical cheeses and are from 1 to 3 inches apart. Cheese of the same kind made in the United States is known as Domestic Swiss, and that made in the region of Lake Constance is called Algau Emmenthaler.

There is a slight difference in manipulation of the milk in making Emmenthaler cheese in this country as compared with Switzerland. In the latter country the evening's and morning's milk is mixed and made into cheese, while in the United States it is popularly believed that the evening's milk must be made into cheese immediately after milking, as is done with the morning's milk.

However, there is a growing tendency to make the cheese from milk delivered once a day or from milk that has been slightly ripened, as it is believed that the quality of the cheese is thereby improved.

Swiss cheese is made both with home-made rennet and with commercial rennet. When homemade rennet is employed usually no additional cultures are used. In some cases the homemade rennet is inoculated with a pure culture starter of lactobacillus bulgaricus. With modern methods it has been found desirable to use the following pure cultures: (1) The lactobacillus bulgaricus to check undesirable fermentation and to aid in controlling the ripening; (2) the use of an eye and flavor culture to aid in the development of eyes and flavor. These pure cultures are sent out by the Bureau of Dairy Industry of the United States Department of Agriculture or by State agencies.

It has been found that by clarifying the milk a much better quality of cheese can be produced, both in regard to eye

formation and in improving the body of the cheese. Clarification tends to reduce the number and to increase the size of the eyes. It is estimated that fully two-thirds of the factories of Wisconsin now clarify their milk for the manufacture of wheel and block Swiss.

In making the cheese in Switzerland the evening's milk is skimmed; the morning's milk is heated to 108° or 110° F., and the cream from the evening's milk is added and both thoroughly mixed. The evening's milk cooled with a little saffron to color it, is then added, and the whole is mixed. The milk is then brought to a temperature of 90° in summer and 95° in winter, and sufficient rennet is added to coagulate the milk in 30 or 40 minutes. The whole process is carried through in a huge copper kettle holding 300 gallons. The rennet used is obtained by soaking the calf's stomach in whey for 24 hours. When the milk has thickened to almost the desired point for cutting, which is practically the same as for ordinary American or Cheddar Cheese, the thin surface layer is scooped off and turned wrong side up. This is supposed to aid in incorporating the layer of cream into the cheese. The curd is then cut very coarse by means of a so-called harp. The cheesemaker, with a wooden scoop in each hand, then draws the mass of curd toward him that lying on the bottom of the kettle being brought to the surface. At this point the cheesemaker and an assistant commence stirring the curd with the harp, a breaker having first been fitted to the inside of the kettle to interrupt the current of the whey and curd. The harps are given a circular motion and cut the curd very fine—about the size of wheat kernels.

After this stage is reached heating is commenced. In Switzerland until recently all the heating was done over an open fire, the kettle being swung on a large crane; most of the factories have the same method at the present time. In this country the same method was followed in the early days of the industry, but at the present time inclosed fireplaces, into which the kettle can be swung and doors closed to retain the heat, are largely employed. This takes away much of the discomfort of the operation. In a few instances the kettle is set in cement and an iron car containing the fire is run under it. The most modern factories use steam, which appears to be the most satisfactory way. When the heating is begun the contents of the kettle are brought rapidly to the desired temperature, which may be from 126° to 140° F., the higher temperature often being necessary to get the curd sufficiently firm. In the meanwhile the stirring continues for about one hour, with slight interruptions near the end of the process, when the curd has become so firm that it will not mat together. The end of the cooking is determined by the firmness of the curd, which is judged by matting a small cake with pressure by the hands and noting the ease with which the cake breaks when heating the edge.

When the curd is sufficiently firm, the contents of the kettle are rotated rapidly and allowed to come to a standstill as the momentum is lost. This brings all the curd into a cone-shaped pile in the center of the kettle. One edge of a heavy linen cloth resembling burlap is wrapped around a piece of hoop iron, and by this means the cloth is slipped under the pile of curd. The mass of curd is then raised from the whey by means of a rope and pulley and lowered into a cheese hoop on the draining table. These hoops are from 4 to 6 inches deep and vary greatly in diameter. The cloth is folded over the cheese, a large follower is put on top, and the press is allowed to come down on the cheese. The press is usually a log swung at one end and operated by a double lever. Pressure is continued for the first time just long enough for the curd mass to retain its shape. The hoop is then removed, the cheese turned over, and a dry cloth substituted. The cheese is allowed to remain in the press about 24 hours, during which time it is turned and a dry cloth substituted six or more times.

At the end of the pressing, the curd should be a homogeneous mass without holes. The cheese is then removed to the salting board, covered with a layer of salt, and occasionally turned. In a day or two it is put into the salting tank in a brine strong enough to float an egg; it remains there at the discretion of the cheesemaker for from one to four days. Often no brine tank is used with Emmenthaler cheese.

The cheese is then taken to the curing cellar. In the best factories two or more cellars with different temperatures are available, and the cheeses are placed in them according to their development. If it appears that the cheese may develop too fast and have too many and too large eyes, it is placed in a cool cellar; if the reverse is true, a warm cellar is selected. The cellars vary in temperature from 55° to 65° F., though in extreme cases 70° or a little higher may be used. While the cheeses are in the ripening cellar, which in Switzerland may be from 6 to 10 months or longer, and in the United States three to six months, they should be turned and washed every other day for the first two or three months and less often subsequently. At the same time a little coarse salt is sprinkled on the surface. In a few hours this salt has dissolved, and the brine is spread over the surface with a long-handled brush.

The cheeses are very large, about 6 inches in thickness and sometimes as much as 4 feet in diameter, and weigh from 60 to 220 pounds. In shipping, a number of them are placed in a tub which may contain 1,000 pounds of cheese. Sometimes Emmenthaler cheese is made up in the form of blocks instead of in the shape of millstones. The blocks are about 28 inches long and 8 inches square in the other dimensions and weigh usually from 25 to 28 pounds.

Gorgonzola Cheese

This variety, known also as Stracchino di Gorgonzola, is a rennet, Italian cheese made from whole milk of cows. The interior of the cheese is mottled or veined with a penicillium much like Roquefort, and for that reason the cheese has been grouped with the Roquefort and Stilton varieties. As seen upon the markets in this country the surface of the cheese is covered with a thin coat resembling clay, said to be prepared by mixing barite or gypsum, lard or tallow, and coloring matter. The cheeses are cylindrical in shape, about 12 inches in diameter and 6 inches in height, and as marketed are wrapped in paper and packed with straw in wicker baskets.

The milk used in making this cheese is warmed to a temperature of about 75° F. and coagulated rapidly with rennet, the time required being usually from 15 to 20 minutes. The curd is then cut very fine, inclosed in a cloth and drained, after which it is put into hoops 12 inches in diameter and 10 inches high. It was formerly the custom to allow the curd from the evening's milk to drain overnight and to mix it with the fresh, warm curd from the morning's milk prepared in the same way. The curd from the evening's milk and that from the morning's milk, crumbled very fine, were put into hoops in layers with moldy bread crumbs interspersed among the layers. The cheese is turned frequently for four or five days, the cloths being changed occasionally, and is salted from the outside, the process requiring about two weeks. It is then transferred to the curing rooms, where a low temperature is usually maintained. At an early stage in the process of ripening, the cheese is usually punched with an instrument about 6 inches long, tapering from a sharp point to a diameter of about one-eighth inch at the base. About 150 holes are made in each cheese. This favors the development of the penicillium throughout the interior of the cheese. Well-made cheese may be kept for a year or longer. In the region where it is made, much of the cheese is consumed while in a fresh condition.

Limburg Cheese

This is a soft, rennet cheese made from cows' milk which may contain all the butterfat or may be partly or entirely skimmed. The best Limburg is undoubtedly made from the whole milk. This cheese has a very strong and characteristic odor and taste, weighs about 2 pounds, and is about 6 by 6 by 3 inches in size.

Limburg cheese originated in the Province of Lüttich, Belgium, in the neighborhood of Hervé, and was marketed in Limburg, Belgium. Its manufacture has spread to Germany and Austria, where it is very popular, and to the United States, where large quantities are made, mostly in New York and Wisconsin.

Sweet milk, without any coloring matter, is set at a temperature of from 91° to 96° F. with sufficient rennet to coagulate the milk in about 40 minutes. In foreign countries a kettle is used, but in the United States an ordinary rectangular cheese vat is found to be more satisfactory. The curd is cut or broken into cubes of about one-third of an inch and is stirred for a short time without additional heating. It is then dipped into rectangular forms 28 inches long, 5½ inches broad, and about 8 inches deep. These forms are kept on a draining board, where the whey drains out freely. When the cheese has been in the forms, with frequent turnings, for a sufficient length of time to retain its shape, it is removed to the salting table, where the surface is rubbed daily with salt. When the surface of the cheese commences to get slippery the cheese is put into a ripening cellar having a temperature of about 60° F. While in the cellar the surface of each cheese is frequently rubbed thoroughly. To ripen requires one or two months. When ripe the cheese is wrapped in paper, then in tin foil, and put into boxes, each containing about 50 cheeses.

Contrary to the popular belief, no Limburg is imported into this country at the present time. This type of cheese is made so cheaply and of such good quality in this country that the foreign make has been crowded out of the market.

Loaf or Process Cheese

It is defined as the clean, sound, heated product made by comminuting and blending, with the aid of heat and water and with or without the addition of salt, one or more lots of cheese into a homogeneous plastic mass.

At present it is estimated that one-half of all cheese made in this country is marketed as loaf or process cheese. American Cheddar, Swiss, Brick, Limburg, and even Camembert have been handled in this manner.

In the preparation of this product, cheese of different degrees of ripeness and of inferior quality with respect to flavor and texture may be used. Well-cured Canadian, well-cured Emmenthaler, or cultured Swiss cheese is often used to impart a typical flavor. It is stated that as much as 20 per cent white American cheese is often blended with Swiss cheese in order to give the finished product the proper texture.

The method of manufacture consists in cleaning the surface of the cheese, grinding it, and then adding a small quantity of an emulsifier, such as sodium citrate, sodium phosphate, or rochelle salts, dissolved in water, and finally heating the mixture in jacketed containers with constant agitation until the cheese has reached the proper degree of consistency. It is then put into suitable containers either directly or by specially designed machinery. From 1 to 2 per cent of emulsifiers are often used. Considerable skill is required in selecting the best kind of cheese to use as well as in regulating the manner and duration of the cooking. Ordinarily the cheese is gradually heated and stirred until a temperature of 140° to 160° F. is reached. The stirring is continued at this temperature for a longer or shorter period according to the nature and kind of cheese.

In the initial heating there is at first a slight separation of fat. This is followed by physical changes in the character of the curd so that the cheese becomes plastic and stringy. Upon further heating this plastic state is gradually broken down and a homogeneous mass with but slight plastic qualities is developed. When the cheese has reached this creamy condition and while still very hot, it is weighed and run into tin-foil-lined containers. Such packages render the cheese remarkably free from subsequent mold development.

Most of the process cheese manufactured in this country is made in a few large plants. At the present time there are no regulations as to the kind or quality of cheese that may be used in blending and no statement on the package as to whether or not emulsifiers are used.

Münster Cheese

Münster is a rennet cheese of the whole milk of cows, made in the vicinity of Münster, in the western part of Germany near the Vosges Mountains. Similar cheese made in the neighboring portion of France is called Géromé, and Münster cheese made near Colmar and Strassburg is sometimes given the names of those two cities.

The milk is set at about 90° F., with sufficient rennet to coagulate it in 30 minutes. The curd is then broken up and allowed to stand from 30 to 45 minutes without stirring, when it is dipped with a sieve, which gives slight pressure to the curd and holds back the small particles. After removing the whey the curd is scooped into forms or hoops, and caraway or anise seed is usually added. The hoops are made in two parts, the lower being 4 inches high and 7 inches in diameter, with holes in the bottom for draining, and the upper of the same dimensions. The whole resembles an ordinary cheese hoop with bandages. The hoop is lined with cheesecloth. After the curd has been in the hoop for 12 hours the upper part of the latter may be removed, the cheese turned, and the cloth removed. The cheese is now put into the upper portion of the hoop and turned frequently for from four to six days. In

the meantime the temperature is held at 68° F. After salt has been rubbed on the surface daily for three days the cheese is taken to the cellar, which has a temperature of from 51° to 55° F., where it is allowed to ripen for two or three months.

Neufchâtel Cheese

This is a soft rennet chese made extensively from either whole or skim milk of cows. Bondon, Malakoff, Petît Carré, and Petit Suisse are essentially the same as Neufchâtel but have slightly different shapes.

Neufchâtel cheese is made in the same manner as cream cheese, except that a little less rennet is used, perhaps 1 ounce of commercial liquid rennet to 1,000 pounds. Either whole milk or partly skimmed milk is used. Rennet is added to it at ordinary temperatures, and the curd when sufficiently firm is broken up, put into molds, and subjected to pressure. After being salted, the cheese is cured for from 8 to 15 days in a so-called drying room and then ripened in a cellar at a temperature of about 55° F. During the process of ripening the cheese becomes covered at first with a whitish mold and later with a blue mold in which red spots appear. After about one month it is ready for sale.

Parmesan Cheese

The milk, which has been skimmed to a greater or less extent, is heated in copper kettles to a temperature varying, according to the acidity of the milk, from 90° to 100° F. The kettle is then removed from the fire, rennet added, and the kettle covered and allowed to stand for 20 minutes to one hour, when the curd is cut very fine and cooked, with stirring, to 115° or 125° F. for from 15 to 45 minutes. The curd is removed from the kettle by means of a cloth, and after draining for a short time is put into hoops about 10 inches high and 18 inches or more in diameter, and lined with coarse cloth before filling. Pressure is then applied for 24 hours, the cheese being turned frequently and the cloths changed. The salting, which is begun in from one to three days after removing from the press, is continued for a considerable length of time, often 40 days. The cheeses are then transferred to a cool, well-ventilated room, where they may be stored for years, the surface being rubbed with oil from time to time. The exterior of the cheese is dark green or black, due to coloring matter rubbed on the surface. A greenish color in the interior has been attributed to the contamination with copper from the vessels in which the milk is allowed to stand before skimming.

Parmesan cheese when well made may be broken and grated easily and may be kept for an indefinite number of years. It is grated and used largely for soups and with macaroni. A considerable quantity of this cheese is imported into this country and sells for a very high price.

Roquefort Cheese

This is a soft, rennet cheese made from the milk of sheep. It is also stated from good authority that as much as 2.46 per cent of cows' milk and 0.18 per cent of goats' milk are mixed with the sheep's milk. There are, however, numerous imitations, such as Gex and Septmoncel, made from cows' milk, which resemble Roquefort. One of the most striking characteristics of this cheese is the mottled or marbled appearance of the interior, due to the development of a penicillium, which is the principal ripening agent.

Part of the milk is heated to 122° to 140° F. When this milk is mixed with the remainder the resulting temperature should be 76° to 82°, which is the setting temperature for the cheese. In from one to two hours after the addition of rennet the curd is cut until the particles are about the size of walnuts. The whey is dipped off, and the curd is put into hoops which are about 8½ inches in diameter and 3½ inches in height. The hoops usually are filled in three layers, a layer of moldy bread crumbs between each. The bread used for this purpose is prepared from wheat and barley flour, with the addition of whey and a little vinegar. It is thoroughly baked and kept in a moist place from four to six weeks, during which time it becomes permeated with a growth of the mold. The crust is removed, and the interior is crumbled dried, ground very fine, and sifted. The cheese is not subjected to pressure. It is turned usually one hour after putting into hoops and is not wrapped in cloths.

Formerly the manufacture of the cheese up to this stage was carried on by the shepherds themselves, but in recent years centralized factories have been established, and much of the milk is collected and there made into cheese. The cheese is then taken to the caves. These are for the most part natural caverns which exist in large numbers in the region of Roquefort. The temperature in these caves is 40° to 45° F., and the air circulates very freely through them. Recently artificial caves have been constructed and used. When the cheeses reach the caves they are salted, which serves to check the growth of the mold on the surface. One or two days later they are rubbed vigorously with a cloth and are afterward subjected to thorough scraping with knives, a process formerly done by hand, but now performed much more satisfactorily and economically by machinery. The salting, scraping, or brushing seems to check the development of mold on the surface. In order to favor the growth of mold in the interior, the cheese is pierced by machinery with from 20 to 60 small needles, which process permits the free access of air. The cheese may be sold after from 30 to 40 days or may remain in the caves as long as five months, depending upon the degree of ripening desired. During the process of ripening by scraping and evaporation the cheese loses from 16 to 20% of the original weight. When ripened, it weighs 4½ or 5 pounds.

Stilton Cheese

This is a hard, rennet cheese, the best of which is made from cows' milk to which a portion of cream has been added. The cheese is about 7 inches in diameter, 9 inches high, and weighs 12 or 15 pounds. It has a very characteristic wrinkled or ridged skin or rind, which is probably caused by the drying of molds and bacteria on the surface. When cut it shows blue or green portions of mold which give its characteristic piquant flavor. The cheese belongs to the same group as the Roquefort of France and the Gorgonzola of Italy.

The morning's milk is put into a tin vat, the cream from the night's milk is added, and the whole is brought to a temperature of 80° F., when the rennet is added. It is claimed by some cheesemakers that the curd should be softer when broken up or cut than the curd for Cheddar cheese, whereas others believe that it should become very firm before it is disturbed, one or two hours being allowed for setting. When sufficiently firm, the curd is dipped into cloths which are placed in tin strainers. After draining for one hour, the cloths containing the curd are packed closely together in a large tub and allowed to remain for 12 hours, when they are again tightened and packed for 18 hours. The curd is ground up coarse, and salt is added, 1 pound to 60 pounds of curd. It is then put into tin hoops 8 inches in diameter and 10 inches deep. The cheeses remain in the hoops for six days, when they are bandaged for 12 days, or until they become firm, and are then placed in the curing room at 65° F. Ripened Stilton cheese of late is often ground up and put into jars holding from 1 to 2½ pounds.

Sherbet Using Ice Cream Mix

Cane Sugar	25.0 lb.
Corn Sugar	7.0 lb.
Agar	0.2 lb.
(3.2 ounces or 90.6 grams)	
Gum Tragacanth or High-grade India Gum	0.2 lb.
(3.2 ounces or 90.6 grams)	
Ice Cream Mix, without Sugar or Gelatin	10.0 lb.
Water, Fruit, Fruit Acid, Flavor, and Color	57.6 lb.

Overrun—25 to 30%—Total yield 13.5 gallons.

The mixture should be prepared by first weighing most of the water or all of the milk, if any is used, leaving out enough water to dissolve the agar and to allow for fruit juices, etc. The sugars should be thoroughly mixed with the powdered gum tragacanth or high-grade india gum and slowly poured into the water while the water is being agitated rapidly. Powdered agar is preferable to granular or shreds because it can be more readily dissolved. The powdered agar should be poured into 50 times its weight of boiling water while the water is being agitated rapidly. The water with agar should continue to boil for about five minutes when the agar will be completely dissolved. The hot agar solution should be added to the mix as if it were a hot gelatin solution. The gelatinization strength of agar is reduced by boiling in acid solutions, but it is only slowly altered by boiling in water, so it is important that fruit acid should be added to the mix after the agar. All

other ingredients used should be added to the mix at this time and the total weight brought up to the required amount with water, making allowance for the fruit and fruit acids or juices which are usually added at the freezer.

There is no necessity of aging water ices or sherbets made with agar and gum as stabilizers because the action of each takes place within a few minutes. Evidence of a weak gel formation should be readily observed at once if sufficient agar has been used, since agar solutions set at 40° to 42° C. and since the temperature of the cold mixes is much lower.

Sherbets

Sugar	13.5 lb.
Sheragum	2¾ to 3 oz.
Flavor, Water, Acid, Color	

and mix to make 5 gallons of mixture.

1. Directions if not pasteurizing:

Mix well 3 oz. or slightly less of Sheragum with all of the sugar of the mix. Add this to the cold water in the vat, agitating all of the time. Add the flavor and mix thoroughly. If the flavor contains a high sugar content, cut down on the amount of sugar added. The amount of sugar given is satisfactory when orange or lemon sherbets are made. This mixture requires no aging, but if aged overnight will give a smoother product.

Freeze with cold refrigerant and when the mixture has started to thicken slightly add the acid (3–4 oz. of 50% citric acid). When the mix is a little stiffer, add 2 quarts of regular mix. Draw when frozen or when the overrun reaches 25 to 30%.

The regular formula used by the plant may be used. The only things to watch are—that the gum is mixed well with a large quantity of sugar and added slowly to the cold water, or milk if milk is used. Do not add the acid until the mixture is being frozen. The mix may be added any time. We always add the mix at the freezer because if the mixture is very acid, it may curdle the mix.

2. Directions if product is pasteurized:

The same rule is followed, but that 2½ oz. of Sheragum will be sufficient in this case. Acid, color and flavor are not pasteurized. Since heating brings out a little flavor from the gum, the gum and enough sugar to carry it should be left out until the mixture is cooled.

Sherbet Using Milk

Cane Sugar	25.0 lb.
Corn Sugar	7.0 lb.
Agar	0.2 lb.
(3.2 ounces or 90.6 grams)	
Gum Tragacanth or High-grade India Gum	0.2 lb.
(3.2 ounces or 90.6 grams)	
Whole Milk	50.0 lb.
Water, Fruit, Fruit Acid, Flavor, and Color	17.6 lb.

Overrun 25 to 30%—Total yield 13.5 gallons.

Ice Cream Powder

Dried Milk Powder	51
Sugar Powder	52
Sod. Carbonate	2
Cream of Tartar	4.4
Vanillin	0.06

One pound of above makes 10 lbs. ice cream.

Candy Jellies
Moderately Firm Pectin Jellies for Cast or Slab Work

Ingredients

Water	2½ gal.	
100 Grade Exchange Citrus Pectin	12	oz.
Acetate of Soda (U.S.P.)	1½	oz.
Citric Acid (crystals or powdered)	2¼	oz.
Glucose (43° Bé.)	20	lb.
Granulated Sugar	20	lb.
Color and Flavor	as desired	

Directions

(1) Put 2½ gallons of water in a kettle and heat hot (170° F.). (Open fire or steam-jacketed kettle may be used.)

(2) *Thoroughly mix* 12 ounces of 100 Grade Exchange Citrus Pectin with about 6 pounds of granulated sugar.

(3) Add the Pectin-Sugar mixture to the warm water as it is being stirred with a paddle. Continue to stir and heat to boiling. Boil vigorously for a moment.

(4) Combine the acetate of soda and citric acid. Dissolve in a small portion of hot water.

(5) Add the acetate of soda-citric acid solution to the kettle and then the 20 pounds of glucose. Heat to boiling again.

(6) Add the remainder of the sugar (14 pounds) and cook to 222°–224° F., or to a good "sheet." (This temperature corresponds to 75–78% total soluble solids at sea level. It is sufficient to cook the batch to 10°–12° F. above the boiling point of water at your factory.)

(7) Add the color and flavor, then cast into starch at once. This formula will produce about 48 to 50 pounds of candy. The finished piece may be crystallized, sanded, iced, or coated with chocolate.

Note: Cooking the batch to 224° F. is recommended for slab work.

Refined Corn Sugar may be substituted for all or a part of the cane or beet sugar given in the above formula.

Tart and Moderately Firm Pectin Jellies for Cast or Slab Work
(Especially for Fruit Flavors)

Ingredients

Water	2½ gal.	
100 Grade Exchange Citrus Pectin	12	oz.
Acetate of Soda (U.S.P.)	3	oz.
Citric Acid (crystals or powdered)	4	oz.
Glucose (43° Bé.)	20	lb.
Granulated Sugar	20	lb.
Color and Flavor	as desired	

Directions

(1) Put 2½ gallons of water in a kettle and heat hot (170° F.). (Open fire or steam-jacketed kettle may be used.)

(2) *Thoroughly mix* 12 ounces of 100 Grade Exchange Citrus Pectin with about 6 pounds of granulated sugar.

(3) Add the Pectin-Sugar mixture to the warm water as it is being stirred with a paddle. Continue to stir and heat to boiling. Boil vigorously for a moment.

(4) Combine the acetate of soda and citric acid. Dissolve in a small portion of hot water.

(5) Add the acetate of soda-citric acid solution to the kettle and then the 20 pounds of glucose. Heat to boiling again.

(6) Add the remainder of the sugar (14 pounds) and cook to 222°–224° F., or to a good "sheet." (This temperature corresponds to 75–78% total soluble solids at sea level. It is sufficient to cook the batch to 10°–12° F. above the boiling point of water at your factory.)

(7) Add the color and flavor, then cast into starch at once. This formula will produce about 48 to 50 pounds of candy.

The finished piece may be crystallized, sanded, iced, or coated with chocolate.

Note: Cooking the batch to 224° F. is recommended for slab work.

Refined Corn Sugar may be substituted for all or a part of the cane or beet sugar given in the above formula.

Firm Pectin Jellies for Cast or Slab Work

Ingredients

Water	3	gal.
100 Grade Exchange Citrus Pectin	15	oz.
Acetate of Soda (U.S.P.)	1½	oz.
Citric Acid (crystals or powdered)	2	oz.
Glucose (43° Bé.)	20	lb.
Granulated Sugar	20	lb.
Color and Flavor	as desired	

Directions

(1) Put 3 gallons of water in a kettle and heat hot (170° F.). (Open fire or steam-jacketed kettle may be used.)

(2) *Thoroughly mix* 15 ounces of 100 Grade Exchange Citrus Pectin with about 8 pounds of granulated sugar.

(3) Add the Pectin-Sugar mixture to the warm water as it is being stirred with a paddle. Continue to stir and heat to boiling. Boil vigorously for a moment.

(4) Combine the acetate of soda and citric acid. Dissolve in a small portion of hot water.

(5) Add the acetate of soda-citric acid solution to the kettle and then the 20 pounds of glucose. Heat to boiling again.

(6) Add the remainder of the sugar (12 pounds) and cook to 222°–224° F., or to a good "sheet." (This temperature corresponds to 75–78% total soluble solids at sea level. It is sufficient to cook the batch to 10°–12° F. above the boiling point of water at your factory.)

(7) Add the color and flavor, then cast into starch at once. This formula will produce about 48 to 50 pounds of candy. The finished piece may be crystallized, sanded, iced, or coated with chocolate.

Note: Cooking the batch to 224° F. is recommended for slab work.

Refined Corn Sugar may be substituted for all or a part of the cane or beet sugar given in the above formula.

Thickening of Jams, Preserves and Other Fruit Pastes

For many specific uses, particularly in baking and for soda fountain use, true fruit as well as imitation fruit jams, preserves and pastes must be thickened. This thickening is necessary to prevent leakage in pies and pastries and too rapid flow when used as coatings and dressings. Here Galagum fills a long felt want with a resultant lowering of costs in addition.

The method for making 100 pounds of finished jam or preserves is as follows: Mix thoroughly 7 ounces of Galagum with 35 ounces of cane sugar. The usual amount of sugar and fruit is boiled together in a steam-jacketed kettle. Start the stirring paddle when boiling begins and add VERY SLOWLY the above mentioned mixture of Galagum and sugar. Heat up to 221° F. and then turn off heat. Continue stirring until cool. If desired the jam may be worked on the cooling table, mixing it occasionally. The use of Galagum in this process increases the bulk or volume more than 5%.

Imitation Jellies

The corn syrup imitation jelly is made as follows: The 8 pints of water is brought to a boil. Add slowly with stirring the 70 grams of Aacagum, which has been previously mixed with the 7 oz. of Cerelose. Bring to a boil and cook for one minute. Now add the certified food color which has been dissolved in a little warm water. Then add the 7 lb. of warm corn syrup. Stir until completely mixed and at no time need the temperature be higher than 200° F. Transfer the jelly to pail, allow to cool down about 150° F. Then add with stirring the 35 grams of phosphoric acid and fruit flavor. The jelly will set in several hours or allow to set all night.

The imitation cane sugar jelly is made exactly the same way as the corn syrup jelly with the exception that you mix the 70 grams of Aacagum with about 10% of the weight of cane sugar. This mixture will aid the Aacagum considerably in going into solution when added to the hot water.

The phosphoric acid used in the above formulae was made by diluting 85% phosphoric acid with an equal volume of water. The fruit flavors used were of the fruit oil type and were dissolved in Glycopon XS.

Jelly (Non-Sweating)

Agar-Agar or Pectin	0.752–1%
Sodium Alginate	0.5–1%
Sugar	15–20%

| Water | 78–83% |
| Citric Acid | 0.03–0.04% |

Guava Jelly

Preparation of Juice:

Wash Guavas, and slice into small pieces with a sharp knife. For each pound of fruit add 2 pints of water and boil until soft (about 25 minutes), allow to stand until cold. Pour into cheese cloth bag and allow to drain pressing to extract all juice. This juice is then drained without pressing through a clean flannel jelly bag.

Making the Jelly:

Bring the juice to a boil, and then add the sugar. Continue boiling until the jellying point has been reached, which is indicated by the flaking or sheeting from the spoon. The jellying point of the guava is 108° C. or 226½° F.

Kumquat Jelly

Kumquats	1 lb.
Sugar	1 lb.
Water	1½ pt.

Wash kumquats, treated with soda, and then cut in halves. For each pound of fruit taken add 1½ pints water. Boil for 15 minutes, then the kettle is covered and set aside for 15 hours. After again boiling for 5 minutes, remove from the stove, and allow to drain. Let this stand for one hour, then pour into a flannel jelly bag, press to obtain all possible juice, drip through a bag to remove particles of fruit. The juice is then placed in a kettle and brought to a boil, at which time there is added 1 lb. sugar for each pound fruit taken. The jellying point is determined by dipping a spoon into the boiling solution, and then holding it above kettle allowing the syrup to drop. When it drops in flakes or sheets from the spoon pour immediately into clean, sterilized jelly glasses. When jelly is cold pour hot paraffin over it and store it away.

Fig Preserves

Figs	6 qt.
Sugar	2 qt.
Water	3 qt.

Add one cup soda to 6 quarts boiling water. Plunge figs into hot soda solution and allow to remain until white, milky fluid is extracted (about 15 minutes) or until water is cold enough to plunge hand into comfortably. Put figs through two cold water baths to rinse well.

Cooking. Drain figs thoroughly and add gradually to the syrup you have made by boiling the sugar and water together 10 minutes and skimming. Cook rapidly until figs are clear and tender (about to hours).

Fig Jam

Select very ripe figs, wash and drain. To every gallon of peeled figs add 2 quarts sugar, mash and cook to the proper consistency. When nearing the finishing point be careful not to scorch. If using a thermometer, cook to 222° F. or 106° C.

Grapefruit Preserves

Grapefruit Peel	1 lb.
Sugar	¾ lb.
Water	1 pt.
Slices of lemon	2

Preparation: Select bright fruit with a thick peel, wash carefully. Cut peel into strips or shapes. To 1 lb. of fruit add 2 pints of water and the lemon. Boil for 15 minutes, change the water and boil again. Repeat the process as often as is necessary to remove as much of the bitter of the peel as is desired. Remove the peel and the lemon from the water and drop them into a boiling syrup made by adding ¾ sugar to 1 pint water for each pound of peel taken and boiling until the sugar is dissolved. After the peel is added boil until the peel is transparent and the syrup sufficiently heavy.

Peach Preserves

Peeled Sliced Cling Stone Peaches	10 lb.
Sugar	7 lb.
Water	3 pt.
Peach Kernels	10

Bring sugar and water to a boil, add the peaches and kernels. Cook until the fruit is clear when lifted from the syrup. Pack in sterilized containers and seal.

Orange Marmalade

Oranges	3 lb.
Lemons	3
Water	1½ pt.
Sugar	3 lb.

Wash, remove the peel and seeds, cutting one half of the peel into very thin strips, and add it to the pulp and balance of the peel, which has first had the yellow portion grated off and has been passed through a food chopper with the pulp. Cover with water and let stand overnight. Boil for 10 minutes the next morning, allow to stand for 12 hours, add the sugar and again stand overnight. Cook it rapidly next morning until the jelly test can be obtained (about 222° F.). Cool to 176° F. pour into sterilized glasses, and seal with paraffine.

Gelatin Dessert Powder

Gelatin Powder (best grade)	80
Sugar Powder	450
Tartaric Acid Powder	10

Curry Powder (Spicing)
A.

Coriander Seed	16	oz.
White Pepper	1	oz.
Cayenne Pepper	½	oz.
Turmeric	1½	oz.
Ginger	1	oz.
Mace	½	oz.
Clove	½	oz.
Fennel	½	oz.
Celery Seed	½	oz.
Cardamom	½	oz.
Slippery Elm	4	oz.

B.
Indian Curry Powder

Coriander Seed	5	oz.
Turmeric	5	oz.
Cardamom	40	oz.
Cayenne Pepper	10	oz.
Fenugreek Seed	4	oz.

The above ingredients are mixed and allow to dry in a warm oven to drive off the moisture. It is then ground very fine and packed in tins.

Spiced Chocolate—I

Cacao	2500 g.
Sugar	2500 g.
Powdered Cinnamon	36 g.
Powdered Cloves	19 g.
Powdered Cardamom Seed	8 g.

Spiced Chocolate—II

Cacao	4000 g.
Starch Flour	130 g.
Powdered Cloves	70 g.
Sugar	4000 g.
Powdered Cinnamon	125 g.
Powdered Cardamom Seed	33 g.
Peru Balsam	6 g.

Coffee Chocolate

Cacao	2000 g.
Sugar	2000 g.
Ground Coffee	500 g.

Marshmallow

Soak together:

Gelatine	3½	oz.
Cold Water	13	oz.

Then heat to 140° F. and add

Hot Water	24	oz.
Invert Sugar	16	oz.
Icing Sugar	104	oz.
Vanilla	1	oz.

Beat stiff and use while warm.

Marshmallow and Meringue Powders
Formula No. 1

Dried Egg Albumen	25 lb.
Carob Bean (first grade powder)	25 lb.
Corn Starch	40 lb.
Skimmed Milk Powder	5 lb.
Powdered Alum	5 lh
Vanillin	to suit

Mix the above well and run through a fine mesh sifter.

Formula No. 2

Dried Egg Albumen	25 lb.
Carob Bean (first grade powder)	25 lb.
Tapioca Starch	10 lb.
Cane Sugar (powdered)	35 lb.
Skimmed Milk Powder	5 lb.
Vanillin	to suit

Mix the above well and run through a fine mesh sifter.

Formula No. 3

Dried Egg Albumen	25 lb.
Carob Bean (first grade powder)	25 lb.
Corn Starch	25 lb.
Corn Sugar (powdered)	20 lb.
Skimmed Milk Powder	5 lb.

Mix the above well and run through a fine mesh sifter.

The above meringue formulae are to be used as follows:

Take 5 oz. of meringue powder to 1 quart cold water and 3 lb. cane sugar. Put the cold water into a clean kettle, then add to it the sugar and meringue powder. Beat in the machine until the required stiffness is obtained. For marshmallow whip take 2 oz. of meringue powder, 1 quart cold water, 3 lb. of cane sugar and whip to the desired stiffness.

Now dissolve thoroughly 2½ oz. of Gelatin in ½ pint hot water. Add this slowly to the beaten meringue, and continue to beat up until the desired consistency is attained.

———◆———

Pineapple Icing

Pineapple (grated or crushed) 1 lb. Thicken to proper consistency with icing sugar. Heat to 110° C. and apply while warm.

———◆———

Lemon Icing

Hot Water	16	oz.
Sugar	120	oz.
Lemon Grating or Juice	2	oz.
Glucose Syrup	4	oz.

———◆———

Orange Icing

Hot Water	16	oz.
Sugar	120	oz.
Orange Grating or Juice	2	oz.
Glucose	4	oz.

———◆———

Maraschino Icing

Hot Water	16	oz.
Maraschino Juice	6	oz.
Chopped Cherries (to suit)		
Sugar	120	oz.
Glucose Syrup	4	oz.

———◆———

Coffee Icing

Fresh Made Coffee	16	oz.
Sugar	96	oz.
Invert Sugar	8	oz.
Caramel Color	⅛	oz.

———◆———

Vanilla Icing

Hot Water	16	oz.
Glucose	4	oz.
Sugar	112	oz.
Vanilla	½	oz.
Egg Whites	3	oz.

———◆———

Chocolate Icing

Hot Water	16	oz.
Sugar	96	oz.
Melted Butter	4	oz.
Melted Chocolate	16	oz.
Inverted Sugar	8	oz.

Home Made Icing

Beat stiff:

Egg Whites	32	oz.
Salt	¼	oz.
Sugar	16	oz.
Vanilla (to suit)		

Boil together to 236–240° F.

Sugar	104	oz.
Glucose	8	oz.
Water	2	oz.

Add cooked syrup to beaten egg whites and beat until stiff. Add chopped fruits, nuts as desired.

———◆———

Light Meringue Icing

Beat until stiff:

Egg Whites	32	oz.
Salt	⅛	oz.
Vanilla	¼	oz.

Boil to 240° F.

Sugar	96	oz.
Glucose	8	oz.
Water	32	oz.

Add syrup to beaten whites, and beat up until desired consistency is reached.

———◆———

Royal Icing

Beat light:

Egg White	16	oz.
Icing Sugar	96	oz.
Juice of Lemon	1	oz.
Cream of Tartar	⅛	oz.
Vanilla	¼	oz.

———◆———

Cocoa Icing

Beat together until smooth and glossy:

Plastic Cocoanut Butter	16	oz.
Invert Sugar	20	oz.
Water	12	oz.
Cocoa	20	oz.
Icing Sugar	88	oz.
Milk Powder	4¾	oz.
Salt	⅛	oz.
Vanilla	½	oz.

———◆———

Chocolate Pudding Dessert

Corn Starch	23	parts
Tapioca Starch	9	parts
Cocoa Powder	18	parts
Cane Sugar	50	parts
Vanilla Flavor to suit.		

The above powders are very carefully mixed. Four ounces when carefully cooked up with a pint of milk will make a delicious pint of chocolate pudding.

Chocolate Fudge

Bring to a boil:

Chocolate	16 oz.
Butter	4 oz.
Sugar	16 oz.
Milk	16 oz.
Glucose	6 oz.

Cool to 120° F.

Then add and mix smooth

Vanilla	1 oz.
Sugar Icing	72 oz.
Egg Whites	2 oz.

Mix smooth.

Butterscotch Fudge

Cook to 235° F.:

Brown Sugar	64	oz.
Milk	32	oz.
Butter	8	oz.
Glucose	1½	oz.

Cool to 120° F.

Then add

Milk	16	oz.
Lemon Juice	1	oz.
Salt	⅛	oz.
Butter	8	oz.
Icing Sugar	128	oz.
Burnt Sugar	¼	oz.

Use Warm.

Green Tomato Mince-Meat

Green Tomatoes	1	peck
Raisins	2	lb.
Brown Sugar	2½	lb.
Suet or Cocoanut	½	lb.
Ground Cinnamon	2	tsp.
Nutmeg	2	tsp.
Cloves	2	tsp.
Vinegar	½	cup
Salt	2	tsp.

Chop tomatoes fine and drain. Cover with cold water, heat through and drain again. Add chopped raisins and other ingredients. Cook 30 minutes. Pack into sterilized jars and process 15 minutes.

BEVERAGES AND FLAVORS

Almond Extract

Oil Bitter Almonds F.P.A. 1¼ fl. oz.		
Alcohol	3	pt.
Water	5	pt.

Almond Flavor

Oil Bitter Almonds	1 fl. oz.
Alcohol	40 fl. oz.
Water	59 fl. oz.

Imitation Almond Flavor

Benzaldehyde (F.F.C.)	1.3
Alcohol	16
Glycerol	24
Water	128

Anise Flavor

Oil Anise	3 fl. oz.
Alcohol	75 fl. oz.
Water	22 fl. oz.

Caraway Flavor

Oil Caraway	2 fl. oz.
Alcohol	70 fl. oz.
Water	28 fl. oz.

Celery Flavor

Oil Celery	4 fl. oz.
Alcohol	70 fl. oz.
Water	26 fl. oz.

Thyme Flavor

Oil Thyme	3 fl. oz.
Alcohol	70 fl. oz.
Water	27 fl. oz.

Cinnamon Flavor

Oil Cinnamon	1 fl. oz.
Alcohol	35 fl. oz.
Water	14 fl. oz.

In making the above flavors the oil should be dissolved in the Alcohol by stirring at room temperature. The water is then added slowly with vigorous stirring. In some cases (where a clear flavor is desired) mix in a weight of magnesium carbonate equal to the weight of the oil used; stir and filter.

Dry Ginger Ale Extract

Solid Extract Jamaica		
Ginger	8	oz.
Oil Ginger	2	drams
Oil Sweet Orange	2	drams
Oil Limes, Distilled	1	dram
Oil Mace	¼	dram
Oil Coriander	¼	dram
Oil Lemenone	¼	dram

Grind the above in a mortar with 4 oz. powdered magnesium carbonate; then add 1 gallon Alcohol slowly while grinding in thoroughly; then add one gallon water slowly and stir thoroughly for 2 hours; add 2 oz. kieselguhr and filter through fine filter paper. The finished product should be aged to develop a finer aroma and taste.

4 oz. of this extract is used per gallon of syrup.

Ginger Ale Extract

Oleo Resin Ginger	15	oz.
Oleo Resin Capsicum	2	oz.
Lemon Extract	5	pt.
Orange Extract	2½	pt.
Alcohol and water	2	gal.

Use to:
1 gallon Simple Syrup.
3 ounces Extract.

Ginger Ale Extract (Belfast)

Oleo Resin Ginger	24	oz.
Oleo Resin Capsicum	5½	oz.
Oil of Lemon (Terpeneless)	36	oz.
Oil of Orange (Terpeneless)	12	oz.
Oil of Cassia	1½	dr.
Oil of Rose, Artificial	½	dr.
Oil of Cloves	1½	dr.
Cologne Spirits	5½	gal.
Water	3	gal.

Use to:
1 gallon Simple Syrup.
2 ounces Extract.

Ginger Ale Extract

Oil of Ginger	4	oz.
Oil of Capsicum	1	oz.
Lemon Extract	16	oz.
Orange Extract	8	oz.
Alcohol	3½	pt.
Water	3½	pt.

Ginger Ale

Jamaica Ginger, fine powder	8 lb.
Capsicum, fine powder	6 oz.
Alcohol	a sufficient quantity

Mix the powders intimately, moisten them with enough alcohol to make them distinctly damp but not wet, set aside for four hours, then pack in a cylindrical percolator and percolate with alcohol until ten pints have been collected; place the percolate in a bottle of at least 2-gallon capacity and add 2 fluid drams of oleoresin ginger, shake and add 2½

pounds of finely powdered pumice stone and agitate frequently for twelve hours, then the next step is most important. Add 14 pints of water in one pint at a time, then shake briskly and add the next, after adding all the water set aside for twenty-four hours, agitating strongly every hour or so, then add:

Oil of Lemon	1½	fl. oz.
Oil of Rose Geranium	3	fl. dr.
Oil of Bergamot	2	fl. dr.
Oil of Cinnamon	3	fl. dr.
Magnesium Carbonate	3	oz.

First rub the magnesia with the oils in a mortar, add 9 fl. oz. of the clear portion of the ginger mixture to which 2 ounces of alcohol have been added and continue trituration, rinsing the mortar out with the ginger mixture, pass the ginger mixture through a double filter and add the mixture of oils through the filter. Finally pass enough water through the filter to make 3 gallons of the finished extract which is to be used 4 fl. oz. to a gallon of syrup. Dilute the syrup, 1 fl. oz. with 6 fl. oz. of carbonated water; bottle.

Note: The ginger ale can be colored a darker color with caramel.

Brewed Ginger Ale

This gives a true flavored ginger ale.

Fifty barrels of hot water are run into the kettle and heated to boiling. Six hundred pounds of granulated sugar are now added, making sure that the same dissolves properly. This having been accomplishd, 75 pounds of powdered ginger, 21 pounds of crystallized citric acid and 8 ounces of powdered capsicum are introduced into the solution, which is permitted to boil for half-hour. Eighteen pounds of good quality hops are now added and the solution boiled for an additional three-quarters of an hour, whereupon it is made up to a volume of, at least, 52 barrels, cooled over the Baudelot cooler and run into a settling tub, where it is permitted to remain overnight.

The following morning the clear supernatant liquid is withdrawn or, to work more economically, the whole solution may be filter-pressed and run into a clean vat or fermenter.

Having reached this stage, the beverage may be treated in one of two different ways. Either five barrels of this solution may be withdrawn, pitched with yeast and permitted to ferment completely and after completed fermentation freed of the yeast by filtration, returned to the main portion of the solution and stored for, at least, ten days. If preferable or more convenient, instead of withdrawing a portion of the solution to be completely fermented and subsequently returning the same, the entire solution can be carefully checked fermented by pitching with the customary amount of yeast and permitting the gravity to decrease no more than 0.8 of one per cent, after which the solution or beverage must be chilled almost to freezing, filtered and run into a clean and sterile vat, where it is to be stored for a period of ten days. The beverage is carbonated and filtered in the usual manner, as practiced in the manufacture of cereal beverages. It is advisable to carbonate twice, after which the beverage is ready for bottling.

The bottled ginger ale may be pasteurized if desired, although this is not necessary. If sold in bulk it is to be racked into freshly pitched packages and can be shipped without any danger of fermentation.

◆

Soluble Ginger Ale Extract

(To be used in the proportion of 4 ounces of extract to 1 gallon of syrup.)

Jamaica Ginger, in fine powder	8 lb.
Capsicum, in fine powder	6 oz.
Alcohol, a sufficient quantity.	

Mix the powders intimately, moisten them with a sufficient quantity of alcohol and set aside for 4 hours. Pack in a cylindrical percolator and percolate with alcohol until 10 pints of percolate have resulted. Place the percolate in a bottle of the capacity of 16 pints, and add to it 2 fluid drams of oleoresin of ginger; shake, add 2½ pounds of finely powdered pumice stone, and agitate thoroughly at intervals of one-half hour for 12 hours. Then add 14 pints of water in quantities of 1 pint at each addition, shaking briskly meanwhile. This part of the operation is most important. Set the mixture aside for 24 hours, agitating it strongly every hour or so during that period. Then take

Oil of Lemon	1½	fl. oz.
Oil of Rose (or geranium)	3	fl. dr.
Oil of Bergamot	2	fl. dr.
Oil of Cinnamon	3	fl. dr.
Magnesium Carbonate	3	fl. oz.

Rub the oils with the magnesia in a large mortar and add 9 ounces of the clear portion of the ginger mixture to which have been previously added 2 ounces of alcohol, and continue trituration, rinsing out the mortar with the ginger mixture. Pass the ginger mixture through a double filter and add through the filter the mixture of oils and magnesia; finally pass enough water through the filter to make the resulting product measure 24 pints, or 3 gallons. If the operator should desire an extract of more or less pugnency he may obtain his desired effect by increasing or decreasing the quantity of powdered capsicum in the formula.

◆

Havana Cigar Flavor

Coumarin, pure, cryst.	1	dr.
Methyl Benzoate	4	dr.
Essence Vanilla, Special	2	pt.
Oil Cascarilla	1	dr.
Oil Valeriana	½	dr.
Acetic Ether, Absolute	5	oz.
Alcohol	1	pt.

◆

Kola Beverage

Fluidextract of Coca	4	fl. oz.
Fluidextract of Kola	2	fl. oz.
Spirit of Orange	1½	fl. oz.
Lime Juice	1½	pints
Ginger Ale Extract	¾	fl. oz.
Cologne Spirit	8	fl. oz.
Sugar	6	lb.
Water	3	pints
Caramel		enough

Mix the fluidextracts, the Cologne spirit and the water, add the spirit of orange and set aside for two days, shaking occasionally. Then filter, add the lime juice and the ginger ale extract and dissolve the sugar in the mixed liquids.

◆

Peppermint Flavor

Oil Peppermint	3	fl. oz.
Alcohol	70	fl. oz.
Water	27	fl. oz.

◆

Orange Extract

Oil Orange	6½	oz.
Alcohol	121½	oz.

Mix, let stand overnight, then filter.

Extract of orange.—Shall be prepared from oil of orange or orange peel, or both, and absolute ethyl alcohol of proper strength, and shall contain not less than 80%, by volume, of ethyl alcohol, and not less than 5%, by volume, of oil of orange.

Terpeneless extract of orange.—Shall be prepared by shaking oil of orange with dilute ethyl alcohol or by dissolving terpeneless oil of orange of proper strength in dilute ethyl alcohol, and shall correspond in flavoring extract to orange extract.

Orange flavor, nonalcoholic.—Shall be a mixture of 20%, by volume, of oil of orange (U.S.P. standard) and 80%, by volume, of cottonseed oil. The cottonseed oil shall be thoroughly refined, winter pressed, sweet, neutral, and free from rancidity. The finished product shall be clear, free from sediment and rancidity.

Orange Oil Emulsion

Gelatin	4 oz.
Water	16 lb.
Cane Sugar	24 lb.
Invert Sugar	60 lb.
Terpeneless Oil Orange	20 oz.
Oil Orange	20 oz.

Dissolve the gelatin in the water, add the cane sugar and heat until dissolved. Then add the invert sugar and mix well; homogenize.

Orange Powder for Soft Drinks

Cane Sugar	80	parts
Dry Orange Powder	20	parts
Citric Acid	⅕	part

Color with an orange certified food color.

The above powders are mixed thoroughly. Four ounces of above powder when mixed with pint of cold water will make a delicious orange drink.

Pure Lemon Extract

Dissolve 5 fluid ounces Lemon Oil in 95 fluid ounces Alcohol; no heating is necessary.

The same proportions of oils of orange, limes, caraway, peppermint, wintergreen, etc., may be used as above to make 5% flavors.

Imitation Lemon Flavor

Citral	5 fl. oz.
Alcohol	96 fl. oz.
Water	189 fl. oz.

Imitation Lemon Flavor

Citral	½	oz.
Alcohol	100	oz.
Glucose 43° Baumé	1	lb.
Water	60	oz.

Lemon Extract

Oil of Lemon, U.S.P.	6½ oz.
Alcohol, 190 proof	121½ oz.

Mix, let stand overnight, then filter.

Lemon Extract (Terpeneless)

Oil of Lemon	30 lb.
Citral	8 oz.
Cologne Spirits	16 gal.

Put in a churn and work 2 hours. Of 11 gallons of water, add gradually about 5 gallons every hour and work for two hours more, then add 3 gallons water and work more. The whole process takes about 10 hours. After 10 hours add 1½ gallons Cologne Spirits. Let stand for 48 hours and filter.

Use to:

1 gallon Simple Syrup.
1 ounce Extract.

East India Lemon Sour Extract

Oil of Lemon	6 oz.
Oil of Limes	2 oz.
Alcohol, 95%	½ gal.
Warm Water	½ gal.
Alum	½ dr.

Add the oils to the alcohol and shake well. Dissolve the alum in the water. Add the water gradually in small quantities, shaking well after each addition. Set aside to settle for 6 hours. A scum will form on top. Separate extract from this with rubber hose. Filter clear through magnesia.

Use to:

1 gallon Simple Syrup.
2½ ounces Lemon Sour Extract.
3 ounces Lemon Sour Acid.
½ ounce Yellow Color.

Lemon Oil Emulsion

1. Gum Arabic	13 oz.
2. Terpeneless Oil of Lemon	20 oz.
3. Oil of Lemon	20 oz.
4. Glycerin	40 oz.
5. Water	to make 10 gal.

Mix one and four, then mix in two and three; to this add five slowly with good stirring. Beat intermittently until homogeneous. Then pass through an homogenizer.

Concentrated extract of lemon.—Shall be prepared from oil of lemon, or lemon peel, or both, and ethyl alcohol of proper strength, and shall contain not less than 20%, by volume, of oil of lemon and not less than 0.8%, by weight, of citral.

Extract of lemon.—Shall be prepared from oil of lemon or lemon peel, or both, and ethyl alcohol of proper strength. It shall contain not less than 80%, by volume, of absolute ethyl alcohol, not less than 5%, by volume, of oil of lemon and not less than 0.2%, by weight, of citral derived solely from the oil of lemon or lemon peel used in its preparation.

Terpeneless extract of lemon.—Shall be prepared by shaking oil of lemon with dilute ethyl alcohol, or by dissolving terpeneless oil of lemon of proper strength in dilute ethyl alcohol, and shall contain not less than 0.2%, by weight, of citral derived solely from oil of lemon.

Lemon flavor, non-alcoholic.—Shall be a mixture of 20%, by volume, of oil of lemon (U.S.P. standard) and 80%, by volume, of cottonseed oil. The cottonseed oil shall be thoroughly refined, winter pressed, sweet, neutral, and free from ¯ancidity. The finished product shall be clear, free from sediment and rancidity.

Lemonade Powder for Soft Drinks

Cane Sugar	86	parts
Dry Lemon Powder	14	parts
Citric Acid	1/10	part

Color with a yellow certified food color. The above powders are mixed and colored. Four ounces of above powder when mixed with pint of cold water will make delicious lemonade.

Pure Vanilla Flavor

Oleoresin Vanilla	4 oz.
Alcohol	2 pt.

Water to make 1 gallon.

Flavoring ingredients must be completely dissolved before water is added. Filter clear after two or three days.

Imitation Vanilla Flavor

1. { Vanillin   2 oz.
   { Coumarin   1/2 oz.
   { Alcohol   32 fl. oz.
   Water to 7 gallons.

2. { Vanillin   2 oz.
   { Coumarin   1 oz.
   { Alcohol   28 fl. oz.
   { Sugar   5 lb.
   Water to 5 gallons.

----

3. { Vanillin   20 oz.
   { Coumarin   4 oz.
   { Alcohol   1½ gal.
   { Water   184 oz.
   Take 1 lb. of above and add water to it slowly with stirring to make 2 gallons.

Artificial Vanilla Flavor

Vanillin	6 dr.
Coumarin	2 dr.
Alcohol	2 pt.
Water	5 pt.
White Sugar Syrup	1 pt.
Glycerin C. P.	1 pt.

Caramel color enough to give the desired shade.

Dissolve the vanillin and coumarin in the alcohol, then add the other materials and let stand for a few days before using. If not clear, filter. The syrup is made by dissolving 12 ounces of sugar in water enough to make a pint of syrup.

Concentrated Vanilla Compound Flavor (Highest Quality)

For dilution with water up to 17 to 1.

Alcohol	60 oz.
Vanillin	6 oz.
Coumarin	2 oz.
Oleoresin Vanilla	4 oz.
Glucose 43° Baumé	3½ lb.
Caramel Color	4 oz.

Balance water to make 1 gal.

The usual procedure on above formulae is to put the Vanillin and Coumarin in a container containing the required amount of Alcohol; stir until completely dissolved. Add to it *slowly* with *stirring* the required amount of water. If caramel color, prune juice, sugar or syrup is to be added; these should be dissolved first in the water.

Where a water-white Vanilla is desired, the solution of Vanillin in Glycopon AAA may be decolorized by the addition of a little tartaric or citric acid.

Non-Alcoholic Vanilla, Lemon and Almond Flavors

Non-Alcoholic Vanilla Flavor

Vanillin	3.2 Gm.
Coumarin	0.19 Gm.
Glycerin	180.00 mils
Syrup	180.00 mils
Water	120.00 mils
Ether	120.00 mils
Color	sufficient

Dissolve the vanillin and the coumarin in the ether. Mix the glycerin, syrup and water, add to this ether solution of the vanillin and couramin. Beat until the ether is entirely volatilized and then add the color.

The Paste type of flavors has been suggested for non-alcoholic lemon and almond. Soak 250 Gm. of gum tragacanth in 4 liters of distilled water for three or four days or until it is softened and has taken up as much water as it will hold. Now forcibly strain it through cheesecloth. Mix 120 mils of this mucilage with 360 mils of glycerin. This will serve as the vehicle for the flavor. For this quantity of paste add gradually and with constant trituration in a mortar 60 mils of oil of lemon.

For almond flavor use 120 mils of the paste and 360 mils of glycerin and to this add gradually and with constant trituration 15 mils of benzaldehyde which must be free from hydrocyanic acid and chlorine.

Compound Vanilla Extract

A. Mexican Vanilla Beans	1	lb.
Bourbon Vanilla Beans	1	lb.
Water	2	gal.
Alcohol	2	gal.
Glycerin	26	oz.
Rock Candy Syrup	2	pt.

Grind or cut the beans small and place in a porcelain jar or clean wooden keg; pour over them the water at a boiling temperature and macerate for twenty-four hours. Then add the alcohol and glycerin and macerate for forty-eight hours; lastly, add the rock candy syrup, stir well and macerate for not less than four weeks.

B. Vanillin	2	oz.
Alcohol	2	pt.

Mix and let stand for twenty-four hours; then add one pint rock candy syrup, and let stand for twenty-four hours longer; add one pint prune juice and let stand for twenty-four hours; then add five pints boiling water and let stand for two weeks. Filter.

To make the extract add one quart of solution (B) to one gallon of solution (A).

Vanilla Extract

Oleoresin Vanilla	4	oz.
Alcohol	4	pints
Simple Syrup	1¼	pints
Water	2¾	pints

Mix by stirring thoroughly. Simple syrup is prepared by dissolving 3½ lb. of sugar in one quart of water.

Pure vanilla extract.—Shall be prepared without added flavoring or coloring, from prime vanilla beans with or without sugar and/or glycerin; shall contain, in 100 cubic centimeters, the soluble matters from not less than 10 grams of vanilla beans; shall contain not less than 40 per cent, by volume, of absolute ethyl alcohol, and show a Wichman lead number not less than 0.70. The strength of the extract in respect to the vanillin and vanilla resins, which shall be derived solely from the beans used, shall be not less than 0.17 per cent vanillin and not less than 0.09 per cent vanilla resins.

Imitation vanilla, artificially flavored and colored.—Shall be a solution of vanillin and coumarin in dilute glycerol with 5 per cent, by volume, of true vanilla extract, colored with caramel. There shall be not less than 0.6 gram of vanillin, 0.1 gram of coumarin, and 35 centimeters of glycerol (U.S.P. standard), and 100 centimeters of the finished product.

Extra concentrated extract of vanilla.—Shall be prepared, without added flavoring or coloring, from prime vanilla beans, with or without glycerin; shall contain, in 100 cubic centimeters, the soluble matters from not less than 100 grams of vanilla beans, and shall contain not less than 30 per cent, by volume, of absolute ethyl alcohol, and when one part by volume, of the product is diluted with nine parts, by volume, of dilute alcohol (40 per cent, by volume) the resulting mixture shall comply with the requirements for vanilla extract except in regard to alcoholic content. The label shall clearly indicate the strength of the product and if the product is not made directly from vanilla beans, the label should contain a statement to that effect.

4X strength, extract of vanilla.—Shall be prepared without added flavoring or coloring, from prime vanilla beans with or without sugar and/or glycerin; shall contain, in 100 cubic centimeters, the soluble matters from not less than 40 grams of vanilla beans; shall contain not less than 35 per cent, by volume, of absolute ethyl alcohol, and when one part, by volume, of the product is diluted with three parts, by volume, of dilute alcohol (40 per cent by volume) the resulting mixture shall comply with the requirements for vanilla extract, except in regard to alcohol content. The label shall clearly indicate the strength of the product and if the product is not made directly from vanilla beans, the label should contain a statement to that effect.

Chocolate Syrup Concentrate

Chocolate Liquor	34	lb.
Granulated Sugar	30	lb.
Nulomoline	6¾	gal.
Water	2¾	gal.

The above formula yields approximately 135 pounds of syrup, or approximately 12 gallons of Chocolate Syrup Concentrate. The method of manufacture is as follows.

Place 2¾ gallons of water in a 25 gallon steam jacketed copper kettle and bring water to a boil. Add 30 pounds of granulated sugar and stir until dissolved. Add 6¾ gallons of Nulomoline and stir until dissolved. Add 34 pounds of melted chocolate liquor, a little at a time with constant stirring, until thoroughly dispersed. Bring the mixture to a boil and boil for 5 minutes. Run syrup through a strainer, and cool to 160° F. Put syrup through a homogenizer at approximately 3000 pounds pressure. The chocolate syrup is run into a holding kettle and then run into cans. When the syrup is colled down to 90° F., put the covers on cans and store cans in a cool place preferably not over 75° F.

It is necessary to keep everything spotlessly clean to avoid mold contamination. This is accomplished by thoroly washing machinery, floors, and etc.

The above syrup can be used as follows.

Cake ingredient, filling between layers and as an icing.

For thin chocolate syrup for fountain use, dilute one gallon can with one gallon simple sugar syrup.

Use as is or dilute as above for Malted Milk and Milk Cocoas.

Could be used as a confection by mixing with chopped nuts and powdered sugar and cutting into squares.

For Ice Cream Frappes, warm syrup and pour over Ice Cream.

Sarsaparilla Extract '

Oil of Wintergreen	4	oz.
Oil of Sassafras	4	oz.
Oil of Anise	1	oz.
Cologne Spirits	5	pt.
Powdered Pumice Stone	4	oz.
Granulated Sugar	8	oz.
Water	2½	pt.
Sugar Color	1	oz.

Dissolve the oils in two pints of the spirits. Each oil must be added separately and well shaken with the spirits before another oil is added. Now put the pumice stone and sugar in a Wedgwood mortar, add the mixture gradually and rub together to a paste. Mix the remainder of the spirits and water together, add the sugar color to these, and dissolve carefully. Mix the whole together gradually, stirring well until all combines, and filter through filter paper.

Use to:

1 gallon Simple Syrup.
1 ounce Extract

Imitation Black Walnut Flavor

Oil of Black Walnut	8	oz.
Alcohol	1½	lb.
Glucose 43° Baumé	1	lb.
Sugar Color	2	oz.

Balance water to make 1 gallon.

Wintergreen Flavor

Methyl Salicylate	3	fl. oz.
Alcohol	70	fl. oz.
Water	27	fl. oz.

Rootbeer Essence

Oil Sassafras, Pure	1	oz.
Oil Anise Russian, Rectified	1	oz.
Oil Lemon, Natural	1	oz.

Methyl Salicylate (Oil Winter-	
green Art.)	18 oz.
Alcohol	6 oz.
Water	11 oz.
Bismarck Brown Color	

Root Beer Oil

Methyl Salicylate	5 oz.
Safrol	8 oz.
Oil Orange	1 oz.
Oil Clove	2 drops
Oil Nutmeg	2 drops
Coumarin	½ oz.
Vanillin	1 oz.
Alcohol	64 oz.
Water q.s.	128 oz.

1 ounce of above flavors 2 gallons.

Fruit Syrup with Pulp

One quart lemon, orange or other fruit pulp; 6½ lb. sugar; 5 pints water; ½–1 oz. citric acid, and 1 oz. Viscogum.

Directions:

Mix thoroughly 1 lb. of sugar with 1 oz. of Viscogum. Bring the 5 pints of water to a boil and add slowly while stirring the mixture of Viscogum and sugar. Then boil vigorously for one minute. If artificial color is desired, it may be added at this point. Now add the balance (5½ lb.) of sugar and cook until completely dissolved. Allow to cool to 180° F. and add the citric acid, previously dissolved in a little water. The fruit pulp is then added and slow stirring is continued until cool. If some additional flavor is desired it is added at this point. If a preservative is indicated then 3.6 grams of Benzoate of Soda is stirred in. The finished syrup is stirred slowly while bottling. It is advisable to shake each bottle the next day before packing for shipment. The pulp will not remain in suspension for long periods.

Essence Sweet Cherry

Heliotropin	60	gr
Solution Jasmin, Concrete		
1: 10 in Glycopon XS	24	mils
Solution Peach Aldehyde, pure		
1: 20 in Glycopon XS	7½	mils
Cyclamic Aldehyde, pure	2	mils
Oil Bitter Almonds,		
F.F.P.A.	16	mils
Vanillin	84	gr.

Fluidextract Rhatany	35	mils
Oil Cloves	2¼	mils
Oil Cinnamon Ceylon	1¼	mils
Cherry Juice	800	mils
Alcohol	800	mils

Essence Wild Cherry Aroma

Heliotropin	40	gm.
Solution Jasmine	24	mils
Peach Aldehyde	7½	mils
Oil Bitter Almond	23	mils
Vanillin	84	gm.
Fl. Extract Rhatany	35	mils
Oil Cloves	2½	mils
Oil Cinnamon	1¼	mils
Cherry Juice	800	mils
Alcohol	800	mils

Oil Cherry Ethereal

Amyl Acetate, Pure	12 pt.
Amyl Butyrate, Pure	8 pt.
Benzaldehyde, free from	
Prussic Acid	12 pt.
Oil Lemon, Handpressed	16 oz.
Oil Sweet Orange, Hand-	
pressed	8 oz.
Oil Cloves, Pure	16 oz.
Oil Cassia, Leadfree	8 oz.
Vegetable Red Coloring.	

Oil of Wild Cherry

Acetic Ether	10 fl. oz.
Benzoic Ether	5 fl. oz.
Oil of Bitter Almonds	5 fl. oz.
Amyl Valerianic Ether	2 fl. oz.
Benzoic Acid	2 fl. oz.
Glycerin	8 fl. oz.
Cologne Spirits	6 pt.

Cherry Acid Solution

Citric Acid	2½ av. lb.
Tartaric Acid	2½ av. lb.
Hot Water	1 gal.

Thoroughly dissolve and add Phosphoric Acid syrupy 2 fluid ounces.

Cherry Compound

Dry Citric Tartaric Acid (½	
Citric and ½ Tartaric)	1¼ lb.
Extract Cherry Concentrated	1 pt.
Vegetable Red Color in liquid	
form	8 oz.

Water, enough to make 1 gallon.

Wild Cherry

Oil of Wild Cherry	½	pt.
Distilled Water	½	gal.
Cologne Spirits	½	gal.
Red Color	¼	fl. oz.

Mix water and Cologne Spirits. Add the oil of Wild Cherry, mix and add the color. Mix well.

Use to:

1 gallon Simple Syrup.
1 ounce Extract.

Essence Concord Grape

Methyl Anthranilate, Pure	10 oz.
Alcohol	100 oz.
Glycerin, Pure	45 oz.
Vegetable Red Liquid	5 oz.

Essence Grape Aroma "Special"

Nerolin	20 gr.
Essence Cognac	10 mils
Sol. Methyl Anthranilate 1 : 10	20 mils
Tinct. Cacao	20 mils
Fluid Ext. Valerian	2 mils
Sol. Benzoic Ether 1 : 10	1 mil
Grape Juice	60 mils
Alcohol	200 mils

Artificial Grape Syrup

Artificial Grape Oil	6	oz.
Tartaric Acid	2¾	lb.
Cream of Tartar	2	oz.
Tannic Acid	15	gm.
Grain Alcohol	3	pt.
Sugar Syrup	7	pt.

Color sufficiently to give the desired shade.

The syrup is made by dissolving 7 pounds granulated sugar in sufficient water to make one gallon.

Artificial Grape Flavor (Powder)

Tartaric Acid	2¾	lb.
Cream of Tartar	2	oz.
Tannic Acid	15	gr.
Granulated Sugar	10	lb.
Concentrated Grape Oil, Artificial	6	oz.

Mix the tannic acid with cream of tartar. (The tannic acid may be omitted if desired.) This should be mixed thoroughly, then mix this with about ½ pound of the acid (fine powdered). Mix well, then work in the remaining acid in lots of ½ pound at a time, thorough mixing being essential. It is best done by sieving several times, mixing well after each sieving. Now work in the sugar the same way, so that the whole forms a perfectly even mixture. Now slowly work in the artificial grape oil, mixing thoroughly. Sufficient color is added to give the required shade when dissolved in water. Mix thoroughly and spread out until dry, then rub again through a sieve and put up in packages.

As the color will vary in strength, it will be necessary to experiment a little to get the exact quantity required to give the desired color when the product is made up into a finished drink.

In the strength given here, a teaspoonful will be sufficient to flavor strongly a quart of water.

French Curacao

Oil Orange	10 oz.
Mace Oil	8 cc.
Cassia Oil	16 cc.
Cloves Oil	8 cc.
Lemon Oil	32 cc.
Rose Oil	1 cc.
Vanillin	1 dr.
Jamaica Rum Essence	2 oz.

Artificial Grape Oil

Benzyl Butyrate	10½	fl. oz.
Methyl Anthranilate	4½	fl. oz.
Methyl Salicylate	½	fl. oz.
Amyl Valerianate	½	fl. oz.
Fluid Extract Valerianate	3	fl. oz.
Port Wine	75	fl. oz.
Alcohol	150	fl. oz.
Grape Juice	50	fl. oz.
Glycerin	25	fl. oz.

Mix the first five with the alcohol, then add the other materials one at a time in the order given, stirring well after each addition. Let stand for 24 hours and filter.

Essence Arrac

Oil Neroli Petale, Extra	15 drops
Essence Jamaica Rum	42 oz.
Extract Vanilla	12 oz.
Essence Cognac Fine Champagne	2 oz. 4 dr.
Essence Raisin Wine	1 oz.

Arrac Aroma Essence

Oil Birch	16 gr.
Oil Cognac	16 gr.
Oil Maraschino	25 gr.
Oil Celery	8 gr.
Rum Essence	250 gr.
Alcohol	250 gr.

Essence Apple Aroma

Oil Apple Ethereal	750 mills
Oil Jasmine Flowers	3 mils
Amyl Valerianate, pure	20 mils
Vanillin	10 gr.
Tinct. Civet 4 oz. to 1 gal.	5 mils
Sol. Peach Aldehyde, pure 1 : 20	1 mil
Alcohol	2000 mils
Apple Cider	1500 mils
Water	750 mils

Essence Apple, Extra

Oil Apple, Ethereal	1500 mils
Peach Flavor	100 mils
Alcohol	5000 mils
Water	3500 mils
Vegetable Liquid Yellow Color	10 mils

Imitation Apple Flavor

Amyl Valerianate	6 oz.
Ether Acetic	3 oz.
Spirits of Nitrous Ether	3 oz.
Amyl Butyrate, Absolute	1 oz.
Aldehyde	½ oz.
Essence of Peach Blossom	½ oz.

Alcohol 95 per cent, enough to make 1 quart.

Cheap Apple Cider

Boiled Cider	2 gal.
Granulated Sugar	25 lb.
Tartaric Acid	¾ gal.
Water	30 gal.

Color to suit with sugar color. Thoroughly mix; let stand three days, then draw off and add one ounce of benzoate of soda to each ten gallons of cider. Keep in a cool place.

Essence Creme de Menthe

Oil Peppermint, Twice Rectified	2 oz.
Menthol	2 dr.
Alcohol	35 oz. 4 dr.

Green Coloring.

Essence Chartreuse

Oil Peppermint, Rectified	1½ dr.
Oil Lemon, Handpressed	2 dr.
Oil Cassia, Leadfree	1 dr.
Oil Cloves Pure	1 dr.
Oil Mace Distilled	1½ dr.
Oil Anise Seed, Russian, Rectified	1 dr.
Oil Angelica Root	40 dr.
Oil Bitter Almonds, F.F.P.A.	½ dr.
Oil Wormwood, American	20 dr.
Oil Neroli Bigrade, Petale, Extra	1 dr.
Oil Cognac, Genuine, White	15 dr.
Glycopon XS	20 oz.

Essence of Peach Blossom

Oil of Peach Blossom	1½ oz.
Peach Aldehyde 100%	2 dr.
Glycopon XS	6 pt.
Water	28 oz.

Essence Tutti Frutti

Essence Benedictine	16 oz.
Essence Maraschino	16 oz.
Essence Curacao	16 oz.
Essence Violet Flowers	16 oz.
Oil Strawbery, Ethereal	32 oz.
Tinct. Vanilla 1 lb. to 1 gal.	32 oz.

Cognac Essence

Oil Bitter Almond	20 drops
Oil Cognac	50 gm.
Violet Flower Essence	25 gm.
Woodruff Essence	50 gm.
Oenanthic Ether	15 gm.
Acetic Ether	120 gm.

Essence Slivovitz

Oil Bitter Almonds, F.F.P.A.	2 mils
Oil Neroli, Artificial	1 mil
Oil Cognac, Genuine, Green	2 mils
Vanillin	5 gm.
Essence Raspberry Aroma	300 mils
Essence Plum	300 mils
Essence Jamaica Rum	25 mils
Essence Raisin Wine	50 mils
Prune Spirit	100 mils
Glycopon XS	100 mils

Essence Kartoffel Schnapps

Essence Rye Whiskey	8 oz.
Essence Nordhaeuser Korn	8 oz.

Essence Nordhaeuser Korn

Carvol	10 oz.
Oil Caraway, Dutch	2 oz.
Oil Coriander, pure	30 drops
Acetic Ether, Absolute	4 dr.
Alcohol	60 oz.
Glycerin, Pure	18 oz.

Essence Prune Juice for Blending

Tinct. St. John's Bread	10 oz.
Extract Vanilla	5 oz.
Prune Juice	28 oz.
Prune Spirit	12 oz. 4 dr.
Essence Rum Kingston	2 oz. 4 dr.
Tinct. Lemosin Oak	30 oz.
Essence Raisin Wine	10 oz.
Essence Cognac Fine Champagne	5 oz.
Essence Figs	2 oz. 4 dr.
Essence Grape Aroma	2 oz.

Pistache Essence

Oil Lemon, Handpressed	4 mils
Oil Bitter Almonds, F.F.P.A.	8 mils
Essence Strawberry Aroma	12 mils
Benzyl Acetate, pure	3 drops
Glycerin, pure	12 mils
Peach Flavor, pure	3 mils
Alcohol	120 mils
Green Color	½ gm.

Essence of Jamaica Rum

Oil of Cassia	1 dr.
Oil of Birch Tar	25 drops
Oil of Ylang Ylang Natural	3 dr.
Oil of Orange Flower Natural	20 drops
Oil of Ceylon Cinnamon	15 drops
Rum Ether Pure	3 pt.
Acetic Ether	2½ oz.
Butyric Ether	1 oz. 1 dr.
Tincture of Saffron 1 lb. to a gal.	4 oz.
Extract of Vanilla Pure	3 oz.
Balsam Peru	2 dr.
Tincture Styrax U.S.P.	2 dr.
Coumarin	5 dr.

Essence Rum New England

Oil Cinnamon, Ceylon	2 dr.
Oil Cloves, Pure	2 dr.
Oil Chamomile, Roman	4 dr.
Rum Ether, Pure	4 pt.
Butyric Ether, Absolute	3 oz.
Extract Vanilla	4 dr.
Acetic Ether, Absolute	3 oz.
Alcohol	8 oz.

Essence Raisin Wine

Extract Vanillin	70 oz.
Essence Raspberry Aroma	2 oz.
Oenanthic Ether, Absolute	4 dr.
Geraniol Pure	2 oz. 2 dr.
Acetic Ether, Glacial	2 oz. 2 dr.
Alcohol	40 oz.
Methyl Anthranilate Pure	20 drops
Water	16 oz.

Essence Whiskey Bourbon

Fusel Oil	1 gal.
Oil Bitter Almond	1½ oz.
Oil Rose Art.	48 min.
Vanilla Extract	32 oz.
Ess. Jamaica Rum	40 oz.
Pineapple Aroma	40 oz.
Acetic Ether	12 oz.

Essence Rock and Rye Whiskey

Oil Corn Fusel	7 oz. 4 dr.
Oil Cognac, Genuine Green	4 dr.
Balsam Peru, True	4 dr.
Essence Jamaica Rum	4 dr.
Vanillin	2 dr.
Acetic Ether Absolute	4 dr.
Coumarin	5 dr.
Essence Raisin Wine	12 oz.
Peach Flavor	4 dr.
Alcohol	35 oz.
Glycerin, Pure	16 oz.

Essence Whiskey "Rye"

Oil Fusel Potato	2 pt.
Oil Fusel Rye	18 pt.
Rum Ether, Pure	20 pt
Oil Coriander, Pure	5 oz.
Oil Bitter Almonds, F.F.P.A.	2 oz. 4 dr.
Alcohol	50 pt.
Tinct. Catechu	1 pt.
Vanillin	2 dr.
Heliotropin	4 dr.
Tinct. Balsam, Peru, True	1 dr.

Essence Whiskey "Scotch"

Guaiacol, pure	4 dr.
Oil Cade, pure	1 oz.
Butyric Ether, pure	4 oz.
Essence Rye Whiskey	2 gal.

Oil Scotch

Oil Corn Fusel	6 oz.
Oil Bitter Almonds	4 dr.
Oil Coriander	4 dr.
Oil Cade	1 oz.

Guaiacol	2 dr.
Butyric Ether	4 oz.
Alcohol	4 oz.

Corn Ether

Glycopon XS	5000 gr.
Acetic Ether	1000 gr.
Fusel Oil	30 gr.
Coriander Oil	4 gr.
Oil Cognac	4 gr.

Oil Gin Holland

Oil Lemon	1 dr.
Oil Anise	1 dr.
Oil Angelica Root	6 dr.
Oil Fusel	4 dr.
Oil Juniper Berries	20 oz.
Oil Rosemary Flavor	6 dr.
Oil Coriander	4 dr.
Alcohol	10 oz.

Oil Gin, Old Tom

Oil Coriander, pure	3 oz. 4 dr.
Oil Angelica Root	3 dr.
Oil Anise, Russian, Rectified	1 oz.
Oil Caraway, Dutch	4 dr.
Oil Juniper Berries, Rectified	7 oz. 4 dr.
Alcohol	1 pt. 8 oz.

Essence Gin, Old Tom

Essence Gin, Holland	1 gal.
Alcohol	1 pt.
Oil Coriander, pure	1 oz.
Oil Calamus	1 oz.

Essence Gin, London Dock

Oil Gin, Old Tom	6 oz.
Oil Gin, Holland	18 oz.
Oil Cassia, Rectified	4 dr.
Alcohol	64 oz.

Gordon Gin Essence

Oil Juniper Berries	16 oz.
Oil Angelica Root	20 cc.
Oil Angelica Seed	20 cc.
Oil Coriander	40 cc.
Oil Lemon	60 cc.
Sweet Orange	20 cc.
Neroli	5 cc.
Geranium Rose	5 cc.
Alcohol to make 1 gal.	
4 oz. of above to make 50 gal.	

Essence Trester Brandy

Oil Cognac, Genuine	4 oz.
Oil Corn Fusel	5 oz.
Methyl Salicylate	3 oz.
Acetic Ether, Absolute	2 lb. 8 oz.
Alcohol	24 pt.
Water	3 pt. 12 oz.

Super Aroma Bourbon 1-5

Oil Fusel Rectified	240 oz.
Ess. Pineapple	½ oz.
Ess. Peach Blossom	½ oz.
Citric Acid Solution 50%	240 oz.
Solution Saccharin Saturated	¼ oz.
Oil Jamaica Rum	13 oz.
Alcohol	133 oz.
Tannic Acid Sol.	1 oz.
	626 oz.

Oil Anisette

Oil Anise Russian, Rectified	465 mils
Oil Sweet Fennel, Rectified	20 mils
Oil Coriander, Pure	10 mils
Oil Star Anise, Leadfree	465 mils
Oil Angelica Root	30 mils
Oil Bitter Almonds, F.F.P.A.	8 mils
Oil Rose, Artificial	2 mils

Anisette Flavor

Oil Star Anise	100 gm.
Oil Anise	50 gm.
Oil Carvol	7 gm.
Oil Lemon	5 gm.
Oil Rose	½ gm.
Oil Neroli	2 gm.
Oil Cardamon	2 gm.

Oil Absinthe, French

Oil Wormwood, American	10 oz.
Oil Star Anise, Leadfree	16 oz.
Oil Anise Russian, Rectified	12 oz.
Oil Fennel, Rectified	6 oz.
Oil Neroli, Artificial	½ dr.
Alcohol	3 oz.
Tinct. Gum Benzoin, Siam 2 lb. to 1 gal.	3 oz.

Oil Alkermes, Cordial

Oil Cinnamon, Ceylon	100 gm.
Oil Cassia, Leadfree	200 gm.
Oil Cloves, Pure	200 gm.
Oil Mace, Distilled	450 gm.
Oil Rose, Genuine	1 gm.
Alcohol	50 gm.

Artificial Oil of Raspberry

Acetic Ether	5 oz.
Formic Ether	1 oz.
Methyl-Salicylic Ether	1 oz.
Nitrous Ether	1 oz.
Oenanthic Ether	1 oz.
Sebacylic Ether	1 oz.
Butyric Ether	1 oz.
Benzoic Ether	1 oz.
Amyl-Butyric Ether	1 oz.
Succinic Acid	1 oz.
Saturated Solution Tartaric Acid in cold Alcohol	5 oz.
Glycerin	4 oz.
Ticture of Orris	100 oz.

Mix the succinic acid with the tincture, add the others and, lastly, the glycerin. One ounce of pure vanilla extract will improve this.

Artificial Oil of Pineapple

Amyl Butyrate	1 oz.
Butyric Ether	4 oz.
Sebacic Ether	1 oz.
Acetic Ether	4 dr.
Amyl Acetate	4 dr.
Pineapple Juice	4 dr.
Glycerine C. P.	4 oz.
Alcohol	50 oz.

Mix, adding glycerin last.

Oil Peach Blossom

Oil Neroli	16 oz.
Oil Cognac Genuine	14 oz.
Oenanthic Ether	14 oz.
Peach Aldehyde 100%	4 oz.
Oil Apple Ethereal	16 oz.
Acetic Ether Absolute	96 oz.
Valerianic Ether Absolute	16 oz.
Alcohol	240 oz.

Artificial Oil of Peach

Ethyl Formate	5 oz.
Ethyl Butyrate	5 oz.
Ethyl Acetate	5 oz.
Ethyl Sebacate	1 oz.
Ethyl Valerianate	5 oz.
Oil of Bitter Almonds	5 oz.
Aldehyde	2 oz.
Glycerin	5 oz.
Amyl Alcohol	2 oz.

Alcohol enough to make up 100 ounces.

Apricot Oil

Oil Neroli Art.	12 oz.
Oil Cognac White	14 oz.
Oenanthic Ether	14 oz.
Peach Aldehyde 100%	4 oz.
Vanillin	64 oz.
Oil Apple Ethereal	16 oz.
Acetic Ether	96 oz.
Valerian Ether Absolute	16 oz.
Alcohol	240 oz.

Oil Benedictine

Oil Sweet Orange, Hand-pressed	72 oz.
Oil Angelica Root	6 oz.
Oil Calamus	3 oz.
Oil Cinnamon, Ceylon	3 oz.
Oil Mace, Distilled	3 oz.
Oil Celery	3 oz.
Alcohol	12 oz.

Oil Pear Ethereal

Benzyl Propionate	1 pt.
Amyl Acetate, pure	11 pt.
Butyric Ether, Absolute	4 pt.

Oil Plum Ethereal

Oil Pineapple. Ethereal	4 pt.
Oil Jamaica Rum	4 pt.
Essence Slivovitz	4 pt.
Essence Peach Blossoms	4 pt.
Alcohol	6 pt.

Oil Neroli Artificial

Ambrettone	2 gr.
Oil Rose Geranium	5 gr.
Infusion Balsam Tolu	8 gr.
Glycopon XS	50 gr.
Phenyl Ethyl Acetate	20 gr.
Orange Oil	40 gr.
Rose Leaf Infusion	75 gr.
Oil Neroli Gen. Bigarde	100 gr.
Geranyl Acetate	100 gr.
Methyl Anthanilate	100 gr.
Inf. Orange Flowers	100 gr.
Linalol	100 gr.
Oil Petit Grain Algerian	150 gr.
Linalyl Acetate	150 gr.

Powdered Flavors

Put about 4 ounces of the powder into a mortar and spray or drop the mixed flavoring materials over it slowly, mixing well. When all have been added, gradually add the remainder of the acid, mixing well after each addition. The color should be dissolved in the flavoring mixture before adding the acid. When well mixed, place in a glass dish and stir often until it has dried out sufficiently to admit of packing. Best put up in glass bottles with closly fitting stoppers, but may be put up in cans. The quantity is sufficient for 45 gallons of liquid.

+

Raspberry

The base as above	1	lb.
Artificial Oil of Raspberry	1½	oz.
Bordeau S. Amaranth		
Color	2 to 5	gr.
Artificial Vanilla Flavor	1	dr.

+

Strawberry

The base as above	1	lb.
Ponceau 3 R Color	2 to 5	gr.
Artificial Oil of Strawberry	1½	oz.
Artificial Vanilla Flavor	1	dr.

+

Cherry

The base as above	1	lb.
Artificial Oil of Cherry	1½	oz.
Bordeau S. Amaranth Color	10	gr.

+

Pineapple

The base as above	1	lb.
Artificial Pineapple Oil	1½	oz.
Napthol Yellow, Color	10	gr.

+

Foam Producers
Soap Bark Foams

Formula A—

Quillaja bark is used in the form of tincture and may be prepared as follows:

Quillaja, fine chips	5½ av. oz.
Alcohol	10 fl. oz.
Water	Sufficient

Mix the drug with 24 fluid ounces of water, boil for 15 minutes. Strain and add enough water through the strainer to make the volume equal to 22 fluid ounces. Mix the liquid when cool with the alcohol, let stand for 12 hours, filter, and to the filtrate add enough water to measure 32 fluid ounces.

If a cheaper preparation is desired, the alcohol may be replaced by water or by glycerin. If the former be used, the preparation must be preserved by the addition of a small amount of salicylic acid solution. Either of the latter is to be preferred to the alcoholic solution, as the alcohol has the tendency to cause premature expulsion of gas from the soda when served.

About one fluid ounce of this preparation is usually sufficient for one gallon of syrup.

Formula B—

Soap Bark (chips)	1	lb.
Boiling Water	10	pt.
Alcohol (95%)	1	pt.

Boil the soap bark in the water for 30 minutes. Allow to cool. Add the alcohol. Pack a small quantity of dry soap bark in a percolator to make a bed and percolate. One-half to 1 ounce of this is used per gallon of syrup.

+

Sapinone Foams

Formula A—

Sapinone	1	lb.
Glycerin	½	gal.
Water	½	gal.

Dissolve the sapinone in ½ gallon of clear water, then add glycerin. Use ½ dram to 1 gallon or 1 ounce to 15 gallons of syrup.

Formula B—

Sapinone	24 av. oz.	
Water	1	gal.

Dissolve sapinone in water by agitation and when dissolved add

Formaldehyde	2 fl. dr.

Use 1 dram to 1 gallon or 1 ounce to 15 gallons of syrup.

+

Plain or Simple Syrup

Granulated Cane Sugar	30	lb.
Water (boiling)	7	qt.

Pour the sugar into the water gradually, stirring meanwhile, and when dissolved, strain through coarse cotton cloth. Do not cover container until thoroughly cooled. This will produce four gallons of syrup. The relative proportions of sugar and water are very important since, if a smaller amount of sugar is employed, fermentation sooner or later will ensue. If too much sugar is used, crystallization will surely follow, resulting in a liquid too thin to keep under ordinary temperature.

+

Beverage Acidulants

Citric Acid Crystals	4	lb.
Boiling Water	4	pt.

When dissolved, filter through filter paper using glass funnel. Keep in glass and avoid contact with metal.

Tartaric Acid Crystals   4 lb.
Boiling Water   4 pt.
Treat the same as above.

Tincture of Grass

Lawn Grass, fresh, cut fine 2 av. oz.
Alcohol   16 fl. oz.

Put the grass in a wide mouth bottle and pour the alcohol upon it. After standing a few days, agitating occasionally, pour off the liquid.

This is a useful preparation for giving a green color to essences, syrup of violets, etc. It can be used with alcohol or water.

Purple Coloring
Tincture of Litmus

Litmus, powder   2½ av. oz.
Water, boiling   16 fl. oz.
Alcohol   3 fl. oz.

Pour the water upon the litmus, stir well, allow to stand for about an hour, stirring occasionally, filter, and to the filtrate add the alcohol.

Brown Red Coloring
Compound Tincture of Cudbear

Cudbear, powder   120 gr.
Caramel   1½ av. oz.
Alcohol, of each   Sufficient
Water, of each   Sufficient

Macerate the cudbear with 12 fluid ounces of a mixture composed of 1 volume of alcohol and 2 of water for 12 hours, agitating frequently, then filter. Add the caramel, previously dissolved in 2 fluid ounces of water, and then pass through the filter enough of the before-mentioned alcohol water mixture to make the whole liquid measure 16 fluid ounces.

This preparation may also be made by dissolving 1½ ounces of caramel in 2 fluid ounces of water, adding 4 fluid ounces of tincture of cudbear and then enough of a mixture composed of 1 volume of alcohol and 2 of water to make the whole measure 16 fluid ounces.

Chlorophyll Coloring

This may be employed in alcoholic solution for coloring preparations of a green tint. It may be purchased or it may be prepared as follows:

Digest leaves of grass, nettles, spinach, or other green herb, in warm water, until soft; pour off the water, and crush the herb to a pulp. Boil this for a short time with a ½ per cent solution of caustic soda, and afterwards precipitate the chlorophyll by means of dilute hydrochloric acid; wash the precipitate thoroughly with water, press and dry it, and use as much for the solution as may be necessary.

ALCOHOLIC LIQUORS

The most important constituent of alcoholic beverages is the alcohol. Its strength depends upon the character of the beverage. If the alcohol is inferior in quality or has an oily taste and odor, the finished product will be unsatisfactory. Be sure to use good alcohol. Sugar is used to sweeten the liqueurs and, in many cases thickens the liqueurs as well, which is desirable.

The colors used should be certified, pure food colors. For brown coloring the most predominant color is burnt sugar color or caramel. Sometimes its taste helps to mellow or round out the taste of liqueurs. Wines and fruit juices also may be used sometimes to bring out the fuller taste.

The quantities of essences or flavoring oils called for in each formula should be carefully measured. It is the essence or oils that gives the alcohol in the finished beverage its characteristic taste and aroma. The skill employed in making these beverages usually decides success or failure. As with all formulas, carelessness, inaccuracy and haste will only result in failure. A formula that imparts good taste and aroma is one always sought for. Good recipes never grow old. They do not change as the science of Chemistry does. And so an old formula when tried and found to be true never grows old.

Some of the liquor formulas in this book may call for substances other than simple oils or simple ingredients. When difficulty arises or should you desire to become more expert in mixing, blending and compounding, call in a reliable, reputable chemist. He will be able to assist you and render valuable service.

Even a freshly prepared mixture of aromatic substances lacks homogeneousness and only after some period of time are the ingredients well mixed and blended. However, storage is necessary in every case to round out taste, flavor and brilliancy—to produce an equilibrium of the reactants present, to give the proper bouquet which characterizes a good product.

When beverages are stored in barrels, the tannin of the wood appears to possess the power of hastening, ageing and improving the taste. Oak barrels are best to use to clear or make liqueur brilliant. Storage is usually sufficient but the clearness can be hastened by the addition of 1 pint of skimmed milk. The clear liquid is then siphoned off later. Where rapid clearing is desired filtration must be resorted to.

Alant Essence

Alant Root	5	gm.
Cinnamon	½	gm.
95% Alcohol	10	kilos

Color: Red.

Essence Aromatic
No. 1

Cardamom	83	gm.
Clove	166	gm.
Mace	166	gm.
Cinnamon	580	gm.
95% Alcohol	10	kilos

No. 2

Curacao Peels	460	gm.
Cloves	83	gm.
Mace	83	gm.
95% Alcohol	10	kilos

Absinthe Essence à la Turine

Oil Angelica	3	gm.
Oil Anise	5	gm.
Oil Fennel	5	gm.
Oil Cardamom	1	gm.
Oil Coriander	5	gm.
Oil Marjoram	3	gm.
Oil Star Anise	6	gm.
Oil Wormwood	3	gm.
95% Alcohol	10	kilos

English Absinthe

Oil Anise	8	gm.
Oil Wormwood	8	gm.
Liqueur Body	11.5	lit.

Color: Green.

Absinthe, Fine

Oil Calamus	1	gm.
Oil Coriander	1.5	gm.
Oil Ginger	1	gm.
Oil Wormwood	1	gm.
Liqueur Body	11.5	lit.

Color: Green.

Swiss Absinthe
No. 1

Oil Angelica	5	gm.
Oil Anise	10	gm.
Oil Fennel	10	gm.
Oil Cardamom	3	gm.
Oil Coriander	10	gm.
Oil Marjoram	10	gm.
Oil Star Anise	12	gm.
Oil Wormwood	15	gm.
95% Alcohol	10	kilos

No. 2

Oil Angelica	8	gm.
Oil Anise	15	gm.
Oil Tincture Arrac No. 5	100	gm.
Oil Fennel	15	gm.
Oil Marjoram	15	gm.
Oil Orange	20	gm.
Oil Wormwood	20	gm.
Oil Lemon	10	gm.
95% Alcohol	10	kilos

Anise Essence

Anise Seed	4	gm.
Oil Star Anise	1	gm.
95% Alcohol	10	kilos

Color: Green.

Anise Liqueur

Oil Anise	4	gm.
Oil Star Anise	4	gm.

Dissolved in 0.25 lit. Alcohol 95%.

Liqueur Body	11.5	lit.

No Color.

Barbado Essence
No. 1

Mace	3	gm.
Cloves	5	gm.
Orange Peel Fresh	100	gm.
Cinnamon	16	gm.
Lemon Peel Fresh	100	gm.
95% Alcohol	10	kilos

Color: Brown.

No. 2

Oil Bergamot	4	gm.
Oil Cloves	1	gm.
Oil Nutmeg	1	gm.
Oil Cinnamon	1	gm.
Oil Lemon	4	gm.
95% Alcohol	10	kilos

Flavoring Bitters, Trinidad

Angostura Bark Genuine	90 gm.
Chamomile	24 gm.
Cardamom	8 gm.
Cinnamon Ceylon	7 gm.
Orange Peel	24 gm.
Raisins	300 gm.
Water	5 kilos
Alcohol	5 kilos

Flavoring Bitters, American

Angostura Bark	18½ gm.
Gentian	7½ gm.
Galgant	17½ gm.
Hazel Root	7½ gm.
Honey	250 gm.
Cardamom	18½ gm.
Catechu	7.6 gm.
Coriander	7½ gm.
Caraway	7½ gm.
Curcuma	100 gm.
Dandelion Root	7½ gm.
Mace Buds	3½ gm.
Nutmeg	7½ gm.
Cloves	1 gm.
Pimento	22 gm.
Orange Peel	30 gm.
Sandalwood Red	30 gm.
Snake Root	7½ gm.
Licorice	7½ gm.
Wormwood	7½ gm.
Cinnamon	7½ gm.
Alcohol 65%	7.2 lit.

Flavoring Bitters, Tropical

Angelica Root	3 gm.
Gentian Root	15 gm.
Galgant Root	15 gm.
Ginger Root	3 gm.
Cardamom Small	20 gm.
Cinnamon	20 gm.
Cloves	3 gm.
Orange Peel Bitter	25 gm.
Sandalwood Red	80 gm.
Tonka Beans	80 gm.
Zedoary Plant	15 gm.

Everything roughly cut and put into 5000 grams of 60% Alcohol. This mixture has to stand 15 days, then filtered. After this add 200 grams Sugar Color, 500 grams Malaga Wine. Let it stand for an additional few days and filter it again.

Anise Brandy

Alcohol 90% by Volume	36 lit.
Anise Oil Essence	30 gm.
Sugar Syrup 65%	4 lit.
Water	60 lit.

Lemon Brandy

Alcohol 90% by Volume	36 lit.
Lemon Essence	50 gm.*
Sugar Syrup 65%	4 lit.
Water	60 lit.

Color Yellow to suit.

Raspberry Brandy

Alcohol 90% by Volume	17 lit.
Cherry Whiskey	3 lit.*
Raspberry Juice	27 lit.
Sugar Syrup 65%	7 lit.
Water	46 lit.

Kummel Brandy

Alcohol 90% by Volume	36 lit.
Coriander Essence	½ lit.*
Sugar Syrup 65%	4 lit.
Water	60 lit.

Cherry Brandy

Alcohol 90% by Volume	16 lit.
Bitter Almond Oil Essence	10 gm.*
Cinnamon Oil Essence	20 gm.*
Clove Oil Essence	10 gm.*
Sugar Syrup 65%	3½ lit.
Water	32½ lit.
Cherry Juice	48 lit.

Clove Brandy

Alcohol 90% by Volume	36 lit.
Clove Oil Essence	100 gm.*
Cinnamon Oil Essence	50 gm.*
Sugar Syrup 65%	4 lit.
Water	57½ lit.
Cherry Juice	2½ lit.

Color: Brown.

Corn Brandy (30% Alcohol)

Alcohol 90% by Volume	33¼ lit.
Coriander Oil Essence	85 gm.*
Rum Essence	¼ lit.
Water	66½ lit.

* In this formula and the others that follow where an essence is used dissolve latter in alcohol first, then add balance of ingredients and then filter.

Peppermint Brandy

Alcohol 90% by Volume	36 lit.
Peppermint Oil Essence	150 gm.*
Sugar Syrup 65%	4 lit.
Water	60 lit.

Filter and clarify with 10 grams Alum. Color green or leave white.

Orange Brandy, White

Alcohol 90% by Volume	36	lit.
Bitter Orange Oil Essence	½	lit.*
Sugar Syrup 65%	4	lit.
Water	59½	lit.

For brown, color with caramel color.

Absinthe Brandy

Alcohol 90% by Volume	36	lit.
Absinthe Essence	½	lit.*
Sugar Syrup 65%	2½	lit.
Water	61	lit.
Color: Green.		

Juniper Brandy

Alcohol 90% by Volume	40	lit.
Juniper Berry Essence	½	lit.*
Sugar Syrup 65%	3	lit.
Water	56½	lit.

Color is white. For brown use caramel color.

Calamus Brandy

Alcohol 90% by Volume	36	lit.
Calamus Essence	½	lit.*
Sugar Syrup 65%	4	lit.
Water	59½	lit.
Color: Brown.		

Bergamot Brandy

Alcohol 90% by Volume	38	lit.
Bergamot Oil Essence	25	gm.*
Sugar Syrup 65%	6	lit.
Water	56	lit.

Peppermint Liqueur

Alcohol 90% by Volume	50	lit.
Peppermint Essence	400	gm.*
Sugar Syrup 65%	30	lit.
Water	20	lit.

Ginger Liqueur

Alcohol 90% by Volume	30	lit.
Ginger Extract	20	lit.*
Sugar Syrup 65%	40	lit.
Water	10	lit.
Color: Brown.		

Anisette

Oil Anise Russian, Rectified	465	mils
Oil Sweet Fennel, Rectified	20	mils
Oil Coriander, Pure	10	mils
Oil Star Anise, Leadfree	465	mils
Oil Angelica Root	30	mils
Oil Bitter Almonds, F.F.P.A.	8	mils
Oil Rose, Artificial	2	mils

Dissolve ½ oz. of above mixture in 22 gallons alcohol. Then add 28 gallons water in which has been dissolved 112 lb. sugar.

Chartreuse

Oil Peppermint, Rectified	1½	dr.
Oil Lemon, Handpressed	2	dr.
Oil Cassia, Leadfree	1	dr.
Oil Cloves, Pure	1	dr.
Oil Mace, Distilled	1½	dr.
Oil Anise Seed, Russian, Rectified	1	dr.
Oil Angelica Root	40	dr.
Oil Bitter Almonds, F.F.P.A.	½	dr.
Oil Wormwood, American	20	dr.
Oil Neroli Bigrade, Petale, Extra	1	dr.
Oil Cognac, Genuine, White	15	dr.
Alcohol	20	oz.

Dissolve 1 oz. of this mixture in 7 gallons alcohol. Then add 9 gallons water in which has been dissolved 3? lb. sugar.

Chartreuse Liqueur

Alcohol by Volume 90%	12½	lit.
Chartreuse Essence	5	gm.
Sugar Syrup 65%	4½	lit.
Water	12½	lit.

Wherever the word essence appears you take one part of the essential oil and mix thoroughly with 7 parts of 95% alcohol and these mixtures or solutions constitute the essences as given in the formulas.

Cognac

Alcohol 90% by Volume	22	lit.
Cognac Essence	500	gm.
Citric Acid	12½	gm.
Rock Candy	1	kilo
Water	28	lit.

Dissolve the Citric Acid in ¼ liter of water. Dissolve the Rock Candy in 1 liter of water. Mix the ingredients thoroughly and allow to remain in the vessel for several weeks.

Cognac

Oil Bitter Almond	20	dr.
Oil Cognac	50	gm.
Violet Flower Essence	25	gm.
Woodruff Essence	50	gm.
Oenanthic Ether	15	gm.
Acetic Ether	120	gm.

Dissolve 1 oz. of above mixture in 36

gallons alcohol. Then add 30 gallons water. Mix. Filter and color with caramel.

Cognac Brandy

Essence Brandy	20 oz.
Extract Vanilla	4 oz.
Tinct. Orrisroot, Florentine (2 lb. to 1 gal.)	2 oz.
Oil Cognac, Genuine	1 oz.
Oil Bitter Almonds, Free from Prussic Acid	2 dr.
Essence Rum, New England	6 dr.
Acetic Ether, Absolute	2 oz. 2 dr.
Nitrous Ether, Absolute	2 oz.
Alcohol	10 oz.

Dissolve 1 oz. of above mixture in 10 gallons alcohol. Then add 10 gallons water. Mix. Filter through magnesium carbonate. Color with caramel.

Lemon Absinthe

Lemon Peels	200	gm.
Peppermint Herb	100	gm.
Wormwood	50	gm.
Alcohol 95%	0.6	lit.
Liqueur Body	11.5	lit.

Color: Green.

Lemon Brandy

Alcohol 90% by Volume	21½	lit.
Lemon Essence	600	gm.
Sugar Syrup	5½	lit.
Water	23	lit.

Collor: Yellow.

Lemon Essence

Alcohol by Volume 95%	2⅓	lit.
Lemon Juice	2⅓	lit.

Mix the alcohol and lemon juice and then filter.

Lemon Liqueur

Alcohol 90% by Volume	34	lit.
Lemon Essence	7	lit.
Sugar Syrup 65%	26	lit.
Corn Syrup	13	lit.
Water	20	lit.

Color: Yellow

Absinthe Brandy (Swiss)

Alcohol 90% by Volume	25	lit.
Absinthe Essence	365	gm.
Water	25	lit.

Color Green to suit.

Absinthe Brandy (French)

Alcohol 90% by Volume	21¾	lit.
Swiss Absinthe Essence	375	gm.
Sugar Syrup	3¼	lit.
Water	25	lit.

Color Green to suit.

Pineapple Brandy

Alcohol 90% by Volume	21¾	lit.
Pineapple Ester (Conc.)	265	gm.
Pineapple Essence from Fresh Fruit	145	gm.
Sugar Solution	3¼	lit.
Water	25	lit.

Italian Orange Brandy

Alcohol 90% by Volume	21½	lit.
Orange Essence	500	gm.
Sugar Solution	8½	lit.
Water	20	lit.

Color Yellow with Tincture of Saffron.

Polish Brandy—"A"

Raisins	280	gm.
Licorice	35	gm.
Cinnamon	25	gm.
Cardamom	25	gm.
Cloves	8	gm.
Galgant	8	gm.
Ammonia Rubber	8	gm.
Anise Seed	8	gm.
Coriander	8	gm.
Alcohol 60%	3	lit.

Extracted for few days, pressed, filtered and mixed with sugar, the last to be dissolved in rose water.

Trester Brandy

Oil Cognac, Genuine		4 oz.
Oil Corn Fusel		5 oz.
Methyl Salicylate		3 oz.
Acetic Ether, Absolute	2 lb.	8 oz.
Alcohol		24 pt.
Water	3 pt.	12 oz.

Filter through magnesium carbonate.

Hamburger Bitters

Alcohol 90% by Volume	21½	lit.
Hamburger Bitter Essence	550	gm.
Sugar Solution	4½	lit.
Water	24	lit.

Color Brown with Caramel.

Hamburger Drops

Alcohol by Volume 90%   21½ lit.
Hamburger Bitter Essence 550   gm.
Sugar Solution   10½ lit.
Water   23 lit.
Color: Brown with Caramel Color.

English Bitter

Benedictine Herb	10	gm.
Gentian Root	20	gm.
Orange Peels	100	gm.
Calamus Root	40	gm.
Lesser Centaury	50	gm.
Orris Root	50	gm.
Wormwood	20	gm.
Cinnamon	10	gm.
Alcohol 95%	0.6	lit.
Liqueur Body	11.5	lit.

Color: Brown.

Spanish Bitter Creme

Spanish Bitter Essence	1.1–1.6	lit.
Liqueur Body	11.5	lit.

Color: Brown.

Stomach Bitter Essence—No. 1

Angelica Root	100	gm.
Gentian Root	100	gm.
Holy Thistle	20	gm.
Buck Bean	80	gm.
Wormwood	80	gm.
Bitter Orange Peel	80	gm.
Lemon Peel	50	gm.
Alcohol	10	kilos

No. 2

Angelica Root	30	gm.
Gentian Root	140	gm.
Holy Thistle	40	gm.
Buckbean	40	gm.
Bitter Orange Peel	200	gm.
Alcohol	10	kilos

Both Bitters Colored Brown Green.

Benedictine

Oil Sweet Orange, Hand-pressed	72	oz.
Oil Angelica Root	6	oz.
Oil Calamus	3	oz.
Oil Cinnamon, Ceylon	3	oz.
Oil Mace, Distilled	3	oz.
Oil Celery	3	oz.
Alcohol	12	oz.

Dissolve 1 oz. of above mixture in 5 gallons alcohol. Then add 5 gallons water to which has been added 24 lb. sugar.

Slivovitz

Oil Bitter Almonds, F.F.P.A.	2	mils
Oil Neroli, Artificial	1	mil
Oil Cognac, Genuine, Green	2	mils
Vanillin	5	gm.
Essence Raspberry Aroma	300	mils
Essence Plum	300	mils
Essence Jamaica Rum	25	mils
Essence Raisin Wine	50	mils
Prune Spirit	100	mils
Alcohol	100	mils

Dissolve 1 oz. of above mixture in 8 gallons alcohol. Then add 8 gallons water. Mix. Filter through magnesium carbonate.

Jamaica Rum

Oil of Cassia	1	dr.
Oil of Birch Tar	25	dr.
Oil of Ylang Ylang Natural	3	dr.
Oil of Orange Flower Natural	20	dr.
Oil of Ceylon Cinnamon	15	dr.
Rum Ether Pure	3	pt.
Acetic Ether	2½	oz.
Butyric Ether	1 oz. 1	dr.
Tincture of Saffron 1 lb. to a gal.	4	oz.
Extract of Vanilla Pure	3	oz.
Balsam Peru	2	dr.
Tincture Styrax U.S.P.	2	dr.
Coumarin	5	dr.

Dissolve 1 oz. of above mixture in 4½ gallons of alcohol. Then add 5½ gallons water. Mix. Filter through magnesium carbonate. Allow to age in barrel.

Goldwasser

Alcohol 90% by Volume	23¼	lit.
Goldwasser Essence	750	gm.
Rose Water	1¼	lit.
Orange Blossom Water	750	gm.
Sugar Solution	5	lit.
Water	20½	lit.

After the mixture has been stored for some time there is added to it a small quantity of genuine Gold Leaf.

Goldwasser Essence

Angelica Oil	4	gm.
Anise Oil	32	gm.
Lemon Oil	290	gm.
Spearmint Oil	32	gm.
Laurel Oil	32	gm.
Lavender Oil	64	gm.
Nutmeg Oil	16	gm.
Balm Oil	20	gm.

Clove Oil	64 gm.
Orange Oil	16 gm.
Rose Oil	16 gm.
Rosemary Oil	32 gm.
Juniper Oil	32 gm.
Orange Blossom Water	750 gm.
Sugar Solution	5 lit.
Water	20½ lit.

After the mixture has been stored for some time there is added to it a small quantity of genuine Gold Leaf.

Whiskey "Rye"

Oil Fusel Potato	2 pt.
Oil Fusel Rye	18 pt.
Rum Ether, Pure	20 pt.
Oil Coriander, Pure	5 oz.
Oil Bitter Almonds, F.F.P.A.	2 oz. 4 dr.
Alcohol	50 pt.
Tinct. Catechu	1 pt.
Vanillin	2 dr.
Heliotropin	4 dr.
Tinct. Balsam, Peru, True	1 dr.

Dissolve 1 oz. of above in 7¼ gallons alcohol. Then add to it 7¾ gallons water. Mix; filter; and color with caramel.

Scotch

Oil Corn Fusel	6 oz.
Oil Bitter Almonds	4 dr.
Oil Coriander	4 dr.
Oil Cade	1 oz.
Guaiacol	2 dr.
Butyric Ether	4 oz.
Alcohol	4 oz.

Dissolve 1 oz. of above mixture in 14 gallons alcohol. Then add 16 gallons water. Mix. Filter through magnesium carbonate. Color with caramel.

Geneva Gin

Alcohol 90% by Volume	22½ lit.
Geneva Essence	150 gm.
Water	27½ lit.

Mix well and store for several weeks.

Gin, Old Tom

Oil Coriander, Pure	3 oz. 4 dr.
Oil Angelica Root	3 dr.
Oil Anise, Russian, Rectified	1 oz.
Oil Caraway, Dutch	4 dr.
Oil Juniper Berries, Rectified	7 oz. 4 dr.
Alcohol	1 pt. 8 oz.

Dissolve 1 oz. of above oil in 4½ gallons alcohol. Then add 5½ gallons water.

Mix. Filter through magnesium carbonate.

Gin, London Dock

Oil Gin, Old Tom	6 oz.
Oil Gin, Holland	18 oz.
Oil Cassia, Rectified	4 dr.
Alcohol	64 oz.

Dissolve 1 oz. of above oil in 3 gallons alcohol. Then add 4 gallons water. Mix. Filter through magnesium carbonate.

Gordon Gin

Oil Juniper Berries	16 oz.
Oil Angelica Root	20 cc.
Oil Angelica Seed	20 cc.
Oil Coriander	40 cc.
Oil Lemon	60 cc.
Sweet Orange Oil	20 cc.
Neroli Oil	5 cc.
Geranium Rose Oil	5 cc.
Alcohol to make 1 gal.	

4 oz. of above is used to 50 gal. 50% alcohol.

Oil Gin Holland

Oil Lemon	1 dr.
Oil Anise	1 dr.
Oil Angelica Root	6 dr.
Oil Fusel	4 dr.
Oil Juniper Berries	20 oz.
Oil Rosemary Flavor	6 dr.
Oil Coriander	4 dr.
Alcohol	10 oz.

Dissolve 1 oz. of above oil in 7 gallons alcohol. Then add 8 gallons water. Mix. Filter through magnesium carbonate.

Holland Gin

Oil Gin	1000 mils
Glycerin C.P.	200 mils
Alcohol	216 oz.

Dissolve 5 oz. of above in 2¼ gallons alcohol. Then add 2¾ gallons water. Mix. Filter through magnesium carbonate.

Bourbon

Oil Bourbon	6 oz.
Alcohol	32 oz.
Sugar Color	20 oz.
Citric Acid Solution	8 oz.
Tannic Acid Solution	1 oz.
	67 oz.
Water	61 oz.
	128 oz.

Filter. Then dissolve 1 oz. of above in ½ gal. alcohol and then add ½ gal. water.

Super Aroma Bourbon

Oil Fusel, Rectified	240	oz.
Ess. Pineapple	½	oz.
Ess. Peach Blossom	½	oz.
Citric Acid Solution 50%	240	oz.
Solution Saccharin Saturated	¼	oz.
Oil Jam. Rum	13	oz.
Alcohol	133	oz.
Tannic Acid Solution	1	oz.

	628¼	oz.

Filter. Then 1 oz. of this will flavor 5 gallons of 50% alcohol.

Cherry Liqueur

Oil Bitter Almond	1	gm.
Vanilla Tincture	2	gm.
Orris Root Tincture	5	gm.
Cinnamon Tincture	0.5	gm.
Liqueur Body	11.5	lit.

Color: Cherry Red.

Cherry Brandy Liqueur

Genuine Cherry Brandy	1	pt.
Cherry Fruit Juice	1½	pt.
Alcohol	2	pt.
Sugar Syrup 65%	2	pt.
Water	2	pt.

Essence for Artificial Cherry Brandy
(1 oz. per gallon)

Oil of Neroli	2	drops
Oil of Cloves	¼	dram
Oil of Cinnamon	¼	dram
Oil of Bitter Almonds	2	oz.
Rum Ether	14	oz.
Wine Brandy	16	oz.
Colorless Cherry Flavor	3	lb.
Genuine Bitter Almond Water	5	lb.

Cherry Liqueur Essence
(2 oz. per gallon)

Vanillin	1½	dram
Oil of Cloves	2	oz.
Oil of Cinnamon	3	oz.
Benzaldehyde	5	oz.
Rum Essence	14	oz.
Alcohol	16	oz.
Cherry Juice	2½	lb.
Cherry Flavor	5	lb.

Kummel Liqueur

Alcohol 90% by Volume	45	lit.
Kummel Essence	1	lit.
Orange Peel Essence	¼	lit.
Sugar Syrup 65%	38¾	lit.
Water	15	lit.

Maraschino Liqueur

Alcohol 90% by Volume	36	lit.
Bitter Almond Oil Essence	115	gm.*
Steroli Oil Essence	200	gm.*
Rose Oil	30	drops
Sugar Syrup 65%	64	lit.

Maraschino

Oil Bitter Almond	3	gm.
Cognac Essence	2	gm.
Raspberry Ether	2	gm.
Oil Neroli	1	gm.
Vanilla Tincture	5	gm.
Liqueur Body	11.5	lit.

No Color.

Chocolate Liqueur

Cocoa Beans Burned	200	gm.
Clove Tincture	5	gm.
Vanilla Tincture	16	gm.
Cinnamon Tincture	5	gm.
Liqueur Body	11.5	lit.

Color: Dark Red.

Sherry Cordial

Alcohol 90% by Volume	35	lit.
Bitter Almond Oil Essence	56	gm.*
Ethyl Acetate	65	gm.*
Sugar Syrup 65%	45	lit.
Water	20	lit.

Creme de Menthe

Oil Peppermint, Twice Rectified	2	oz.
Menthol	2	dr.
Alcohol	35 oz.	4 dr.

Green Coloring.

Dissolve 1 oz. of this mixture in 1½ gallons alcohol. Then add 1½ gallons water in which has been dissolved 5½ lb. sugar.

French Liqueur (Cremes) as below:

Fleur d'Amour (Flower of Love)

Alcohol 90% by Volume	34	lit.
Lemon Oil Essence	½	lit.*
Clove Oil Essence	150	gm.*
Nutmeg Oil Essence	150	gm.*
Sugar Syrup 65%	45	lit.
Water	20½	lit.

Color: Bluish Red.

Flower of Love Essence

Oil Cloves	10 gm.
Oil Nutmeg	10 gm.
Oil Cinnamon	3 gm.
Alcohol 95%	10 kg.

Colored: Light Red.

Arrack

Ethyl Acetate	100	gm.
Black Balsam Peru	130	gm.
Vanilla	16	gm.
Oil of Neroli	5	gm.
Oil of Birch	1	gm.
Ground Horseradish	500	gm.
Onions	125	gm.
Iron Filings	2	kg.
Cocoa	25	gm.
Raisin Stems	1	kg.
Alcohol by Volume 90%	41	lit.
Water	27½	lit.

The above are mixed together and then filtered.

Clove Essence

Cloves	200 gm.
Cinnamon	50 gm.
Alcohol 95%	1 kg.

Color: Red-Brown.

Curacao Essence

1

Bitter Orange Oil	640	gm.
Neroli Oil	27	gm.
Sweet Orange Oil	27	gm.
Cinnamon Oil	13½	gm.

2

Bitter Orange Oil	640	gm.
Neroli Oil	27	gm.
Orange Peel Sweet Oil	27	gm.
Cinnamon Oil	13½	gm.

Creme de Curacao Dutch

Oil Pear	1	gm.
Oil Bitter Orange	1.5	gm.
Raspberry Ether	2	gm.
Oil Neroli	0.4	gm.
Oil Mace	0.4	gm.
Oil Orange	1.5	gm.
Vanilla Tincture	5	gm.
Oil Cinnamon	0.5	gm.
Liqueur Body	11.5	lit.

Collor: Yellow.

Woodruff Essence

Fresh Woodruff	4	kg.
Tonka Beans	100	gm.
Alcohol 95%	10	kg.

Color: Grass-Green.

Fine Caraway Liqueur

Oil Fennel	1	gm.
Oil Caraway	4	gm.
Vanilla Tincture	10	gm.
Oil Cinnamon	1	gm.
Liqueur Body	11.5	lit.

No Color.

Triple Caraway Essence

Oil Anise	2	gm.
Oil Lemon	5	gm.
Oil Coriander	3	gm.
Oil Caraway	150	gm.
Oil Mace	1	gm.

48 gm. of this mixture are to be mixed with 25 lit. of 60% Alcohol, 25 lit. Water, 1 lit. Bourbon Whiskey and 4 kilos Sugar.

Liqueur Body for Cremes and Huiles

No. 1

Sugar Sol. = 437 Grams Sugar in 1 Litre Water.

Sugar Sol. above	57.20 lit.
Alcohol	45.76 lit.
Water	11.40 lit.

No. 2

Sugar Sol. = 393.3 Grams Sugar in 1 Litre Water.

Sugar Sol. above	51.48 lit.
Alcohol	45.76 lit.
Water	28.60 lit.

No. 3

Sugar Sol. = 349.6 Grams Sugar in 1 Litre Water.

Sugar Sol. above	45.76 lit.
Alcohol	48.05 lit.
Water	22.88 lit.

Liqueur Body for Fine Liqueurs

No. 4

Sugar Sol. = 327.7 Grams Sugar in 1 Litre Water.

Sugar Sol. above	42.90 lit.
Alcohol	50.91 lit.
Water	20.59 lit.

No. 5

Sugar Sol. = 305.9 Grams Sugar in 1 Litre Water.

Sugar Sol. above	40.08 lit.
Alcohol	50.33 lit.
Water	24.02 lit.

No. 6

Sugar Sol. = 262.2 Grams Sugar in 1 Litre Water.

Sugar Sol. above	34.32 lit.
Alcohol	50.33 lit.
Water	27.25 lit.

Liqueur Body for Ordinary Liqueur
No. 7

Sugar Sol. = 218.5 Grams Sugar in 1 Litre Water.

Sugar Sol. above	28.60 lit.
Alcohol	53.77 lit.
Water	32.03 lit.

No. 8

Sugar Sol. = 174.8 Grams Sugar in 1 Litre Water.

Sugar Sol. above	22.88 lit.
Alcohol	50.08 lit.
Water	35.46 lit.

For Double Spirits or Whiskey
No. 9

Sugar Sol. = 131 Grams Sugar in 1 Litre Water.

Sugar Sol. above	17.16 lit.
Alcohol	57.20 lit.
Water	40.04 lit.

No. 10

Sugar Sol. = 109.25 Grams Sugar in 1 Litre Water.

Sugar Sol. above	14.30 lit.
Alcohol	58.31 lit.
Water	41.18 lit.

No. 11

Sugar Sol. = 87.4 Grams Sugar in 1 Litre Water.

Sugar Sol. above	11.44 lit.
Alcohol	59.48 lit.
Water	43.42 lit.

For Ordinary Spirits or Whiskey
No. 12

Sugar Sol. = 65.55 Grams Sugar in 1 Litre Water.

Sugar Sol. above	6.86 lit.
Alcohol	60.62 lit.
Water	46.90 lit.

No. 13

Sugar Sol. = 43.7 Grams Sugar in 1 Litre Water.

Sugar Sol. above	5.72 lit.
Alcohol	61.77 lit.
Water	46.90 lit.

Usquebaugh

Oil Anise	1 gm.
Oil Cardamom	0.5 gm.
Oil Lemon	0.5 gm.
Oil Coriander	0.5 gm.
Oil Mace	0.5 gm.
Oil Cloves	0.5 gm.
Oil Cinnamon	0.5 gm.
Liqueur Body	11.5 lit.

Color: Yellow.

Crambambuli

Oil Cardamom	1 gm.
Oil Lemon	1 gm.
Oil Mace	1 gm.
Oil Cloves	0.5 gm.
Oil Orange	1 gm.
Oil Cinnamon	0.5 gm.
Liqueur Body	11.5 lit.

Color: Dark Red.

Mulled Wine Extract

Sugar	47½ kg.
Water	14 lit.
Cherry Juice	8 lit.
Raspberry Juice	2¼ lit.

Cook the above together and then add:

Alcohol 90%	37½ lit.
Clove Essence	⅙ lit.
Cinnamon Essence	⅙ lit.
Moselle Wine	11½ lit.

Color: Dark Cherry.

Rhine Wine Extract

Mix together:

Alcohol 90%	3½ lit.
Strawberry Oil	75 gm.
Orange Oil	50 gm.
Pineapple Essence	20½ lit.
Woodruff Essence	100 gm.
Neroli Oil	48 drops

Color: Slightly Yellow.

Creme de Flauve d'Orange
(Orange Liqueur)

Alcohol by Volume 95%	4¼ lit.
Neroli Oil Essence	132 gm.
Bitter Oil Almond Ess.	175 gm.

| Sugar Syrup 65% | 11¼ lit. |
| Water | 1½ lit. |

Orange Fruit Liqueur

Fresh Orange Juice	1.1 lit.
Alcohol 95%	4.6 lit.
Water	3.4 lit.
Sugar	4.5 kg.
Curacao Tincture	100–200 gm.

The fruit juice stays with alcohol for 8 days, then filter. The clear liquid has to be mixed with a Sugar Solution then add carefully the Curacao Tincture to avoid bitter taste.

Color: Golden Yellow.

Apricot Fruit Liqueur

Apricots	6 kg.
Sugar	4 kg.
Alcohol 95%	3.3 lit.
Water	1.1 lit.
Cinnamon Tincture	50 gm.

Color: Rose Red.

Cherry Fruit Liqueur

Cherries	5.5 kg.
Sugar	3 kg.
Alcohol 95%	4.6 lit.
Water	1.1 lit.
Bitter Almond Tincture	50 gm.

Very sweet dark cherries very ripe, have to be squashed including the pits, in a stone mortar. The mash has to stand a few days in a cool place then press it out and add the sugar and water and heat until it boils. After it cools off add Bitter Almond Tincture and alcohol.

Color: Dark Red.

Date Fruit Liqueur

Dates (Squashed)	4 kg.
Water	4.6 lit.
Alcohol 95%	4.6 lit.
Sugar	4 kg.

Fig Fruit Liqueur

Fresh Figs	5 kilos
Water	2–3 lit.
Alcohol 95%	4–6 lit.
Sugar	2 kilos

Peach Fruit Liqueur

Peaches	6 kg.
Sugar	4.5 kg.
Alcohol 95%	4.6 lit.
Water	1.7 lit.
Bitter Almond Tincture	16 gm.

The fruits skinned and pits removed. Then to be squashed and ex-pressed. To the residue 1.7 lit. water added together with the stamped pits. This mash remains for 2 days then press it. Dissolve sugar in those liquids, add Bitter Almond Tincture. No heating.

Color: Pale Red.

Pineapple Fruit Liqueur

Pineapples	2
Alcohol 95%	4.6 lit.
Water	3.4 lit.
Sugar	4.5 kg.
Vanilla Tincture	50 gm.
Pear Ether	5 gm.

Color: Yellow.

Raspberry Fruit Liqueur

Fresh Pressed Raspberry Juice	10 lit.
Alcohol 95%	8 lit.
Sugar	6 kg.

Raspberry Fruit Liqueur

6 kg. Raspberries (squashed) are to be extracted with 2 lit. Water and 5 lit. Alcohol 95%. Shake daily for 14 days.

Apple Fruit Ether

Ethyl Acetate	10 gm.
Amyl Valeriate	100 gm.
Cold Saturated Malic Acid Solution	10 gm.
Alcohol 95%	1000 gm.

Apricot Fruit Ether

Ethyl Butyrate	100 gm.
Ethyl Valeriate	50 gm.
Oil Bitter Almond	10 gm.
Alcohol 95%	1000 gm.

Cherry Fruit Ether

Ethyl Acetate	50 gm.
Ethyl Benzoate	50 gm.
Oil Bitter Almond	10 gm.
Cold Saturated Solution Benzoic Acid in Alcohol 95%	10 gm.
Alcohol 95%	1000 gm.

Current Fruit Ether

Ethyl Acetate	50 gm.
Ethyl Formate	10 gm.
Ethyl Butyrate	10 gm.
Ethyl Benzoate	10 gm.

Ethyl Oenanthate	10 gm.
Ethyl Salicylate	10 gm.
Ethyl Sebaciate	10 gm.
Amyl Butyrate	10 gm.
Cold Saturated Solution of	
Tartaric Acid	50 gm.
Alcohol 95%	1000 gm.

Grape Fruit Ether

Ethyl Formate	20 gm.
Ethyl Oenanthate	100 gm.
Methyl Salicylate	10 gm.
Cold Saturated Solution of	
Tartaric Acid in Alcohol	50 gm.
Succinic Acid	30 gm.
Alcohol 95%	1000 gm.

Lemon Fruit Ether

Ethyl Acetate	100 gm.
Oil Lemon	100 gm.
Cold Saturated Citric Acid	
Solution	100 gm.
Alcohol 95%	1000 gm.

Melon Fruit Ether

Ethyl Formate	20 gm.
Ethyl Butyrate	40 gm.
Ethyl Valeriate	50 gm.
Ethyl Sebaciate	100 gm.
Alcohol 95%	1000 gm.

Orange Fruit Ether

Ethyl Acetate	50 gm.
Ethyl Formate	10 gm.
Ethyl Butyrate	10 gm.
Ethyl Benzoate	10 gm.
Methyl Salicylate	10 gm.
Amyl Acetate	10 gm.
Orange Flower Oil	100 gm.
Cold Saturated Solution of	
Tartaric Acid in Alcohol	10 gm.
Alcohol 95%	1000 gm.

Peach Fruit Ether

Ethyl Acetate	50 gm.
Ethyl Formate	50 gm.
Ethyl Butyrate	50 gm.
Ethyl Valeriate	50 gm.
Ethyl Sebaciate	10 gm.
Oil Bitter Almond	50 gm.
Alcohol 95%	1000 gm.

Pear Fruit Ether

Ethyl Acetate	50 gm.
Amyl Acetate	100 gm.
Alcohol 95%	1000 gm.

Pineapple Fruit Ether

Ethyl Butyrate	50 gm.
Amyl Butyrate	100 gm.
Alcohol 95%	1000 gm.

Strawberry Fruit Ether

Ethyl Acetate	50 gm.
Ethyl Formate	10 gm.
Ethyl Butyrate	50 gm.
Ethyl Salicylate	10 gm.
Amyl Acetate	30 gm.
Amyl Butyrate	20 gm.
Alcohol 95%	1000 gm.

Weichxel Fruit Ether

Ethyl Acetate	100 gm.
Ethyl Benzoate	50 gm.
Oil Bitter Almond	20 gm.
Cold Saturated Solution	
Malic Acid in Alcohol	10 gm.
Benzoic Acid	30 gm.
Alcohol 95%	1000 gm.

Birch Oil Spirit

| Alcohol 90% | ¼ lit. |
| Oil Birch | 5 gm. |

Bischof Extract

Cardinal Extract	3 lit.
Orange Peels	100 gm.
Bitter Orange Oil	100 gm.

Champagne

Rhine Wine	32 lit.
Whole Lemons and peels	
cut up	4
Raisins	2 kg.
Orange Oil Essence	30 gm.
Oil of Neroli	10 drops
Sugar	8 kg.
Water	2 lit.

Cherry Lemonade

Cherry Juice cooked with	17 lit.
Sugar and	12½ kg.
Tartaric Acid dissolved	
in ¼ lit. water	125 gm.

Lemon Lemonade

Sugar Syrup 65%	45 lit.
Alcohol 90%	4 lit.
Lemon Oil Essence	1½ lit.
Citric Acid dissolved in	
1 lit. water	750 gm.

Orange Lemonade

Sugar Syrup 65%	45	lit.
Alcohol 90%	4	lit.
Citric Acid dissolved in 1 lit. water	750	gm.
Orange Oil Essence	1½	lit.

Raspberry Lemonade

Sugar with	75	kg.
Raspberry Juice	31	lit.
Cherry Juice	10	lit.
Water	7½	lit.
Tartaric Acid	1½	kg.

Heat together juices and sugar; then dissolve acid in water and then mix all together.

Strawberry Lemonade

Sugar Syrup 65%	30	lit.
Alcohol 90%	3	lit.
Strawberry Ether	25	gm.
Citric Acid dissolved in 1 lit. water	750	gm.
Color: Strawberry.		

Claret Essence

Ambergris Tincture	¼	dr.
Ethyl Acetate	3¾	dr.
Carob Tincture	8½	oz.
Cherry Juice	7¼	oz.
Krameria Tincture	4	lb.
Wine Distillate	4	lb.

Claret Lemonade

Clove Tincture	3	dr.
Cinnamon Tincture	5	dr.
Claret Essence	2	oz.
Cherry Juice	5½	oz.
Red Wine	8	oz.

Muscatel Lemonade

Honey Lemonade Essence	½	oz.
Claret Essence	2	oz.
Port Wine Essence	3	oz.
Grape Essence	10½	oz.

Nectar Lemonade

Honey Lemonade Essence	¼	oz.
Rum Essence	1¾	oz.
Port Wine Essence	3	oz.
Currant Essence	3	oz.
Apple Essence	8	oz.

Port Wine Essence

Ambergris Tincture	¼	dr.
Ethyl Acetate	7¾	dr.

Krameria Tincture	1½	oz.
Elder Flower Tincture	2	oz.
St. John's Bread Tincture	3	oz.
Carob Tincture	3	oz.
Cacao Essence	3	oz.
Wine Distillate	3	oz.

White Wine Essence

Cognac Oil	10	dr.
Ethyl Nitrite	22	dr.
Ethyl Acetate	1½	oz.
St. John's Bread Tincture	12½	oz.
Wine Distillate	4½	lb.
Water	4½	lb.

WINE FONDANT FLAVORS

Burgundy Fondant

Ambergris Tincture	½	oz.
Rhatany Tincture	3½	oz.
Cherry Juice	4	oz.
Raspberry Essence	8	oz.
Black Currant Essence	1	lb.
Grape Essence	8	lb.

Claret Fondant

Civet Tincture	½	oz.
Ambergris Tincture	½	oz.
Rhatany Tincture	7	oz.
Black Currant Essence	1	lb.
Cherry Juice	2	lb.
Grape Essence (from dried grapes)	6½	lb.

Madeira Fondant

Pineapple Essence	4	oz.
Brown Cacao Essence	8	oz.
Elder Flower Tincture	8	oz.
Black Currant Essence	12	oz.
Grape Essence	8	lb.

Malaga Fondant

Civet Tincture	5	gm.
Ambergris Tincture	5	gm.
Vanillin	5	gm.
Cherry Water Genuine	9	gm.
Rhatany Tincture	6½	oz.
Black Currant Essence	1	lb.
Carob Tincture	1	lb.
Grape Essence	7½	lb.

Muscatel Essence

Coumarin	¼	oz.
Mace Tincture	1¾	oz.
Elder Flower Essence	6	oz.
Apple Essence	1½	lb.
Grape Essence	8	lb.

Port Fondant

Vanillin	1/4	oz.
Ambergris Tincture	1/4	oz.
Brown Cacao Essence	7 1/2	oz.
Rhatany Tincture	8	oz.
Grape Essence	9	lb.

Rhine Wine Fondant

White Cognac Oil	2 1/2	oz.
Heliotropin	3 1/2	oz.
Ethyl Acetate	10	gm.
Apple Essence	1 lb. 7	oz.
Grape Essence	8 1/2	lb.

Sherry Fondant

Civet Tincture	1/4	oz.
Elder Flower Tincture	2 3/4	oz.
Black Currant Essence	13	oz.
Pineapple Essence	1	lb.
Grape Essence	8	lb.

Tokay Fondant

Civet Tincture	2 1/2	dr.
Pineapple Essence	1/2	lb.
Raspberry Essence	1/2	lb.
Carob Tincture	1	lb.
Grape Essence	8	lb.

Pear Essence

(1 oz. per gallon)

Vanillin	1/2	dr.
Amyl Acetate	1 1/2	dr.
Raspberry Distillate	5	oz.
Bergamot Essence	11	oz.
Orange Flower Water	1	lb.
Wine Brandy	1	lb.

Distilled Water	2 1/2	lb.
Alcohol	4 1/2	lb.

Burgundy Wine Punch Extract

Vanilla Essence	1	oz.
Lemon Juice	1/4	gal.
Rum	1/4	gal.
Arrac	1/4	gal.
Water	2 1/4	gal.
Genuine Burgundy Wine	3	gal.
Sugar Syrup	4	gal.

Claret Punch Extract

Cardamom Tincture	1	oz.
Cinnamon Tincture	3	oz.
Clove Tincture	3	oz.
Lemon Juice	1	lb.
Genuine Rum	1/4	gal.
Sugar Syrup	4	gal.
Dark Claret Wine	4 3/4	gal.

Glowing Wine Punch Extract

Cardamom Tincture	2	oz.
Pineapple Essence	3	oz.
Cinnamon Tincture	5	oz.
Clove Tincture	5	oz.
Genuine Arrac	1/4	gal.
Alcohol	1 1/2	gal.
Cherry Fruit Syrup	4	gal.
Claret Wine	4 1/2	gal.

White Wine Punch Extract from Moselle, Rhine or Chablis Wine

Sweet Orange Juice	1/2	gal.
Genuine Arrac	1	gal.
Sugar Syrup 65%	3 3/4	gal.
Moselle, Rhine or Chablis	5	gal.

Chewing Gum Bases

a. Bubble Gum Base:

Washed Pontianac Gum	425	lb.
Washed Gutta Katian	400	lb.
Washed Gutta Soh	75	lb.
Candelilla Wax	10	lb.

The mixed gums and wax are heated until the total batch contains only 8-9% moisture.

b. Stick Gum Base:

Pontianac Gum	425	lb.
Gutta Katian	400	lb.
Gutta Soh	75	lb.
Candelilla Wax	60	lb.

Chewing Gum
Formula No. 1

Ball Gum:

Base b (above)	22	lb.
Corn Syrup	48	lb.
Sugar	117	lb.
Chicle	3	lb.
Wax	1 1/2	lb.
Caramel Paste	2 1/2	lb.
Flavor	2 2/3	oz.

No. 2

Penny Stick Gum:

Base a (above)	40	lb.
Corn Syrup	40	lb.
Sugar	140	lb.
Flavor	30	oz.

No. 3

Bubble Gum:

Base a (above)	35	lb.
Pontianac Gum	5	lb.
Corn Syrup	45	lb.
Sugar	115	lb.
Flavor	28	oz.

CHAPTER VII

INKS, CARBON PAPER, CRAYONS, ETC.

INKS may be roughly divided into writing, printing and specialty divisions. Each group in turn has many types. Thus a writing ink may be for general records, copying and fountain-pen. Printing inks may be for newspapers, books, magazines, metal foil, "cellophane" etc. Under specialty inks there are inks such as hectograph, stamp-pad, meat-marking, water-proof, invisible, etc.

Crayons, chalks, pencils, etc., likewise vary in compositions depending on the use to which they are put.

Accessories such as carbon paper, stencil paper, hectograph mass and printers roller formulae are included in this chapter.

Ink: Copying and Record

All the ingredients in the standard ink must be of the quality prescribed in the current edition of the United States Pharmacopoeia.

Tannic Acid	23.4 gm.
Gallic Acid Crystals	7.7 gm.
Ferrous Sulfate	30.0 gm.
Hydrochloric Acid, Dilute	25.0 gm.
Phenol (Carbolic Acid)	1.0 gm.
Soluble Blue	3.5 gm.

Water to make 1 liter at 20°C. (68°F.)

Here as in all other formulae, "water" means distilled water, if it can be had. Rain water is second choice.

Dilute hydrochloric acid, U.S.P., is of 10 per cent strength. Concentrated hydrochloric acid as commonly sold is a water solution containing about 36 per cent by weight of hydrochloric acid gas, so as to make the 10 per cent acid, 100 parts by weight of concentrated acid must be diluted with 260 parts by weight of water.

Soluble blue is one of the comparatively few dyes that are not precipitated by the other ingredients of the ink. When buying a supply of it, be careful to say that it is to be used for making ink.

To make the ink, dissolve the tannic and gallic acids in about 400 milliliters of water at a temperature of about 50° C. (122° F.). Dissolve the ferrous sulfate in about 200 milliliters of warm water to which has been added the re-

quired amount of hydrochloric acid. In another 200 milliliters of warm water dissolve the dye. Mix the three solutions and add the phenol. Rinse each of the vessels in which the solutions were made with a small quantity of water, and use the rinsings to make the volume of ink up to 1 liter at room temperature. Be sure the ink is well mixed before it is bottled. If sealed hermetically in a glass bulb, the ink will keep for years with practically no formation of sediment. So when bottling the ink, have good tight corks and fill the bottles almost to the corks.

This ink is primarily for records, and is not like most copying inks. However it will make one good press copy when the writing is fresh, and this will generally suffice.

Writing and Copying Ink

	Fountain Pen Ink	Copying Ink
Tannic Acid	1.55 oz.	3.10 oz.
Gallic Acid	0.50 oz	1.00 oz.
Ferrous Sulfate	2.00 oz.	4.00 oz.
Hydrochloric Acid (dilute)	1.67 oz.	3.34 oz.
Phenol	0.13 oz.	0.13 oz.
Soluble Blue	0.47 oz.	0.47 oz.

Dissolve the Tannin and Gallic Acid in about 3 pints of warm water (of about 130° F.) and add to it the Dilute Hydrochloric Acid (of about 7° Bé.) and then the solution of Ferrous Sulfate and Phenol in about 2 pints of water. Bring up to 1 gallon, mix well and let stand quietly for 4 days. Then decant without stirring up any sediment formed.

---◆---

Ink: Writing

1

Except for the phenol and dye, this ink is half as concentrated as the record and copying ink. It is similar to some of the commercial writing fluids and fountain pen inks. The standard is made in the same way as the preceding ink, and from materials of the same quality. If made with slightly more hydrochloric acid than the formula calls for it will keep longer without depositing sediment, but it will be more corrosive to steel pens.

The standard formula is:

Tannic Aid	11.7 gm.
Gallic Acid Crystals	3.8 gm.
Ferrous Sulfate	15.0 gm.
Hydrochloric Acid, Dilute	12.5 gm.
Phenol (Carbolic Acid)	1.0 gm.
Soluble Blue	3.5 gm.

Water to make 1 liter at 20° C. (68° F.).

---◆---

2
(8 times concentrated)

The ingredients are best dissolved as follows:

2 ounces Ferrous Sulfate { 1⅜ oz. of dil. Hydrochloric
Dissolved in
3 oz. of Water

0.47 oz. of Soluble Blue } 3 oz of Water
0.13 oz. of Phenol
1.55 oz. of Tannic Acid } 6 oz. of Water
0.50 oz. of Gallic Acid

For washing, etc., 2⅜ oz. of Water

Dissolve first the Dye and Phenol; pour into this mixture the acid solution of Iron and then the Tannic-Gallic Acid solution. All solutions should be heated to about 180° F. and the final mixture stirred well for some time and then allowed to cool. Let stand quietly for 2 or 3 days and decant.

---◆---

Writing Ink—Red

Eosine	1 oz.
Gum Arabic	1 oz.
Phenol	½ oz.
Water	1 gal.

---◆---

Writing Ink—Blue Black

Naphthol Blue Black	1	oz.
Gum Arabic	½	oz.
Phenol	¼	oz.
Water	1	gal.

Red Writing Ink

Water, Warm	250	gal.
Crosein Scarlet	15	lb.
Carbolic Acid	1½	lb.

---◆---

Blue Writing Ink

Water, Warm	250	gal.
Methylene Blue	15	lb.
Carbolic Acid	1½	lb.

---◆---

Jet Black Writing Ink

Water, Warm	250	gal.
Nigrosine	15	lb.
Carbolic Acid	1½	lb.

Directions

Dissolve all color in 25 gallons of hot water (about 160° F.), add balance of warm water while mixing. Allow to stand several days then decant without stirring up any sediment.

---◆---

Ink: Red

The standard ink is made by dissolving 5.5 grams of crocein scarlet 3B in 1 liter of water.

---◆---

Concentrated Ink, Powder and Tablets

Concentrated ink that meets all the requirements of the specifications can be made by cutting down the amount of water to a minimum, so as to make a pasty mass or a thick fluid with the solids only partly dissolved. Instead of hydrochloric acid, which is volatile, an equivalent quantity of sulfuric acid is used; that is, 1.77 grams of the usual concentrated acid of 95 per cent strength (66° Baumé).

---◆---

Hectograph Ink

Years before some of the modern duplicating devices had been invented, the hectograph was used for printing small editions of circular letters, etc., and it is still in rather wide use. The original is written with a special ink that contains a large proportion of a dye that has good tinting strength. The letter is then pressed face-downward upon a gelatin-glycerin or a clay-glycerin pad, which absorbs a considerable amount of the ink. From this pad it is possible to print a number of increasingly paler copies upon other sheets of paper. The name, hectograph, ''hundred writing,'' exaggerates somewhat, unless copies so pale as to be barely legible are counted. In experimenting with quite a number of dyes, it was found that the following would give at least 30 copies with unbroken line, and numerous other copies that were easily legible, though there were breaks in the strokes of the pen. Methyl violet gave the most copies, the best red dye was rhodamine B, and emerald green and Victoria blue were the best of their colors.

The ink used in making these tests was prepared according to the formula:

Acetone	8
Glycerin	20
Acetic Acid, Coml. 30%	10
Water	50
Dextrin	2
Dye	10

Hectograph Ink

1

Acetone	8
Glycerin	20
Acetic Acid (28%)	10
Water	50
Dextrin	2
Dye	10

Dissolve dextrin in hot water with stirring; cool and add other liquids and dye.

2

Fuchsin	1	oz.
Alcohol	1	oz.
Glycerin	¼	oz.
Phenol	½	oz.

Hectograph Mass

Good Grade Powdered Glue	2 parts
Water	1 lb.
Glycerin	4 lb.

Proceed as in printers' rollers composition.

Stamp-Pad Ink

1

A solution of dye in water could be used on a stamp pad, but it would soon dry out. A mixture of equal volumes of glycerin and water remains moist under all atmospheric humidities, though the water content of the mixture fluctuates. In each 100 milliliters of the mixture of glycerin and water dissolve 5 grams of dye. The following are used for making the standards of different colors in the specification: water-soluble nigrosine

(black), soluble blue, light green, magenta (red), and acid violet.

2

Glycerin	5 lb. 6 oz.
Water	4 lb. 2 oz.
Warm to 150° F. and add	
Methyl Violet	6⅔ oz.

slowly while stirring. Allow to cool and stand for a few days and filter.

3

Magenta	4 oz.
Acetic Acid	4 oz.
Water	1 qt.
Alcohol	1 pt.
Glucose (43° Bé.)	1 pt.
Glycerin	2 qt.

Add the dye slowly with stirring to the mixture of other ingredients. Warm and stir until dissolved. Allow to stand a few days and filter. For violet and green inks acetic acid may be used as above; for other colors leave out acetic acid.

4

Denatured Alcohol	1 part
Spirit Soluble Aniline Dystuff	1–3 parts
Glycerin	4–5 parts

Mix thoroughly in water bath at 100–130° F. Allow to cool.

Use.—Apply to inking pads or as stenciling ink.

Meat Stamping Inks

1

A. Red

Carmine	16
Ammonium Hydroxide	120
Glycerin	45
Stir until dissolved then stir in	
Dextrin	20

B. Blue

Pure Food Blue Dye	30
Dextrin	20
Glycerin	82
Water	70

2

14 lb. of spirit soluble nigrosine is dissolved in a warmed mixture of Glycerin 28 lb., Glycerin 10 lb., Acetic Acid Glacial 12 lb. Cool and add 136 lb. alcohol.

Recording Inks

For outdoor recording instruments the Weather Bureau uses inks made by dissolving about 10 grams of dye in 1 liter of a mixture of equal volumes of glycerin and water. As this mixture will freeze in some parts of the country, it is sometimes necessary to add a certain proportion of alcohol to the ink.

For recording instruments in the laboratory, the ink needs to contain only enough glycerin to prevent its drying at the tip of the pen. A mixture of 1 volume of glycerin and 3 volumes of water has been found satisfactory.

Almost any water-soluble dye might be used were it not that some of them rather unaccountably make blurred lines on the usual card and paper charts. Dyes that have been found to work well are crocein scarlet, fast crimson, brilliant yellow, emerald green, soluble blue, methylene blue, methyl violet, Bismarck brown, and water-soluble nigrosine.

Finger-Printing Ink

Glycerin	112
Ferric Chloride	10
Colloidal Carbon Black	1
Acetone	90

Blue-Print Ink

For writing on blue prints use the following which bleaches white:

Soda Ash	10 gm.
Water	50

Blue Marking Ink

Shellac	2
Gum Acacia	2
Borax	2
Aniline Dye	sufficient
Ultramarine Blue	sufficient
Water	26

Outdoor Ink

Shellac	12.5
Alcohol	22.5
Cresol Tech.	15
Nigrosine Base	5

Ink, Sympathetic (Invisible)

1

A solution of oxal-molybdic acid yields an "ink" the characters made with which are invisible in the lamp-light, or in weak daylight, but which, exposed to strong sunlight or electric arc light, suddenly appear in deep indigo blue. The acid is prepared by adding to a boiling solution of molybdic acid one of oxalic acid, also boiling, letting cool, and recovering the crystals which form. Dissolve these in cold water to make the "ink." A sheet of paper immersed in the solution and dried in the dark becomes blue when exposed to the sun. If written on with a pen dipped in plain water, the letter will appear white on a blue ground. If the paper be held close to a hot fire, the blue becomes black. Similarly, the blue letters that appear on a white ground, if strongly heated, become permanently brown or black.

2

Make a five or ten per cent solution of cobalt chloride in soft or distilled water. When marks are made with this on paper it is not noticeable when dry at ordinary temperature; on heating the paper, *blue-green* lines will appear.

3

Writing or a drawing made with a ten per cent solution of lead acetate (or sugar of lead) in water will turn *black* if exposed to hydrogen sulfide, or if a weak solution of ammonium sulfide is brushed gently over it.

4

Writing made with a five or ten per cent solution of ammonium or potassium thiocyanate in water will turn a deep *red* if brushed gently or sprayed with a dilute solution of ferric chloride.

5

Cobalt Chloride	3 dr.
Water	4 oz.
Glycerin	1 dr.

6

Linseed Oil	1 dr.
Ammonia Water	20 dr.
Water	100 dr.

This ink leaves no visible stain on the paper, but when it is dipped in water, and while it is wet, the secret can be read. As the paper dries the writing again disappears.

Transfer Ink

1. Ultramarine Blue	50
2. Gum Mastic	30
3. Beeswax	10
4. Petrolatum	10

Melt (3) and (4), work in (1) and mix with melted (2).

Waterproof Drawing Inks

Yellow

Fresh Bleached Shellac	28 gm.
Borax Crystallized	7 gm
Water	1000 cc.

Dissolve the above by warming and stirring; then add with stirring

Erythrosine Yellow	1 oz.

By substituting the following dyes in a like amount the corresponding shades are obtained:

Orange—Brilliant Orange R
Yellow—Chloramine Yellow
Green—Brilliant Milling Green B
Blue—Wool Blue G Extra
Violet—Methyl Violet B
Brown—Benzamine Brown 3GO

Ink Eradicator for Tracing Cloth

Turpentine	17
Pumice Dust	53
Petrolatum	14
Paraffin	16

Typewriter Ribbon Ink

Petroleum Oil	108
Peerless Carbon Black	25–30
Oleic Acid	20
Toner (Oil Soluble)	10

Grind until uniform.

Stencil Paper

A stencil sheet coating composition containing the following substances in substantially the proportions specified:

Aluminum Stearate	2 parts
(45% Solution) Phenol	
Formaldehyde Resin	16 parts
Chlorinated Naphthalene	14 parts
Corn Oil	13 parts

Composition for Printing Rollers

1

Ingredients	Composition "A"	Composition "B"
Glue	10 lb.	32 lb.
Molasses	0 lb.	12 lb.
Sugar	10 lb.	0 lb.
Glycerin	12 lb.	56 lb.
Isinglass	1½ oz.	3 lb.
India Rubber in Naphtha	0 lb.	10 lb.

2

Powdered Hide Glue	1 part
Glycerin	1½ parts

Water	1 part
Sugar	½ part

Add glue and sugar to mixture of water and glycerin and stir well. Allow to stand until glue is thoroughly soaked and then place on water bath and melt. When mass is completely molten and all air bubbles have risen to surface, it is ready to be poured into molds.

Roller, Printers

Glue Highest Grade	20 lb.
Water	20 lb.

Soak ½ hr.

To this add

White Corn Syrup	40 lb.

Cook in double boiler for 2 hrs.

Add

Glycerin	16 lb.
Rezinel No. 2	1 lb.

Agitate with a high speed mixer until uniform and cast on a rubber core.

Flexible Printing Roller

Casein Glue Solution	10
Glycerin	5
Molasses	5
Clovel	1

Mix until uniform and pour into forms.

Carbon Paper

Crystal Violet Base or	
Methyl Violet Base	300 parts

are dissolved in

Red Oil	600 parts

This is introduced into approximately

Sesame Oil	3500–4000 parts

and added to

Carnauba Wax	3500 parts

melted at 105–110° C.

Black Carbon Paper

75% of these materials in proportions suitable for grade desired.

Candelilla Wax
Beeswax
Crude Montan Wax
Mineral Oil

Toners (Oil Soluble)	10%
Peerless Carbon Black	15%

This is ground hot. It is a base formula which may be modified to suit conditions.

Laundry Marking Ink

1

A. Soda Ash   1
   Gum Acacia   1
   Water   10
B. Silver Nitrate   4
   Gum Acacia   4
   Lampblack   2
   Water   40

Wet cloth with solution A and dry.
Write with solution B using a quill pen.

2

Silver Nitrate   6
Gum Acacia   6
Soda Ash   8
Distilled Water   15
Ammonium Hydroxide   8

3

Silver Nitrate   15
Copper Sulfate   35
Gum Arabic   20
Sal Soda   20
Distilled Water   80
Ammonium Hydroxide   50

4

A. Copper Chloride   85
   Sodium Chlorate   106
   Ammonium Chloride   53
   Water   600
B. Aniline Hydrochloride   60
   Glycerin   30
   Gum Acacia   20
   Water   130

Mix 1 part of A with 4 of B and use
immediately as mixture does not keep.
The marking is ''fixed'' by steaming it.

5

Aniline Black   7 gm.
Alcohol   200 cc.
Hydrochloric Acid   12 cc.
Shellac   10 gm.
Alcohol   800 cc.

Dissolve the shellac in alcohol and then
stir in other ingredients.

Marking Ink

A water glass marking ink is made by
cooking together fifty parts by weight of
water glass, 38 to 40 degrees Bé. concentration, and twenty-five parts by weight
of each of water and ground rosin. The
cooking continues until a smooth soap
solution is formed. Before this solution
cools down, twenty parts by weight of
carbon black are added. When the proportions used above are changed, so that
equal parts by weight of water glass,
water and rosin are used, and when this
soap solution is mixed with twenty-five
parts by weight of carbon black and
seventy-five parts by weight of mineral
black, a so-called marking india ink is
obtained. This ink may then be com-
pressed into tablets and dried. When
moistened with a wet brush, the color is
transferred to the same and hence the
ink can be used for marking purposes
with or without stencils.

Marking Ink, Waterproof

A waterproofing marking ink is made
by heating almost to the boiling temperature a mixture of seventy parts by
weight of water, five parts by weight of
ammonia, 0.910 specific gravity, and
twenty-five parts by weight of pulverized, red acaroid resin. The mass is constantly agitated while being heated.
Then sufficient ammonia is added in
small proportions, until the resin is completely dissolved, that is the undissolved
part from the first cooking is brought
into solution. The solution, still in the
hot state, is then passed through a very
fine sieve or through a hair cloth. The
sieved mass is then mixed with one-half
part by weight of acid green, three parts
by weight of bluish or violet-tinted
nigrosin, three parts by weight of sul-
phonated castor oil and 0.1 part by
weight of tri-cresol. In order to make
the ink somewhat thicker in consistency,
a little shellac size or casein solution is
added. If the acaroid resin solution becomes too thick, this is generally due to
the use of too much shellac size or casein
solution or ammonia.

Waterproof Show Card Ink

Hydromalin   13.8 lb.
Carnauba Wax   25   lb.

Heat together for ½ hr. at 120–140°
C. Turn off heat and dissolve with
stirring.

Any oil soluble dye   0.3 lb.

When temperature has fallen to 100°
C., add while stirring vigorously,

Distilled Water, Boiling   178 lb.

Stir until uniform.

Ink for Use on Metals

Copper Sulfate	10 g.
Hydrochloric Acid, Conc.	4 g.
Ammonium Chloride	8 g.
Gum Arabic	4 g.
Lamp Black	2 g.
Water	10 g.

An iron marking black can be obtained by mixing thirty parts by weight of medium hard stearin pitch with twenty-five parts by weight of rosin pitch, forty parts by weight of coal tar light oil and five parts by weight of carbon black. The two pitches are first melted together, the molten mass removed from the flame and then very carefully mixed with the light oil or crude benzol. Great care must be taken to avoid the mass running over or the benzol or light oil catching fire. Then the carbon black is added after first being passed through a fine screen. This ink is very well suited for marking metal containers and sheet metal and in fact for all purposes where the ink does not penetrate into material and hence must possess a marked tendency to adhere firmly to the surface of the same.

Ink for Zinc

Copper Acetate	1
Ammonium Chloride	1
Water	15
Lampblack	½

Copper Sulfate	1
Potassium Chlorate	1
Water	36

Black Stencil Ink

Paris Paste is thinned down with water and rapid stirring to the consistency desired.

If a waterproof ink is desired the water is replaced by a rubless wax emulsion or borax shellac solution.

Mimeograph Moistening Compound

Powdered Soap	8 oz.
Castile Soap	5 oz.
Glycerin	4 oz.
Water to make	1 gal.

Mimeograph Ink Base

1. Lampblack (Best Grade)	10.5
2. Violet Toner	1.1
3. Aluminum Hydrate Light	3.8
4. Long Varnish	1.1
5. Castor Oil	65.5
6. Lanolin	18.0

Mix (1), (2) and (3) dry and add (4) and (5) and continue mixing until uniform; add (6) and mix until thoroughly incorporated. Then grind on a four roll mill. This base ink is too heavy for direct use and is thinned down with castor oil to suit.

Mimeograph Ink

Lampblack (Best Grade)	6.4
Violet Toner	0.6
Aluminum Hydrate (Light)	2.2
Long Varnish	0.6
Castor Oil	78.5
Lanolin	11.7

Follow same procedure as for mimeograph ink base.

Embossing Ink

Glycerin, 5 parts; Silicate of Soda (water glass), 2 parts; water, 8 parts; Red Ink, or other ink, enough to color. The coloring is only to guide the pen when writing. The glycerin and silicate keep the writing moist and tacky so that it will retain the gold or silver powder dusted on. Thin writing produces the best results. Dust the gold or other powder on the writing and flick off the surplus powder with the finger, from the back of the paper. An electric iron (as used for ironing clothes) is the best thing to heat the writing and bring out the embossed effect. Have the iron hot, but not too hot or it will burn the paper and spoil the embossing powder. Turn the switch off when iron is hot enough and on again as needed, if you are doing considerable work. Experience will guide you in the amount of heat necessary. Hold the back of the paper upon which the embossing is to appear, over the iron, pressing down lightly with some metallic object like a table knife or fork or a nail file. As soon as the writing or imprinting raises, remove from heat. The raised letters will be smooth and stand up when the work is correctly done. After the work has cooled off, wipe off any surplus powder with a clean rag.

Printing Inks

Printing Inks may be divided into three classes—typographic, lithographic and rotographic. They consist principally of a pigment, vehicle and drier.

Typographic Inks

Typographic Inks are printed from a raised surface. They dry principally by oxidation and penetration. Magazine and book inks dry largely by oxidation. Representative formulae would be as follows:

Black

Carbon Black	20 lb.
No. O Lithographic Varnish	30 lb.
Rosin Oil	30 lb.
Cobalt Drier	10 lb.
Stearine Pitch	5 lb.

Yellow

Chrome Yellow	75 lb.
No. O Lithographic Varnish	25 lb.
Lead-Manganese Drier	2 lb.

Red

Lithol Red	45 lb.
No. O Lithographic Varnish	50 lb.
Drier	5 lb.

Besides these pigments, formulae contain many other colors, depending upon their use and desired shade. News inks, which come under the typographic class, dry principally by penetration, assisted in some cases by oxidation. The following would be representative formulae:

News Inks

Black

Carbon Black	12 lb.
Mineral Oil	85 lb.
Methyl Violet	1 lb.
Stearine Pitch	2 lb.

Blue

Peacock Blue	15 lb.
White Extender	7 lb.
No. 2 Lithographic Varnish	20 lb.
Mineral Oil	58 lb.

Red

Lithol Red	12 lb.
White Extender	10 lb.
Mineral Oil	25 lb.
No. O Lithographic Varnish	25 lb.
Rosin Oil	27 lb.

Lithographic Inks

The lithographic process depends upon the fact that oil or greasy substances and water will not mix. Most present day lithographic printing is done from grained zinc or aluminum plates. The original designs or characters are made onto the plates by the artist actually drawing or painting the original onto the grained plate or by transferring the designs from another print by transfer ink or by a photo litho process, whereby the design or negative is developed on the metal plate after it has been sensitized with an albumen coating.

This coating which has no affinity for water, allows the ink to transfer from a rubber roll to the plate and then to the paper. Lithographic inks, in composition, are very similar to typographic inks. Generally a heavier lithographic varnish is used as a vehicle. The only essential difference in pigments is that they must not bleed in water or weak acids to any great degree.

Vehicles. — The vehicles in printing inks are, as already mentioned. Lithographic varnish is nothing more than a heat bodied linseed oil. It may range in viscosity anywhere from 2 poises to 500 poises. Rosin oils and mineral oils may be either of high or low viscosity. Although the above oils are most commonly used in typographic inks, china wood oil, perilla oil and fish oil are also used.

Other ingredients may be found in inks such as waxes, resins and sometimes solvents.

Driers. — Driers are made from lead, manganese and cobalt compounds. These are dispersed in various oils and varnishes. Generally lead and manganese driers are used in light colors while cobalt is used in the darker colors. The kind of driers used are also dependent on the application.

Rotographic Inks

Rotographic inks are printed from an etched surface. They dry almost completely by evaporation. Generally solvents such as Toluene, Xylene and High Flash Naphtha are used to dissolve the resins which, together, make up the vehicles. Practically any resin soluble in the above mentioned solvents may be used. A formula would contain approximately

Pigment	33⅓ lb.
Resin	33⅓ lb.
Solvent	33⅓ lb.

Until recently only black and brown pigments were used, but at present rotographic inks may be made in other colors.

Printing inks are made by wetting and dispersing solid pigment colors in a suitable liquid medium. The vehicle used is usually a combination of oils and varnishes together with small amounts of driers, wax and grease compounds. The ink is manufactured by first mixing the ingredients in a change can or kneading mixer and then ground on steel roller mills.

In formulating a printing ink, only those pigments should be used that will meet the requirements of the printed matter, such as permanency to light, alkali proof, etc., and the method of printing used (either typographic, planographic or intaglio). The skillful blending of these pigments in a formula produces practically any desired color in the chromatic scale.

The specific gravity and oil absorption of the pigments will govern the ratio of pigment to vehicle. The type of vehicle will vary according to the body, tack, penetration, hardness of printed films, and drying properties that is desired to give to the ink. These in turn are governed by the method of printing used, type of press, size of the form, and nature of the stock the ink is printed on.

The final test of the suitability of a printing ink is its ability to work well on the printing press, print perfectly and to adhere properly to the printing surface.

The commercially available pigments, the properties of each and typical formulae containing these pigments are listed in the following:

Offset Tin Printing Yellow

No. 1 Transparent Lithographic Varnish	20 lb.
No. 00 Transparent Lithographic Varnish	2 lb.
No. 2 Transparent Lithographic Varnish	4 lb.
No. 3 Transparent Lithographic Varnish	2 lb.
C. P. Medium Chrome Yellow Dry	55 lb.
Gloss White Dry	15 lb.
Offset Ink Wax Compound	1 lb.

on last pass over mill and add

No. 7 Lithographic Varnish	1 lb.

------◆------

Process Transparent Yellow

Tartrazine Yellow Lake Dry	4 lb. 12 oz.
No. 0 Lithographic Varnish	2 lb. 8 oz.
Cobalt Linoleate Liquid Drier	3 oz.
No. 00 Lithographic Varnish	1 lb.
Lead Manganese Paste Drier	6 oz.
Paraffin Wax	2 oz.

Kerosene Oil	6 oz.
Amber Petrolatum	3 oz.

------◆------

Cadmium Yellow

Cadmium Yellow Light Dry	15 lb.
No. 1 Lithographic Varnish	4 lb.
No. 3 Lithographic Varnish	8 oz.
Lead Manganese Drier	4 oz.
Wax Compound	4 oz.
Aluminum Hydrate Dry	1 lb.
No. 0 Lithographic Varnish	1 lb.

------◆------

Opaque Orange Ink

Orange Mineral Powder, Dry	30 lb.
No. 0 Lithographic Varnish	6 lb.
No. 1 Lithographic Varnish	12 lb.
No. 3 Lithographic Varnish	3 lb.
Persian Orange, Dry	8 lb.
Alumina Hydrate, Dry	13 lb.
No. 6 Lithographic Varnish	1 lb. 8 oz.

------◆------

Transparent Orange

Persian Orange Dry	7 lb. 8 oz.
No. 0 Lithographic Varnish	6 lb.
Wool Grease	12 oz.
Cobalt Linoleate Liquid Drier	8 oz.
Lead Manganese Paste Drier	4 oz.

------◆------

Gloss Die Stamping Red

Gloss Stamping Varnish	33 lb.
No. 1 Burnt Plate Oil	2 lb.
Plate Paste, Drier	6 lb.
Blanc Fixe Dry	27 lb.
Paris White (Whiting) Dry	28 lb.
Calcium Lithol Toner Red Dry	4 lb.

------◆------

Cylinder Press Red

Sodium Lithol Toner Dry	8 lb.
Barium Lithol Toner Dry	5 lb.
Gloss White, Dry	5 lb.
Magnesium Carbonate, Dry	5 lb.
No. 0 Lithographic Varnish	20 lb.
Boiled Linseed Oil	2 lb.
Lead Manganese Paste Drier	2 lb. 8 oz.
Cobalt Linoleate Liquid Drier	1 lb.

------◆------

Job Press Bright Red

No. 1 Lithographic Varnish	4 lb.
No. 0 Lithographic Varnish	5 lb.
Lead-Manganese Paste Drier	1 lb. 8 oz.
Barium Red for Lake C Dry	4 lb.
Gloss White Dry	7 lb.
Aluminum Hydrate Dry	3 lb.
Cobalt Drier	4 oz.

Cylinder Press Red Ink

No. 1 Lithographic Varnish	35 lb.
No. 00 Lithographic Varnish	12 lb.
Neutral Wool Grease	4 lb.
Paste Drier (Manganese Resinate Lead Acetate)	4 lb.
Gloss White, Dry	22 lb.
Barium Lake for Red C, Dry	23 lb.

Offset Process Red

No. 1 Lithographic Varnish	32 lb.
No. 3 Lithographic Varnish	4 lb.
Barium Red for Lake C, Dry	44 lb.
Aluminum Hydrate, Dry	8 lb.
Blanc Fixe Dry	8 lb.
Offset Ink Wax Compound	2 lb.
Paste Drier (Lead Acetate Manganese Borate)	2 lb.

Label Red

No. 0 Lithographic Varnish	5 lb.
No. 1 Lithographic Varnish	8 lb.
Medium Bodied Rosin and Mineral Oil Varnish	6 lb.
Para Red Dark Dry	6 lb.
Para Red Light Dry	2 lb.
Aluminum Hydrate Dry	8 lb.
Wool Grease	1 lb. 8 oz.
Cobalt Linoleate Liquid Drier	2 lb.
Wax Compound	8 oz.
Barium Sulfate Dry	10 lb.

Madder Lake Ink

Madder Lake, Dry	5 lb. 8 oz.
No. 0 Lithographic Varnish	3 lb.
No. 2 Lithographic Varnish	5 lb.
Lead-Manganese Paste Drier	8 oz.
Cobalt Linoleate Drier	12 oz.

Process Red

Phloxine Toner Red, Dry	12 lb.
Alumina Hydrate, Dry	10 lb.
No. 0 Lithographic Varnish	7 lb.
No. 1 Lithographic Varnish	14 lb.
No. 6 Lithographic Varnish	1 lb.
Wax Compound	3 lb.

Job Press Blue

Bronze Blue, Dry	9 lb.
Permanent Violet, Dry	1 lb. 8 oz.
No. 0 Lithographic Varnish	9 lb. 8 oz.
Lead Manganese Paste Drier	4 oz.
No. 6 Lithographic Varnish	4 oz.
No. 1 Lithographic Varnish	2 lb. 8 oz.
Barium Sulfate, Dry	6 lb.
Petrolatum	4 oz.

Label Blue

Bronze Blue, Dry	8 lb.
No. 0 Lithographic Varnish	2 lb. 4 oz.
No. 1 Lithographic Varnish	1 lb.
Mineral Ink Oil	4 lb. 8 oz.
Barium Sulfate, Dry	3 lb. 8 oz.
Aluminum Hydrate, Dry	6 oz.
Permanent Violet, Dry	4 oz.
Wool Grease	6 oz.
Cobalt Linoleate Liquid Drier	4 oz.
Lead Manganese Paste Drier	12 oz.

Lichtdruck or Photogelatin Blue

No. 1 Lithographic Varnish	44 lb.
No. 3 Lithographic Varnish	3 lb.
Milori Blue, Dry	50 lb.
Mutton Tallow	1 lb.

Steel Plate Blue

Bronze Blue, Dry	52 lb.
Barytes, Dry	14 lb.
No. 0½ Plate Oil	27 lb.
No. 1 Plate Oil	7 lb.

Process Blue

Peacock Blue, Dry	8 lb. 8 oz.
No. 0 Lithographic Varnish	4 lb.
No. 1 Lithographic Varnish	3 lb.
Cobalt Linoleate Liquid Drier	1 lb.
Wax Compound	8 oz.

Glassine and Cellophane Blue

Spec. Hard Grip Varnish (No. 1 Lithographic Varnish and Amberol)	25 lb.
Cobalt Linoleate Drier	8 lb.
Beeswax (Melted into Drier)	2 lb.
Red Shade Reflex Alkali Blue Ink	60 lb.
No. 00000 Lithographic Varnish	5 lb.

Blue Lake Ink

Aluminum Hydrate, Dry	3 lb.
Magnesium Carbonate, Dry	3 lb.
Permanent Blue Toner, Dry	2 lb. 8 oz.
No. 0 Lithographic Varnish	9 lb.
Cobalt Linoleate Liquid Drier	1 lb.
Lead Manganese Paste Drier	12 oz.
Wax Compound	8 oz.

Ultramarine Blue Ink

Ultramarine Blue, Dry	15 lb.
Aluminum Hydrate, Dry	4 lb.
No. 1 Lithographic Varnish	8 lb.

No. 2 Lithographic Varnish 2 lb.
No. 3 Lithographic Varnish 8 oz.
Cobalt Linoleate Liquid Drier 2 oz.

Job Green

Milori Green, Dry 8 lb. 8 oz.
No. 0 Lithographic Varnish 1 lb.
No. 1 Lithographic Varnish 4 lb.
No. 2 Lithographic Varnish 6 lb.
Copal Green Varnish 4 lb.
Primrose Yellow, Dry 22 lb.
Aluminum Hydrate, Dry 1 lb.

Milori Green Ink

Milori Green, Dry 12 lb.
No. 1 Lithographic Varnish 3 lb.
No. 0 Lithographic Varnish 2 lb.
Copal Green Varnish 1 lb.
No. 00 Lithographic Varnish 8 oz.

Light Green Lake

Green Lake Light, Dry 7 lb. 8 oz.
No. 1 Lithographic Varnish 8 lb.
No. 2 Lithographic Varnish 1 lb.
Quinoline Yellow Lake, Dry 3 lb.
No. 6 Lithographic Varnish 12 oz.
Cobalt Linoleate Liquid
Drier 1 lb. 4 oz.
Lead Manganese Paste
Drier 1 lb. 8 oz.
Wool Grease 12 oz.

Heavy Job Black

Carbon Black 8 lb.
Bronze Blue, Dry 5 lb.
Alkali Blue Toner 3 lb.
No. 1 Lithographic Varnish 5 lb.
No. 3 Lithographic Varnish 10 lb.
No. 5 Lithographic Varnish 4 lb.
Gloss Varnish 3 lb.
Cobalt Drier 2 lb. 8 oz.
Lead Manganese Paste Drier 3 lb.

Bond Ledger or Job Black

No. 3 Lithographic Varnish 16 lb.
Gloss Varnish (Lithographic
Varnish and Amberol) 19 lb.
Amber Petrolatum 3 lb.
Highgrade Carbon Black, Dry 22 lb.
Blue for Black in Ink Form 14 lb.
Paste Drier (Manganese
Resinate Lead Actate) 12 lb.
Cobalt Linoleate, Liquid Drier 14 lb.

Halftone Black for Coated Stock

Boiled Linseed Oil 16 lb.
No. 00 Lithographic Varnish 12 lb.

No. 3 Lithographic Varnish 12 lb.
Concentrated Cobalt
Linoleate Drier 8 lb.
Soft Wax Non-offset Com-
pound (see below) 12 lb.
Red Shade Reflex Alkali
Blue, Ink 10 lb.
High Grade Carbon Black,
Dry 18 lb.
Blue for Black in Ink Form 12 lb.

Web Press Black for Newsprint

Heavy Body Mineral Ink Oil 33 lb.
Second Run Rosin Oil 22 lb.
Rosin Varnish (60 parts Mineral
Oil and 40 parts Rosin) 34 lb.
News-grade Carbon Black,
Dry 10 lb.
Blue Toner (10% Methylene
Blue in Oleic Acid) 1 lb.

Lithographic Black

No. 3 Lithographic Varnish 24 lb.
No. 1 Lithographic Varnish 24 lb.
No. 7 Lithographic Varnish 1 lb.
Red Shade Reflex Alkali Blue
Ink 9 lb.
Finest Grade Carbon Black,
Dry 32 lb.
Concentrated Cobalt Drier 10 lb.

Copper Plate Black

No. 1 Burnt Plate Oil 26 lb.
No. 2 Burnt Plate Oil 4 lb.
Hard Black (Bone Black)
Dry 37 lb.
Soft Black (Bone Black)
Dry 16 lb.
Plate Paste Drier 10 lb.
Prussian Blue, Dry 7 lb.

Bookbinder's Black

No. 0 Lithographic Varnish 15 lb.
Gloss Copal or Kauri Varnish 25 lb.
Concentrated Cobalt
Linoleate Drier 10 lb.
High Grade Carbon Black
Dry 25 lb.
Bronze Blue in Ink Form 15 lb.
Reflex Alkali Blue, Red
Shade, Ink 10 lb.

Wax Offset Compound

1. Beeswax 22
2. Petrolatum Amber 20
3. Mutton Tallow 5
4. Paraffin Oil 22

5. Kerosene 10
6. Naphtha (High Flash) 4

Melt (1), (2), (3) and (4) and stir until dissolved. Turn off heat and work in (5) and (6).

Soft Wax Non-Offset Compound

No. 1 Lithographic Varnish	35
Soft Cup Grease	35
Paraffin Wax	10
Beeswax	20

Non-Offset Compound

No. 1 Lithographic Varnish	35
Soft Cup Grease	35
Paraffin Wax	10
Beeswax	20

Melt together; cool and run in mill.

1. Ordinary Composition for Transfers.

	Parts by Weight
Rosin	100
Beeswax	30
Gold Bronze or Pigment	30

2. Indelible Marking Composition — Blacks

	Parts by Weight
Stearic Acid	100
Induline Base	150

3. Indelible Marking Composition — Colors.

	Parts by Weight
Cumar Light	100
No. 4 Litho Varnish	25
Mineral Oil	8
Cobalt Drier	2½
Permanent Pigment	30

4. Permanent Marking Composition.

	Parts by Weight
Cumar Light	100
Processed Rapeseed Oil	50
Bronze or Pigment	35

5. Water Soluble Transfer Composition.

1. Printing Compound.
    a. Glycerin 100 by wt.
    b. Gum Arabic 40 by wt.
    c. Color Dye or Pigment) 25 by wt.
2. Dusting Material.
    a. Gum Tragacanth Powder

6. Embroidery Composition for Transfers.

	Parts by Weight
Cumar	16
Rosin	4
Canauba Wax	4
Stearic Acid	2
Ultramarine Blue	8
Titanox Ground	31.2
Litho Varnish Ground	8.8

7. Leather Composition for Transfers.

	Parts by Weight
Shellac—Orange or White	100
Venice Turpentine	50
Pigment	40

8. Indelible Transfer Ink.

	Parts by Weight
Cumar	100
Varnoline	10
No. 4 Litho Varnish	10
Turkey Red Oil	20
Dyestuff (Induline Base)	20
Permanent Pigment	30

9. Flexible Marking Composition.

	Parts by Weight
Light Cumar	100
Processed Rapeseed Oil	55
Rubber Latex	30
Vermilion	45

10. Fugitive Transfer Composition.

	Parts by Weight
Rosin	100
Beeswax	10
Cobalt Drier	1
Gold Bronze	25

11. Water Fugitive Transfer Composition.

	Parts by Weight
Mutton Tallow	1
Cocoa Butter	1
Paraffin	4
Rosin	6
Sufficient quantity—Pigment	

Drawing Crayons
Black

Kaolin	24	lb.
Carbon Black	22	lb.
Garnet Shellac	12	lb.
Denatured Alcohol	1	gal.
Turpentine	½	gal.

Dissolve shellac in alcohol; add turpentine and then mix in solids and grind to smooth paste. Mould and dry slowly.

Blue

Soapstone	34	lb.
Chinese Blue	14	lb.
Garnet Shellac	12	lb.
Denatured Alcohol	1	gal.
Turpentine	½	gal.

Method—as under Black.

Light Green—Dark Green

Paraffin Wax—138-140° F.	100	lb.
Stearic Acid—double pressed	10	lb.
Carnauba Wax	4½	lb.
Precipitated Chalk	¾	lb.
Chrome Green Light or Dark	12	lb.

Orange or Yellow

Paraffin Wax—138-140° F.	100	lb.
Stearic Acid—double pressed	4	lb.
Carnauba Wax	4½	lb.
Precipitated Chalk	¾	lb.
Chrome Yellow or Orange	18	lb.

Purple

Paraffin Wax—138-140° F.	100	lb.
Stearic Acid—double pressed	12	lb.
Carnauba Wax	4½	lb.
Precipitated Chalk	¾	lb.
Purple Lake—reddish	8	lb.

Light Red—Dark Red

Paraffin Wax—138-140° F.	100	lb.
Stearic Acid—double pressed	12	lb.
Carnauba Wax	4½	lb.
Precipitated Chalk	¾	lb.
Para Red on a hydrate base	21	lb.

Bluish shade for dark red.
Light shade for light red.

Light Blue

Paraffin Wax—138-140° F.	100	lb.
Carnauba Wax	4½	lb.
Precipitated Chalk	¾	lb.
Ultramarine Blue	21	lb.

Dark Blue

Paraffin Wax—138-140° F.	100	lb.
Stearic Acid—double pressed	10	lb.
Carnauba Wax	4½	lb.
Precipitated Chalk	¾	lb.
Prussian Blue	10	lb.

Dark Brown

Paraffin Wax—138-140° F.	100	lb.
Carnauba Wax	4	lb.
Precipitated Chalk	¾	lb.
Burnt Turkey Umber	25	lb.

Light Brown

Paraffin Wax—138-140° F.	100	lb.
Carnauba Wax	4	lb.
Precipitated Chalk	¾	lb.
Burnt Sienna	25	lb.

Black

Paraffin Wax—138-140° F.	100	lb.
Carnauba Wax	4	lb.
Precipitated Chalk	¾	lb.
Lamp Black	12	lb.

When stearic acid is used in the foregoing formulas it is always best to add the color to the melted stearic acid before adding the other materials.

------◆------

Animal Marking Crayon

Tallow	180
Rosin	5
Rozolin	2

Melt together and add while stirring a mineral pigment such as Prussian Blue, Red Iron Oxide, etc. Cast in glass or metal tubes.

------◆------

Blackboard Crayon

Calcium Carbonate (precipitated)	60	lb.
Kaolin Clay	40	lb.
Saponified Oleic Acid	5	lb.
Caustic Soda	¾	lb.

The Oleic Acid and Caustic Soda are mixed, warm, in a separate kettle and added to the clay mix along with enough water to bring to about the consistency of putty. The mixing is done in a standard type dough mixer or other clay mixing equipment.

------◆------

Lithographic Crayon

1

Sodium Stearate	7
Beeswax	6
Carbon Black	1

2

Beeswax	30
Tallow	25
Soap	20
Shellac	15
Lamp Black	6

Heat in enamelled pot to melt together. Then heat strongly until vapor ignites. Allow to burn for a while and smother flame with cover of pot. Take out a sample and test for elasticity. If not satisfactory ignite again in same way.

Marking Crayons

Ceresin	40
Carnauba Wax	35
Paraffin Wax	20
Beeswax	5
Talc	50
Chrome Green or Other Pigment	15

Crayon, Tailors' Marking

Carnauba Wax	11
Stearic Acid	2
Ceraflux	76
Ozokerite	6
Terra Alba	5

Wax Crayons

The manufacture of wax crayons follows very closely that of the molded candle, both in procedure and materials and an attempt to go into details would be endless and rather futile. A finely divided dry color is usually more suitable as the coloring medium and usually more dependable. The dry color is added to the wax combination after the wax is melted in a steam jacketed aluminum kettle. Mechanical agitation is continued until the kettle has been emptied in order to prevent any tendency of the color to settle to the bottom. The wax should be maintained as nearly to the melting point as practicable and rapid cooling is perhaps more important here than in candles. A good starting point on the wax combination would be as follows:

Double Pressed Saponified	
Stearic Acid	40 lb.
Paraffin	45 lb.
Beeswax	10 lb.
Carnauba Wax	5 lb.
Dry color to suit.	

The above proportions may be changed to create a harder or softer crayon and Candelilla Wax may be added or substituted for the Beeswax. Care should be taken not to make the crayon too hard as a tendency of the points to crack or flake will be noted.

Wax Drawing Pastels
Black

Hard Soap	80
Beeswax Crude	60
Spermaceti Crude	28
Carbon Black	14
Burnt Umber	5
Prussian Blue	4

Melt waxes and soap, mix in pigments and grind until smooth; pour hot in molds; and plunge into cold water to "set."

Red

Hard Soap	28
Saponified Japan Wax	28
Spermaceti	16
Carnauba Wax	2
Beeswax Crude	8
Orange Chrome Yellow	12
Method—as under Black.	

Tailors' Chalk
Yellow

Chalk (Powd.)	28
Soapstone	18
Pipe Clay	10
Yellow Ochre	7
Lemon Chrome Yellow	1½

Make into a paste with water and mold.

White

French Chalk	20
Pipe Clay	20
White Curd Soap	6

Make into a stiff paste with water and dry.

Black

Soapstone	56
Bone Black	8
Yellow Soap	6
Gum Arabic	2
Glycerin	1

Dissolve gum in water, add glycerin, mix in pigments; grind to a smooth paste with water and mold.

Warehouse Chalk

Gypsum	40
Soapstone	55
Carbon Black	6
Petrolatum	1

Mix to a uniform paste with a thin glue solution and mold.

Spotting Pencil

(For restoring color on fabrics, etc.)

Stearic Acid	50 parts
Japan Wax	50 parts

Required amount of oil dyes for shade.

Place material in a steam-jacketed vessel, preferably; melt slowly and agitate until thoroughly mixed. Pour into forms desired to cool.

Use

Stains or spots removed previously on fabrics and on last of original shade these spotting pencils can be used advantageously in restoring original shade.

------◆------

Colored Pencil Leads

Ammonium Hydroxide	2
Shellac	3
Venice Turpentine	1
Prussian Blue or other pigment	6
Clay or Chalk	4

The pigment is ground to a fine paste with water; the shellac is dissolved in the ammonia. The Venice turpentine is rendered fluid by short heating. The clay is worked to a smooth slurry with water and pressed through muslin and dried and powdered. Mix everything together in a mill until the consistency is that of a thick dough. This is then fed into a pressing machine of the macaroni type with openings of the size required. The extruded leads are placed in a drying oven for drying.

------◆------

Red Indelible Lead

Rosin Soap	60 gm.
Water	6 kg.

Dissolve with heat and add

Shellac	40 gm.

Stir in

Ponceau-Creosot	2 kg.

and

Albumen	40 gm.
Gum Tragacanth	40 gm
Water	120 gm.

Emulsifiable Transfer Ink

Diglycol Stearate	20 oz.
Ethyl Cellulose	5 oz.
Sodium Abietate	10 oz.
Pigment	10 oz.

------◆------

Ink Remover

For cleaning dry printing ink from printers' rolls and type.

Denatured Alcohol	2½ gal.
Commercial Toluol	1¼ gal.
Heavy Naphtha	3¾ qt.
Creosote Oil	1¼ gal.

Mill in

Cinnabar Powd.	2 kg.
Kaolin Powd.	2 kg.

Extrude through press and dry.

------◆------

Blue Copying Pencil

Aniline Blue (Water Soluble)

Powder	2 kg.
Water	4 kg.

Dissolve by heating; then cool and add Gum Tragacanth Powder   20 gm. and stir until dispersed; now add

Milori Blue (Powder)	4⁄7 kg.
Kaolin (Powder)	3¾ kg.

Make acid with sulfuric acid; allow to stand overnight and neutralize with soda ash. Extrude the leads and dry for a few days. Rub off crystals which have formed on leads, by means of a damp rag. Dry in an oven and clean off crystals again in same way. Repeat until no more crystals form on drying.

Redissolve in a similar amount of water to which has been added the following filtered solution.

Sugar	80 gm.
Albumen	20 gm.
Water	120 gm.

then add with stirring

Indigo-Carmin	500 gm.

and heat on a water bath until of a doughy consistency.

The Milori Blue and Kaolin should first be mixed together with water to form a slurry and ground wet and dried and powdered. To this is added and thoroughly mixed in

Sulfuric Acid	½ kg.

The finished lead is waxed or greased to protect it from atmospheric moisture.

Billiard Chalk

Formula No. 1

a.	Calcium Carbonate, Precipitated	115 g.
	Gypsum, Calcined	35 g.
	Pigment Powder (Blue, Green)	50 g.
b.	Borax Water (2%)	about 180-200 g.

to make a pasty liquid

This paste is poured into slightly oiled molds.

No. 2

Calcium Carbonate	100 g.
Gypsum	30 g.
Borax Water (2%)	115-130 g.

As above.

CHAPTER VIII

LEATHER, SKINS, FURS, ETC.

THE manufacture of leather from raw skins is a lengthy and laborious process. Different types of leathers require different treatments. After the leather is finally tanned it must be given a finish suitable for the use to which it will be put. Many typical formulae are given in this chapter varying from the treatment of cow-hides to lizard skins.

Fur skin tannings are similarly described as well as methods for dyeing and bleaching.

Home Tanning of Leather and Fur Skins

Preparation of the hide or skin for tanning may be started as soon as it has been taken off the animal, drained, and cooled from the body heat. Overnight will be long enough. If tanning is not to be started at once or if there are more hides than can be handled at one time, the hides may be thoroughly salted and kept for from three to five months. The hides must never be allowed to freeze or heat during storage or tanning. Some tanners state that salting before tanning is helpful. It can do no harm to salt a hide for a few days before it is prepared for tanning.

The directions here given have been prepared for a single heavy cow, steer, or bull hide weighing from 40 to 70 pounds or for an equivalent weight in smaller skins, such as calf or kip skins. The heavy hides are best suited for sole, harness, or belting leather. Lighter hides weighing from 20 to 40 pounds should be used for lace leather.

Preliminary Operations

Before it is tanned a hide or skin must be put through the following preliminary operations. As soon as the hide or skin has been put through these processes, start the tanning, following the directions given for the particular kind of leather desired.

Slaking Lime

Put from 6 to 8 pounds of burnt or caustic lime in a clean half barrel, wooden tub, or bucket, with a capacity of at least 5 gallons. Use only good-quality lime, free from dirt and stones; never use air-slaked lime. To the lime add about 1 quart of water. As the lime begins to slake add more water, a little at a time, to keep the lime moist. Do not pour in enough water to quench the slaking. When the lime appears to be slaked, stir in 2 gallons of clean water. Do all this just as in making whitewash. Slake the lime on the day before the soaking of the hide is begun, and keep the limewater covered with boards or sacks until ready to use it.

If available, fresh hydrated lime, not air-slaked, may be used instead of the burnt or caustic lime. In this case use from 8 to 10 pounds in 4 or 5 gallons of water.

Soaking and Cleaning

If the hide has been salted, shake it vigorously to remove most of the salt. Spread it out, hair side down, and trim off the tail, head, ears, all ragged edges, and shanks.

Place the hide, hair side up, lengthwise, over a smooth log or board, and, with a sharp knife, split it from neck to tail, straight down the backbone line, into two half hides, or "sides." It will be more convenient in the later handling, especially when the hide is large, to then split each side lengthwise through the "break," just above the flanks, into two strips, making the strip with the backbone edge about twice as wide as the belly strip. Thus a whole hide will give two sides or four strips. If desired, small skins need not be split. In these directions "side" means side, strip, or skin, as the case may be.

Fill a 50-gallon barrel with clean, cool water. Place the sides, flesh side out, over short sticks or pieces of rope and hang them in the barrel of water. Let them soak for two or three hours. Stir them about frequently to soften, loosen, and wash out the blood, dirt, manure, and salt. The sticks or pieces of rope may be held in place by tying a loop of cord on each end and catching the loops over nails in the outside of the barrel near the top.

After soaking for about three hours take out the sides, one at a time, and place them, hair side up, over a "beam."

A ready-made beam can be bought. A fairly satisfactory one may be made from a very smooth slab, log, or thick planed board, from 1 to 2 feet wide and 6 to 8 feet long. The slab or log is inclined, with one end resting on the ground and the other extending over a box or trestle so as to be about waist high.

With the side lying hair side up over the beam, scrub off all dirt and manure, using if necessary a stiff brush. Wash off with several bucketfuls of clean water.

Turn the side over, flesh side up, and scrape or cut off any remaining flesh. Work over the entire flesh side with the back edge of a drawing or butcher knife, held firmly against the hide, while pushing away from the body. Wash off with one or two bucketfuls of clean water. This working over should always be done.

Refill the soak barrel with clean, cool water and hang the sides in it as before. Pull them up and stir them about frequently until they are soft and flexible. Usually a green or fresh hide needs to be soaked for not more than from 12 to 24 hours and a green salted hide for not more than from 24 to 48 hours.

When the sides are properly softened—that is, about like a fresh hide or skin—throw them over the beam and thoroughly scrape off all remaining flesh and fat. It is of the greatest importance to remove all this material. When it can not be scraped off, cut it off, but be careful not to cut into the hide itself. Even should there appear to be no flesh to take off and nothing seems to be removed, it is necessary to thoroughly work over the flesh side in this way with the back of a knife. Finally wash off with a bucketful of clean water.

The side must be soft, pliable, and clean all over before being put into the lime, which is the next step.

Liming

Wash out the soak barrel. Pour in all of the slaked lime; nearly fill the barrel with clean, cool water; and stir thoroughly. Place the sides, hair side out, again over the short sticks or pieces of rope, and hang them in the barrel so that they are completely covered by the limewater. See that the sides have as few folds or wrinkles as possible and also be sure that no air is trapped under them. Keep the barrel covered with boards or bags. Pull up the sides and stir the limewater three or four times each day until the hair will come off easily. This takes from 6 to 10 days in summer and possibly as many as 16 days in winter.

When thoroughly limed, the hair can be rubbed off readily with the hand. Early in the liming process it will be possible to pull out the hair, but the hide must be left in the limewater until the hair comes off by rubbing over with the hand. For harness and belting leathers leave the hide in the limewater for from 3 to 5 days after this condition has been reached.

Unhairing

When limed, throw the side, hair side up, over the beam, and, with the back edge of a drawing or butcher knife, held nearly flat against the side, push off the hair from all parts. If the side is sufficiently limed, a curdy or cheesy layer of skin rubs off with the hair. If this layer does not rub off, the side must be returned to the limewater. After removing the hair, put the side back in the limewater again for another day, until any fine hairs that may remain can be easily scraped off. Now thoroughly work over the grain or hair side with a dull-edged tool to "scud" or work out as much lime, grease, and dirt as possible.

Fleshing

Turn the side over and "scud" it again, being sure to remove all fleshy matter. Shave down to the hide itself, but be careful not to cut into it. Remove the flesh by scraping and by using a very sharp knife, with a motion like that of shaving the face.

Now proceed as directed under "Bark-tanned sole and harness leather," "Chrome-tanned leather" or "Alum-tanned lace leather," depending upon the kind of leather desired.

Wastes from Liming

The lime, limewater, sludge, nd flesh-ings from the liming process may be used as fertilizer, being particularly good for acid soils. The hair, as it is scraped from the hide, may be collected separately, and, after being rinsed several times, may be used in plastering. If desired, it can be thoroughly washed with many changes of water until absolutely clean and, after being dried out in a warm place, can be used for padding, upholstering, insulation of pipes, etc.

Bark-tanned Sole and Harness Leather

Deliming

After the sides have been put through the unhairing and fleshing operations, rinse them with clean water. Wash the sides in cool, clean water for from six to eight hours, changing the water frequently.

Buy 5 ounces of U. S. P. lactic acid (or 16 ounces of tannery 22 per cent lactic acid). Nearly fill a clean 40 to 50 gallon barrel with clean, cool water, and stir in the lactic acid, mixing thoroughly with a paddle. Hang the sides in the barrel and leave them there for 24 hours, pulling them up and stirring frequently.

Take out the sides, work over or "scud" them thoroughly, as directed under "Unhairing," and hang them in a barrel of cold water. Change the water several times, and finally leave them in the water overnight.

If lactic acid can not be obtained, use a gallon of vinegar instead.

Tanning

The sides are now ready for the actual tanning. From 15 to 20 days before this stage will be reached weigh out from 30 to 40 pounds of good-quality, finely-ground oak or hemlock bark and pour onto it about 20 gallons of boiling water.

Finely-ground bark, with no particles larger than a grain of corn, will give the best results. Simply chopping the bark into coarse pieces will not do. Do not let the tan liquor come in contact with iron vessels. Use the purest water available. Rain water is best.

Let this bark infusion stand in a covered vessel until ready to use it. Stir it occasionally. When ready to start tanning, strain off the bark liquor through a clean, coarse sack into the tanning barrel. Fill the barrel about three-quarters full with water, rinsing the bark with this water so as to get out as much tannin as possible. Add 2 quarts of vinegar. Stir well. Place the sides, from the deliming, over sticks, and hang them in this bark liquor with as few folds and wrinkles as possible. Move the sides about and change their position often in order to get an even color.

Just as soon as the sides have been hung in the bark liquor, again soak from 30 to 40 pounds of ground bark in about 20 gallons of hot water. Let this second bark liquor stand until the sides have become evenly colored, or for from 10 to 15 days. Take out of the tanning barrel 5 gallons of liquor and pour in about one-quarter of the second bark liquor. Also add about 2 quarts more of vinegar and stir it in well. Five days later add another fourth of the tan liquor only (no vinegar). Do this every 5 days until the second bark liquor is used up.

The progress of the tanning varies somewhat with conditions and can best be followed by inspecting a small sliver cut from the edge of the hide. About 35 days after the actual tanning has been started a fresh cut should show two dark or brown narrow streaks about as wide as a heavy pencil line coming in from each surface of the hide.

At this stage weigh out about 40 pounds of fine bark and just moisten it with hot water. Do not add more water than the bark will soak up. Pull the sides out of the bark liquor and dump in the moistened bark, keeping in the barrel as much of the old tan liquor as possible. Mix thoroughly and while mixing hang the sides back in the barrel. Actually bury them in the bark. All parts of the sides must be kept well down in the bark mixture. Leave the sides in this bark for about six weeks, moving them about once in a while.

At the end of six weeks pull the sides out. A cutting should show that the tanning has spread nearer to the center. Pour out about half the liquor. Stir the bark in the barrel, hang the sides back, and fill the barrel with fresh, finely ground bark. Leave the sides in for about two months, shaking the barrel from time to time and adding bark and water as needed to keep the sides completely covered.

At the end of this time the hide should be evenly colored all the way through, without any white or raw streak in the center of a cut edge. If it is not struck through, it must be left longer in the wet bark, and more bark may be needed.

For harness, strap, and belting leather the sides may be taken out of the bark liquor at this stage, but for sole leather they must be left for two months longer. When fully tanned through, the sides are ready for oiling and finishing.

Oiling and Finishing

Harness and belting leather.—Take the sides from the tan liquor, rinse them off with water; and scour the grain or hair side thoroughly with plenty of warm water and a stiff brush. Then go over the sides with a "slicker," pressing the slicker firmly against the leather while pushing it away from the body. "Slick" out on the grain or hair side in all directions. For harness, belting, and the like, this scouring and slicking out must be thoroughly done.

A slicker can be made from a piece of copper or brass about one-fourth inch thick, 6 inches long, and 4 inches wide. One long edge of the slicker is mounted in a wooden handle and the other long edge is finished smooth and well rounded. A piece of hardwood, about 6 inches square, 1½ inches thick at the head, and shaved down wedge-shape to a thin edge, will also serve as a slicker.

While the sides are still damp, but not very wet, go over the grain or hair side with a liberal coating of neat's foot or cod oil. Hang up the sides and let them dry out slowly. When dry, take them down and dampen well by dipping in water or by rolling them up in wet sacking or burlap.

When uniformly damp and limber, evenly brush or mop over the grain or hair side a thick coating of warm dubbin. The dubbin is made by melting together about equal parts of cod oil and tallow or neat's-foot oil and tallow. This dubbin when cool must be soft and pasty, but not liquid.

Hang up the sides again and leave until thoroughly dried. When dry, scrape off the excess tallow by working over with the slicker. If more grease in the leather is desired, dampen again and apply another coating of the dubbin, giving a light application also to the flesh side. When again dry, remove the tallow and thoroughly work over all parts of the leather with the slicker. Rubbing over with sawdust will help to take up any surface oiliness.

If it is desired to blacken the leather, this must be done before greasing. A black dye solution can be made by dis-solving one-half ounce of water-soluble nigrosine in 1¼ pints of water, with the addition, if handy, of several drops of ammonia. Evenly mop or brush this solution over the dampened but ungreased leather and then grease as directed in the preceding paragraph.

Sole leather.—Take the sides from the tan liquor and rinse them thoroughly with clean water. Hang them up until they are only damp and then apply a good coating of neat's foot or cod oil to the grain or hair side. Again hang them up until they are thoroughly dry.

When repairing shoes with this leather it is advisable, after cutting out the piece for soling, to dampen and hammer it down well, and then, after putting it on the shoe, to make it waterproof and more serviceable by setting the shoe in a shallow pan of melted grease or oil and letting it stand for about 15 minutes. The grease or oil must be no hotter than the hand can bear. Rubber heels should not be put in oil or grease. The soles of shoes with rubber heels may be waterproofed in the same way, using a pie pan for the oil or grease and placing the heels outside the pan. Any good oil or grease will do. The following formulas have been found satisfactory:

Formula 1:	Ounces
Neutral Wool Grease	8
Dark Petrolatum	4
Paraffin Wax	4
Formula 2:	
Petrolatum	16
Beeswax	2
Formula 3:	
Petrolatum	8
Paraffin Wax	4
Wool Grease	4
Crude Turpentine Gum (gum Thus)	2
Formula 4:	
Tallow	12
Cod Oil	4

Chrome-tanned Leather

For many purposes chrome-tanned leather is considered to be as good as the more generally known bark or vegetable-tanned leather. The chrome process, which takes only a few weeks as against as many months for the bark-tanning process, derives its name from the use of chemicals containing chromium or "chrome." It is a chemical process requiring great care. It is felt, however, that by following exactly the directions here given, never disregarding details which may seem unimportant, a serviceable leather can be produced in a comparatively short time. The saving in time seems sufficient to justify a trial of this process.

Deliming

After the sides have been put through the unhairing and fleshing operations rinse them off with clean water.

If sole, belting, or harness leather is to be tanned, soak and wash the sides in cool water for about six hours before putting them into the lactic acid. Change the water four or five times.

If strap, upper, or thin leather is to be tanned, put the limed white sides into a wooden or fiber tub of clean, lukewarm (about 90° F.) water and let them stay there for from four to eight hours before putting them into the lactic acid. Stir the sides about occasionally. Be sure that the water is not too hot. It never should be so hot that it is uncomfortably warm to the hand.

For each large hide or skin buy 5 ounces of U. S. P. lactic acid (or 16 ounces of tannery 22 per cent lactic acid). Nearly fill a clean 40 to 50 gallon barrel with clean, cool water, and stir in the lactic acid, mixing thoroughly with a paddle. Hang the sides in the barrel, and leave them there for 24 hours, plunging them up and down occasionally. For light skins, weighing less than 15 pounds, use only 2 ounces of U. S. P. lactic acid in about 20 gallons of water.

If lactic acid can not be obtained, use 1 pint of vinegar for every ounce of lactic acid. An effort should be made to get the lactic acid, however, for vinegar will not be as satisfactory, especially for the medium and smaller skins.

After deliming, work over both sides of the side as directed under "Unhairing."

For sole, belting, and harness leathers, hang the sides in a barrel of cool water overnight. Then proceed as directed under "Tanning."

For thin, softer leathers from small skins, do not soak the sides in water overnight. Simply rinse them off with water and proceed as directed under "Tanning."

Tanning

The tanning solution should be made up at least two days before it is to be used—that is, not later than when the sides are taken from the limewater for the last time.

Remember that this is a chemical process and all materials must be of good quality and accurately weighed, and that the specified quantities of water must be carefully measured.

The following chemicals are required: Chrome alum (chromium potassium sulfate crystals); soda crystals (crystallized sodium carbonate); and common salt (sodium chloride).

For each hide or skin weighing more than 30 pounds use the following quantities for the stock chromium solution:

Dissolve 3½ pounds of soda crystals (crystallized sodium carbonate) and 6 pounds of common salt (sodium chloride) in 3 gallons of warm, clean water in a wooden or fiber bucket. The soda crystals must be clear or glasslike. *Do not use the white crusted lumps.*

At the same time dissolve, in a large tub or half barrel, 12 pounds of chrome alum (chromium potassium sulfate crystals) in 9 gallons of cool, clean water. This will take some time to dissolve and will need frequent stirring. Here again it is important to use only the very dark, hard, glossy, purple or plum-colored crystals of chrome alum, not the lighter, crumbly, dull lavender ones.

When the chemicals are dissolved, which can be told by feeling around in the tubs with a paddle, pour the soda-salt solution slowly in a thin stream into the chrome-alum solution, stirring constantly. Take at least 10 minutes to pour in the soda solution. This should give one solution of about 12 gallons which is the *stock chrome solution.* Keep this solution well covered in a wooden or fiber bucket, tub, or half barrel.

To start tanning, pour one-third (4 gallons) of the stock chrome solution into a clean 50-gallon barrel and add about 30 gallons of clean, cool water; that is, fill the barrel about two-thirds full. Thoroughly mix the solution in the barrel and hang in it the sides from the deliming. Work the sides about and stir the solution frequently, especially the first two or three days. This helps to give the sides an even color. It should be done every hour or so throughout the first day. Keep the sides as smooth as possible.

After three days, temporarily remove the sides from the barrel. Add one-half of the remaining stock chrome solution, thoroughly mixing it with that in the barrel, and again hang in the sides. Move the sides about and stir the solution three or four times each day.

Three days later, once more temporarily remove the sides. Pour into the barrel the rest of the stock chrome solution, thoroughly mixing it with that in the barrel and again hang in the sides. Move the sides about and stir frequently as before.

After the sides have been in this solution for three or four days, cut off a small piece of the thickest part of the side, usually in the neck, and examine the freshly cut edge of the piece. If the cut edge seems to be evenly colored greenish or bluish all the way through, the tanning is about finished. Boil the small piece in water for a few minutes. If it curls up and becomes hard or rubbery the tanning is not completed and the sides must be left in the tanning solution for a few days longer, or until a small piece when boiled in water is changed little if at all.

The foregoing quantities and directions have been given for a medium or large hide. For smaller hides and skins the quantities of chemicals and water can be reduced. For each hide or skin weighing less than 30 pounds, or for two or three small skins together weighing not more than 30 pounds, the quantities of chemicals may be cut in half, giving the following solutions:

For the soda-salt solution, dissolve 1¾ pounds of soda crystals (crystallized sodium carbonate) and 3 pounds of common salt (sodium chloride) in 1½ gallons of clean water.

For the chrome-alum solution, dissolve 6 pounds of chrome alum (chromium potassium sulfate crystals) in 4½ gallons of cool, clean water.

When the chemicals are dissolved pour the soda-salt solution slowly into the chrome-alum solution as already described. This will give one solution of about 6 gallons which is the *stock chrome solution*. For the lighter skins tan with this solution, exactly as directed for medium and large hides, adding one-third, that is, 2 gallons, of this stock chrome solution each time, and begin to tan in about 15 gallons instead of 30 gallons of water. Follow the directions already given as to stirring, number of days, and testing to determine when tanning is completed. Very small, thin skins probably will not take as long to tan as will the large hides. The boiling-water test is very reliable for showing when the hide is tanned.

Washing and Neutralizing

When the sides are tanned, take them out of the tanning solution and put them in a barrel of clean water. The barrel in which the tanning was done can be used after it has been thoroughly washed.

When emptying the tanning barrel be sure to carefully dispose of the tanning solution. Although not poisonous to the touch, it probably would be fatal to farm animals should they drink it, and it is harmful to the soil.

Wash the sides in about four changes of water. For medium and large hides, dissolve 2 pounds of borax in about 40 gallons of clean water and soak the sides in this solution overnight. For hides and skins weighing less than 25 pounds, use 1 pound of borax in about 20 gallons of water. Move the sides about in the borax solution as often as feasible. After soaking overnight in the borax solution, remove the sides and wash them for an entire day, changing the water five or six times. Take the sides out, let the water drain off, and proceed as directed under "Dyeing black," or, if it is not desired to blacken the leather, proceed as directed under "Oiling and finishing."

Dyeing Black

Water-soluble nigrosine.—One of the simplest and best means of dyeing leather black is the use of nigrosine. Make up the dye solution in the proportion of one-half ounce of water-soluble nigrosine dissolved in 1¼ pints of water. Be sure to get water-soluble nigrosine. Evenly mop or brush this solution over the damp leather after draining as already directed and then proceed as directed under "Oiling and finishing."

Iron liquor and sumac.—If water-soluble nigrosine can not be obtained, a fairly good black may be secured with iron liquor and sumac. To make the iron liquor, mix clean iron filings or turnings with one-half gallon of good vinegar and let the mixture stand for several days. See that there are always some undissolved filings or turnings in the vinegar. For a medium or large hide put from 10 to 15 pounds of dried crumbled sumac leaves in a barrel containing from 35 to 40 gallons of warm water. Stir well and when cool hang in it the wet, chrome-tanned sides. Leave the sides in this solution for about two days, pulling them up and mixing the solution frequently. Take out the sides, rinse off all bits of sumac, and evenly mop or brush over with the iron liquor. Rinse off the excess of iron liquor and put the sides back in the sumac overnight. If not black enough the next morning, mop over again with iron liquor, rinse, and return to the sumac solution for a day. Take the sides out of the sumac, rinse well, and scrub thoroughly with warm water. Finally, wash the sides for a few hours in several changes of water.

While both of these formulas for dyeing have been given, it is recommended that water-soluble nigrosine be used whenever possible, as the iron liquor and sumac formula is somewhat troublesome and may produce a cracky grain. After blackening, proceed as directed under ''Oiling and finishing.''

Oiling and Finishing

Thin leather.—Let the wet tanned leather from the dyeing, or, if not dyed, from the neutralizing, dry out slowly. While it is still very damp go over the grain or hair side with a liberal coating of neat's foot or cod oil. While still damp tack the sides out on a wall or tie them in frames being sure to pull them out tight and smooth, and leave them until dry. When dry take down and dampen well by dipping in warm water or by rolling them up in wet sacking or burlap. When uniformly damp and limber go over the sides with a ''slicker,'' pressing the slicker firmly against the leather, while pushing it away from the body. ''Slick'' out on the grain or hair side in all directions.

After slicking it may be necessary to ''stake'' the leather. This is done by pulling the damp leather vigorously back and forth over the edge of a small smooth board about 3 feet long, 6 inches wide, and 1 inch thick, fastened upright and braced to the floor or ground. The top end of the board must be shaved down to a wedge shape, with the edge not more than one-eighth inch thick and the corners well rounded. Pull the sides, flesh side down, backward and forward over this edge, exactly as a cloth is worked back and forth in polishing shoes.

Let the sides dry out thoroughly again. If not sufficiently soft and pliable, dampen them with water, apply more oil, and slick and stake as before. The more time given to slicking and staking, the smoother and more pliable the leather will be.

Thick leather.—Thick leather from the larger hides is oiled and finished in a slightly different manner. For harness and strap leather, let the tanned sides, dyed if desired, dry down. While they are still quite damp slick over the grain or hair side thoroughly and apply a liberal coating of neat's foot or cod oil. Tack on a wall or tie in a frame, stretching the leather out tight and smooth, and leave until dry. Take the sides down, dampen them with warm water until limber and pliable, and apply to the grain side a thick coating of warm dubbin. This dubbin is made by melting together about equal parts of cod oil and tallow or neat's foot oil and tallow. When cool it must be soft and pasty, but not liquid. If too nearly liquid, add more tallow. Hang up the sides again and leave them until thoroughly dried. When dry, scrape off the excess tallow by working over with the slicker. If more grease in the leather is desired, dampen again and apply another coating of the dubbin. When again dry, slick off the tallow and thoroughly work over all parts of the leather with the slicker. Rubbing over with sawdust helps to take up surface oiliness.

Chrome-tanned leather is stretchy, so that in cutting the leather for use in harness, straps, reins, and similar articles it is best to first take out most of the stretch.

Chrome leather for shoe soles must be heavily greased, or, in other words, waterproofed, unless it is to be worn in extremely dry regions. Waterproofing may be done after preparing the shoes by setting them in a shallow pan of oil or grease so that just the soles are covered by the grease. The soles should be dry before they are set in the melted grease. Melted paraffin wax will do, although it makes the soles stiff. The simple formulas given are satisfactory for waterproofing chrome sole leather.

Alum-tanned Lace Leather
Deliming

After the sides have been put through the unhairing and fleshing operations, rinse them off with cool, clean water for from six to eight hours, changing the water frequently.

Buy 5 ounces of U.S.P. lactic acid (or 16 ounces of tannery 22 per cent lactic acid). Nearly fill a clean 40 to 50 gallon barrel with clean, cool water and stir in the lactic acid, mixing thoroughly with a paddle. Hang the sides in the barrel and leave them there for 24 hours, pulling them up and stirring them about frequently. Take out the sides, work over or "scud" thoroughly, as directed under "Unhairing," and hang them in a barrel of cool water. Change the water several times, and finally leave them in the water overnight.

If lactic acid can not be obtained, use a gallon of vinegar instead.

Tanning

While the sides are being delimed, thoroughly wash out the barrel in which the hide was limed. Put in it 15 gallons of clean water and 12 pounds of ammonia alum or potash alum and stir frequently until it is completely dissolved.

Dissolve 3 pounds of washing soda (crystallized sodium carbonate) and 6 pounds of salt in 5 gallons of cold, clean water in a wooden bucket. The soda crystals must be clear and glasslike. *Do not use white crusted lumps.*

Pour the soda solution into the alum solution in the barrel very, very slowly, stirring the solution in the barrel constantly. Take at least 10 minutes to pour in the soda solution in a small stream. If the soda is poured in rapidly the solution will become milky and it will not tan. The solution should be cool, and enough water to nearly fill the barrel should be added.

Hang each well-washed side from the deliming in the alum-soda solution. Pull up the sides and stir the solution six or eight times each day. Do not put the bare hands in the liquor if they are cut or cracked or have sores on them.

After six or seven days remove the sides from the alum-soda solution and rinse well for about quarter of an hour in clean, cold water.

Oiling and Finishing

Let the sides drain and dry out slowly. While still very damp go over the grain or hair side with a liberal coating of neat's-foot or cod oil. After the oil has gone in and the sides have dried a little more, but are still slightly damp, begin to work them over a "stake." The time to start staking is important. The sides must not be too damp; neither must they be too dry. When light spots or light streaks appear on folding it is time to begin staking. Alum-tanned leather must be thoroughly and frequently staked.

Staking is done by pulling the damp leather vigorously back and forth over the edge of a small, smooth board, as described. The sides must be staked thoroughly all over in order to make them pliable and soft, and the staking must be continued at intervals until the leather is dry.

When dry, evenly dampen the sides by dipping them in water or by leaving them overnight covered with wet burlap or sacks. Apply to the grain or hair side a thick coating of warm dubbin. This dubbin is made by melting together about equal parts of neat's-foot oil and tallow or cod oil and tallow. When cool, the dubbin must be soft and pasty but not liquid. If too nearly liquid, add more tallow. Leave the greased sides, preferably in a warm place, until dry. Scrape off the excess tallow and again stake the sides. If the leather is too hard and stiff, dampen it evenly with water before staking.

After staking, go over the sides with a "slicker," pressing the slicker firmly against the leather, while pushing it away from the body. Slick out on the grain or hair side in all directions.

Alum-tanned leather almost invariably dries out the first time hard and stiff. It must be dampened again and restaked while drying. In some cases this must be done repeatedly and another application of dubbin may be necessary. By repeated dampening, staking, and slicking the leather can be made as soft and pliable as desired.

Leather Rolls, Coating for

Red Lead	2.5 oz.
Clovel	2.5 oz.
Lampblack	2 oz.
Glycerin	2.5 oz.
Gelatin	1.5 lb.
Acetic Acid	1 gal.

Patent Leather Softening Emulsion

Castor Oil	4 parts
Casein	4 parts
Methylated Spirits	1 part
Benzol	1 part
Water	50 parts
Preservative	A trace

Leather Finishes

Unpigmented finishes, known as seasonings are applied in dilute solutions to the grain side, leaving a very thin flexible film, sufficiently hard to take a polish when the leather is glazed. That is when the leather is rubbed on glass or agate.

Egg Albumen Finish (for light colored leather)

Egg Albumen	1.5 parts
Milk	4.5 parts
Water	94.0 parts

The above are thoroughly mixed together. This film becomes insoluble to water when exposed to light and air over a period of time. A much more rapid method of rendering it insoluble is by ironing the skin or by treating it with a dilute solution of a metallic salt which does not react with the tannin of the skin.

Note: In making the above mixture, care must be taken not to exceed 130° F. otherwise the albumen will coagulate.

Blood Albumen Finish (for glazed black leather)

Blood Albumen	10 to 18%
Nigrosine	1%
Glycerin	½%
Milk	10%

Water to make 100%

The skin is also ironed to render the film insoluble.

Temperature of mixing should not exceed 130° F.

Casein Finish

Only lactic casein should be used, and not rennet casein.

Casein	2 %
Borax	0.35%
Water	90 %
Milk	10 %

The casein is added to the warm milk and water at about 130° F. and the borax is stirred in afterwards. Formaldehyde is added as a fixative. The formaldehyde (less than 10%) must be added cold, very slowly in a thin stream with constant agitation to the cold casein solution, otherwise it will cause the casein solution to gel.

Nitrobenzene is added as a preservative.

Cellulose Finish for Patent Leather Splits

After the usual rolling and smoothing processes, the splits are brushed free from dust. They are then given two priming coats and a final gloss finish.

Priming Coat:

Celluloid	100 gm.
Amyl Acetate	100 gm.
Ethyl Acetate	50 gm.
Acetone Alcohol	300 gm.
Fusel Oil	300 gm.
Solvent Naphtha	100 gm.
Alcohol	100 gm.
Castor Oil	125 gm.
Mineral Dye (Umber)	50 gm.

The celluloid is dissolved in the mixture of amyl acetate, ethyl acetate and acetone alcohol. The dye is dissolved in the castor oil and a little of the solvents. It is then milled and added to the dissolved celluloid together with the rest of the solvents. The mixture is blended in a mill and applied to the splits by brush and dried at 35° C. When dry, the leather is pressed and a second coat of primer is given. When dry, the flesh side of the splits is wetted down and the grain side pressed with a grain-patterned plate. It is then sprayed with the final gloss finish.

Gloss Finish:

Celluloid	100 gm.
Amyl Acetate	100 gm.
Ethyl Acetate	150 gm.
Acetone Alcohol	300 gm.
Fusel Oil	200 gm.
Solvent Naphtha	200 gm.
Alcohol	200 gm.
Castor Oil	100 gm.

Solution of above is effected similar to the priming coat.

Leather Finish

Dissolve

1 oz. Nigrosine sol. in spirit in a mixture of

3 gills spirit shellac solution and

¾ gill acetine by heating on the water bath, allow to cool and filter.

Spirit Shellac Solution

is prepared by dissolving

8 oz. shellac in

1 gallon methylated spirit by heating on the water bath, filter, and allow to cool.

The leather is brushed over once or twice with this solution and after drying polished with a cloth with or without the application of cream.

Leather Finish

A typical example of wax pigment finish—a russet finish—is as follows:

Boil 40 lbs. grey carnauba wax with 4 lbs. caustic soda and 5 gallons of water for at least 8 hours, making to original volume with water, until saponifcation is complete; often a further boiling is necessary. Then add the following pigments:

Venetian Red	3 lb.
Raw Umber	11 lb.
Brown Acid Dye	2 lb.

and more water as required.

Artificial Leather Dope

1

High grade for hand finishing.

Pyroxylin (30–40 second viscosity)	8 oz.
Butyl Acetate	1 qt.
Amyl Acetate	1 pt.
Butanol	1 pt.
Tuluol or solvent Naphtha	2 qt.
Acetanilid	1 oz.
Camphor	2 oz.

2

Cheaper grade

Pyroxylin	26 oz.
Ethyl Acetate	2 pt.
Methyl Acetate	1 pt.
Denatured Alcohol	1 pt.
Benzol	4 pt.
Camphor	2 oz.

Castor or Rapeseed Oil to be used as plasticizer for both of above.

Pyroxylin artificial leather is made from a cotton fabric, upon which has been built up a plurality of coats of mixtures of oils pyroxylin and plasticizers together with pigments to give the desired color. When the desired thickness has been attained the material is run through an embossing machine where, under proper conditions, the desired grain effect is impressed into the fabric. If a hard finish is desired a nitrocellulose coating with a minimum of oil is applied as a final measure. But since from 3 to 30 coats are applied it is probably economical to use low grade dope for the intermediate coats and a high grade one for the first two coats and the last two or three coats. The dope itself is applied by a blunt knife operated by a machine. For this reason they are rather viscous. The manipulation of the solvent formulae to give the desired qualities together with cheapness is a very specialized art and each manufacturer cherishes what he conceives to be the best and cheapest formula. To avoid blushing when using cheap low boiling solvents use forced drying under heated drying tunnels at a temperature of 150° to 200° F.

Split Leather

Split leather is technically treated the same as cotton cloth, but has the added advantage of it being possible to correctly call it "leather" and a compensating cost from splitting, with that of only requiring three coats whereas 6 to 30 coats are used on cotton. Because of the irregular shade of the hide the dope is applied by hand with a 2½" by 6" swab and since it is brushed it is necessary to use high boiling point solvents and, in the case of black or patent leather, each coat is pumiced smooth to remove all flow and brush marks.

Leather Stuffing

Ozokerite	6
Paraffin Wax	8
Rosin Oil	40
Mineral Oil	48

Coloring Leather Black

Make a thin paste of Colloid Carbon Black and water and rub into the leather. When dry, coat with a bright drying wax emulsion or shellac solution. This gives a permanent non-fading black.

Belt Dressing Stick

Rosin	65 lb.
Tallow	6 lb.
Stearic Acid	1 lb.
Scale Wax	20 lb.
Castor Oil	2.0 lb.
Rosin Oil	0.5 lb.
Lanolin	4.2 lb.

Coloring Belt Edges

Brown

Bismarck Brown	1 oz.
Water	1 pt.
Borax Shellac Water Solution	1 pt.

Black

Nigrosine Crystals	1 oz.
Water	1 pt.
Borax Shellac Water Solution	1 pt.

Leather, Applying Basic Dyes to

Before dyeing with basic dyes, tanned leather is treated for 30 min. with a liquor containing as much Copper Sulfate as the dye to be afterwards applied, whereby the depth of shade obtained subsequently is 4–5 times that similarly obtained on non-treated leather, whilst exaggerated grain defects and a tendency for the dyed flesh side of the leather to be loose to rubbing (evident in leather not fixed after tanning) are avoided. The Cu treatment colors the tanned leather from a pale yellow to brown, but insufficiently to affect the shade obtained with the basic dye, and enables acid dyes to be satisfactorily replaced by basic dyes.

Black Leather Dye

Deodorized Kerosene	20
Benzol or Toluol	80
Pylam 77 Black Dye	4
Pylam 512 Black Dye	3

Brown Leather Dye

1

Deodorized Kerosene	20
Benzol or Toluol	80
Pylam 123Y Dye	4
Sudan 5BA Dye	2

2

Spirit Brown	2
Sudan Brown 5BA	0.5
Oil Black	0.5
Toluol	50
Methanol	50

Tan Leather Dye

Spirit Yellow Dye	1
Spirit Brown Dye	0.2
Alcohol	100

Blue Leather Dye

Oil Blue Green Dye	1
Spirit Blue Dye	2
Spirit Brown Dye	0.5
Alcohol	75
Ethyl Acetate	25

Leather Soles, Impregnant for

Crepe Rubber	15
Rosin	30
Linseed Oil	35
Turpentine	17
Paraffin	3

Keep melted with occasional stirring until rubber has dissolved.

Leather "Nourisher"

For leggings, boots, base-ball gloves, etc.

Menhaden Oil	39
Tallow	60
Clovel	1

Protecting Leather during Manufacture

Shoes, bags, novelties, etc., made of leather are soiled readily while being handled in various "putting together" operations.

To avoid this they are dipped or sprayed with following and dried

Rubber Latex	20
Carnauba Wax Emulsion	10
Water	40

After articles are finished the deposited film is easily stripped off.

Leather Preservatives

A. Neatsfoot Oil	
(20° Cold Test)	20
Castor Oil	20

B. Lanolin Anhydrous	40
Neatsfoot Oil	
(20° Cold Test)	60

C. Neatsfoot Oil	
(20° Cold Test)	50
Lanolin Anhydrous	35
Japan Wax	20
Soap Chips	8
Water	90

Imitation Leather Dressing

A transparent dressing for imitation leather may be made as follows:

½ second dope solution (nitrocellulose approximately 30%)	19 lb.
Wood alcohol	33 gal.
Castor Oil	2 qt.
Amyl Acetate	13 gal.

Should a colored dressing be desired a

proper dye may be added to the above solution to obtain desired shade.

White Shoe Dressing

Pipe Clay	450 gm.
Spanish Whiting	225 gm.
Flake White	180 gm.
Precipitated Chalk	115 gm.
Powdered Tragacanth	8 gm.
Phenol	4 gm.
Water to make a paste.	

White Shoe Dressings (Liquid)
A

Titanium-Lithopone Mixture	40 gm.
Shellac (Bleached)	6 gm.
Ammonium Hydroxide (0.880)	½ c.c.
Water	50 c.c.
Alcohol	50 c.c.

B

Lithopone	40 gm.
Bright Drying Wax Emulsion	80 c.c.
Triethanolamine	2 c.c.
Water	30 c.c.

White Shoe Dressing Cake

Water	95
Gum Arabic	5
Preservative	1

Allow to soak a few hours and stir till dissolved.

''Wetting Out'' Agent	3
Glycerin	1

Whiting or Lithopone sufficient to make a stiff paste.
Mold in blocks or cylinders and dry.

White Shoe Cleaners, Paste
(For use in tubes)

Soap Flakes	10
Proflex	5
Water	35
White Pigment	150

White Shoe Cleaners, Liquid
A

Soda Ash	1
Rochelle Salts	2
Titanox C	40
Water	57

B

Soda Ash	0.5
Soap Flakes	3

Lithopone	40
Water	53
Gum Arabic (50% Sol.)	4

Paste Shoe Polish

Carnauba Wax	25 gm.
Ceresin	10 gm.
Triethanolamine	1 c.c.

Melt together and stir; add following slowly

Turpentine	30 c.c.
Deodorized Kerosene	5 c.c.
Gasoline	30 c.c.

For black polish add 1% Oil Black Dye
For brown polish add Brown Iron Oxide sufficient to color.
The above mix is poured into cans at 35° C. and allowed to set.

Shoe Polish, Paste

Carnauba Wax	20
Paraffin Wax	12

Heat to 200° F. and add to this slowly with good stirring while heating on a steam table

Turpentine	65
Carbon Black No. 1	2.5
Oil Soluble Black Dye	0.5

Stir until uniform.

Non-Caking Shoe Dressings

White shoe polishes, especially, have a tendency toward cake formation of the pigments. This can be overcome by grinding the pigment with Aquaresin G.M. The latter forms a thin film around each particle of pigment. While this does not prevent settling, it does prevent formation of a hard cake and slight shaking distributes the pigment thoroughly.

Shoe Polish and Preservative

Carnauba Wax	2 parts by wt.
Beeswax	2 parts by wt.
Neatsfoot Oil	1 part by wt.

Heat by hot water bath (not over fire) till melted, and then add turpentine until a soft paste is obtained when the mixture is cold. This should be applied to the clean, dry leather with a rag or a piece of waste, and rubbed hard until no more polish is absorbed. Polish with a clean cloth. A higher polish will be obtained by reduction of the proportion of oil, but the leather will not be so well preserved.

Shoe Cream

1. Trihydroxyethylamine

Stearate	25 lb.
Beeswax	10 lb.
Candelilla Wax	30 lb.
Carnauba Wax	40 lb.
Turpentine	20 lb.
2. Water	500 lb.

Heat (1) to 200° F. and in a separate pot heat (2) to 200° F. Run (1) into (2) slowly while stirring vigorously until cold. This gives a beautiful light cream. If a colored cream is desired dissolve some oil soluble dye in the wax mixture while it is melting.

Shoe Cream, Neutral

Hydrowax Cream	50

Heat to 200° F. and to it add following solution warmed to 150° F. and stir until smooth.

Turpentine	29
Water	24
Proflex	3
Soap Flakes	1

Shoe Polish

1.
Carnauba Wax	55 parts
Crude Montan Wax	55 parts

are melted at 105–110° C.

Nigrosine Base	10 parts

dissolved in

Stearic Acid	20 parts

added, then

Ceresine	150 parts

and finally

Turpentine Oil	900 parts

The mass is filled at 45° C. (105° F.).

2.
Carnauba Wax	65 parts
Crude Montan Wax	40 parts
Dyestuff Soluble in Oil	30 parts
Paraffin	110 parts
Ozokerite	10 parts
Turpentine Oil	760 parts

3.
Carnauba Wax	65 parts
Crude Montan Wax	40 parts
Dyestuff Soluble in Oil	30 parts
Paraffin	40 parts
Ceresine	75 parts
Turpentine Oil	760 parts

It is recommended to use only stearic acid or crude Montan wax for dissolving the bases, as oleine or mixtures of crude Montan wax with oleine do not give such fine surfaces.

Saponified Water-Wax, Shoe Polish

Mixture 1

Carnauba Wax	8 parts
Montan Wax	8 parts
Paraffin Wax	4 parts

These are saponified in a hot solution of:

Potash	3 parts
Water	50 parts

Replace any evaporation with additional warm water.

Mixture 2

No. 1 Polish Black	4 parts
Water	25 parts

These should be milled together in a color mill until thoroughly dispersed.

While Mixture No. 1 is hot, add Mixture No. 2 slowly and with constant stirring. As it cools, the mass will slowly set to a paste. Before it is too stiff for flowing pour into suitable containers and set aside until cold.

Liquid Shoe Blacking

Nigrosine Base	8
Rozolin	17

Warm and stir until dissolved. Cool and add

Alcohol	24
Acetone	22
Benzol	42

Black Shoe Cream

Montan Wax Crude	15
Carnauba Wax Refined	15
Rosin	3
Caustic Potash	6
Soap Flakes	1
Water	156
Nigrosine (Water Soluble)	4

French Shoe Dressing

Shellac	10 gm.
Ammonia 26°	1 c.c.

Heat and stir until dissolved. Dissolve in this

Nigrosine	0.1 gm.

Add to this

Hydrowax Liquid N	100 c.c.

or other bright drying carnauba wax emulsion.

Black Water Soluble Dye	0.5 gm.

When this is swabbed on shoes it dries with a lustre without rubbing.

Shoe Dye

Shellac	12.7 kg.
Borax	3.2 kg.
Water	82.0 kg.
Carnauba Wax	6.3 kg.
Marseilles Soap	1.5 kg.
Potassium Carbonate	0.3 kg.
Nigrosin	12.0 kg.
Water	32.0 kg.

The shellac solution in borax and water is made first, the carnauba wax is emulsified in the soap, carbonate solution as above and the nigrosin and water added to it, it is then added to the shellac soln. with rapid agitation. Some ammonia may be added to prevent lumps.

Cold Polishing Dyes for Dressing Shoes

Carnauba Wax	7.5 kg.
Marseilles Soap	1.0 kg.
Potassium Carbonate	1.5 kg.
Water	79.0 kg.

Melt the carnauba wax, and add the heated mixture of the other ingredients. Stir rapidly, and add 11 kg. nigrosine previously dissolved in a small amount of the soap soln.

Waterproof Boot Dressing

Spermaceti	3 oz.
Raw India Rubber	6 dr.
Tallow	8 oz.
Hog's Lard	2 oz.
Amber Varnish	5 oz.

Calf Leather Cleaner and Polish

Water	20	gal.
Potassium Oleate	7.5	lb.
Trisodium Phosphate	0.5	lb.
Beeswax, Yellow	6	lb.
Carnauba Wax	6	lb.
Turpentine	4.5	gal.
Pine Oil	0.5	gal.
Terpineol	0.25	gal.

Greasy Leather Cleaner

Water	10	gal.
Castile Soap	0.75	lb.
Trichlorethylene Soap	3.5	lb.
Methyl Acetone	0.5	gal.
Lemongrass Oil	0.15	lb.

Dissolve soap in water by heating and stirring: cool and stir in other ingredients.

Leather Cleaner

Castile Soap (Powd.)	6
Water	160

Boil until dissolved: cool and add

Ammonium Hydroxide	6
Glycerin	14
Ethylene Dichloride	7

Powdered Glove Cleaner

Cream of Tartar Powd.	480
Soap Bark	160
Whiting	96
Oil Birch Tar	12

Suede Cleaner

Precipitated Chalk or Whiting	12	lb.
Quilaya Bark	20	lb.
Cream of Tartar Powder	60	lb.
Oil Birch Tar	1½	oz.

Leather Finish Remover

Ethyl Acetate	60
Butyl "Cellosolve"	20
Butyl Acetate	20

or

Ammonium Hydroxide	20%

Leather Stain Remover

A solution for removing stains from the flesh side of leather is composed of the following:

Water	250 cc.
Oxalic Acid	3 gr.

Removing Mildew from Leather

Make a thick paste of bicarbonate of soda and rub into the leather, and stand in the sun for a day. This will kill the mildew, though the leather will doubtless require painting with a new leather finish.

Waterproofing Shoes

Natural Wool Grease	8 oz.
Dark Petrolatum	4 oz.
Paraffin Wax	4 oz.

Melt the ingredients together by warming them carefully and stirring thoroughly. Apply grease when it is warm but never hotter than the hand can bear.

Bleach for Furs

Water	3 gal.
Hydrogen Peroxide	3 oz.
Potassium Persulfate	6 oz.
Sodium Pyrophosphate	6 oz.

CHAPTER IX

LUBRICANTS, OILS, ETC.

TO lengthen the life of moving parts which are in contact, a lubricant of some sort is essential. The lubricant should spread itself in a thin continuous film to prevent the parts from rubbing against each other and thus eventually wearing down.

The type of lubricant varies with temperature, speed, nature of the surfaces and many other factors. Because of this there is no universally satisfactory lubricant. In one case a very light oil (e.g. for clocks or sewing machines) is used while a very heavy oil (e.g. in tractors) is necessary in other cases. Greases and fatty compositions as well as waxes are used in heavy duty applications. In some cases mineral matter such as graphite or talc is included for special purposes. In high speed cutting or grinding operations, where much heat is generated locally, "cutting-oils" which contain water (introduced for its cooling effect) are used.

Boring Oil

A.	1. Oleic Acid	15
	2. Thin Mineral Oil	75
	3. Caustic Soda (40° Bé.)	5
	4. Alcohol	5

Warm 1 and 3 with stirring until uniform and while mixing vigorously run into it 2 and 4.

B.	Turkey Brown Oil	30
	Thin Mineral Oil	50
	Caustic Soda (20° Bé.)	10
	Alcohol	10

C.	Rosin Oil	18
	Thin Mineral Oil	74
	Caustic Soda (40° Bé.)	5
	Isopropyl Alcohol	5

D.	Naphthenic Acid	25
	Red Oil	25
	Thin Mineral Oil	100
	Caustic Soda (24° Bé.)	25
	Alcohol	25

E.	Rosin Oil	10
	Red Oil	10
	Thin Mineral Oil	70
	Caustic Soda (36° Bé.)	5
	Methanol	5

The above are mixed with water for use.

Cutting Oil

1. Mineral Oil	280 lb.
2. Miscibol	32 lb.
3. Oleic Acid	24 lb.
4. Water	15 lb.
5. Denatured Alcohol	10 lb.

Mix 1, 2 and 3 mechanically until dissolved. Heating speeds solution. Stir 4 into this and then add 5 with stirring. This produces a clear, stable, "soluble" oil.

If 70 parts water are added slowly, with stirring, to 10 parts of the above, a beautiful white stable emulsion results. The amount of water may be larger or smaller as needs require. This emulsion is useful as a lubricant, cutting oil, polish or agricultural spray.

Cutting Oil Emulsions

The term "cutting oil" is applied to soluble lubricating oils which are used as machine lubricants. In lathe and speed-tool operations the first requirement is a cooling medium which will carry heat away from the cutting edge. In addition, a certain amount of true lubricant is advantageous, and both of these requirements are satisfied by a dilute oil emulsion. With the proper oil and emulsifying agent, the corrosive action of the water is likewise decreased and rusting of steel prevented. In practice a soluble oil is used to produce a 5 to 25 per cent oil emulsion, and this is flowed over the cutting edge and continuously recirculated.

One of the most important requirements of a soluble oil for cutting is its dependability. It should not separate when left in open containers and it should always emulsify in water with only the simplest stirring methods. The resulting emulsion should also remain stable and uniform, a 5 per cent emulsion not separating oil in 24 hours. Soluble oils fulfilling these qualifications can be made with Triethanolamine. This agent, for one thing, permits the use of oils of high lubricating value which are otherwise difficultly emulsifiable. In addition it yields emulsions of such high dispersion and uniformity that lower concentrations of oil in water than are customary can be used with equal lubricating effect.

Another interesting application for soluble oils is in the lubrication of textile machinery. The elimination of ordinary oil spots from fabrics is usually an expensive hand operation. On the other hand, when the spot is caused by a soluble oil, it may be completely and readily removed in the regular scouring operation. If a stiffer lubricant, more of the texture of a grease, is desired, this can be made of any consistency by stirring thoroughly up to 20 per cent water into one of the soluble oils. Another way of making a soluble grease consists in melting 10 per cent of stearic acid into a lubricating oil, and then emulsifying this with an equal weight of hot water containing 4 per cent Triethanolamine.

Rosin Soluble Cutting Oil

Rosin	7.5 lb.
100 visc. Spindle Oil	2.5 gal.
Oleic Acid	6.0 lb.
100 visc. Spindle Oil	5.5 gal.
32° Bé. Caustic Soda	4.0 lb.
Alcohol	2.1 lb.
Yield	10 gal.

Heat the rosin with the first portion of the spindle oil at a temperature of about 212° F. until the former is melted, then add the other ingredients in the order listed. The alcohol should be added when the batch has been cooled to room temperatures.

Drawing Oil

A. Rosin Oil	28
Caustic Potash (38° Bé.)	10
Thin Mineral Oil	64
B. Degras	40
Rosin	29
Rosin Oil	21
Caustic Soda (40° Bé.)	10
C. Tallow	10
Thin Mineral Oil	10
Japan Wax	1
Caustic Soda (40° Bé.)	4.2

Oil for Leather

Rosin Oil	10
Degras	10
Mineral Oil	82

Penetrating Oil

1

Pine Oil	30
Blown Rape Seed Oil	30
Carbon Tetrachloride	10
Kerosene	100
Light Paraffin Oil	70

2

For freeing rusted bolts, screws, etc.

Kerosene	20
Mineral Oil, Light	70
Secondary Butyl Alcohol	10

Slushing Oil (for foreign shipment)

Neutral 28° Paraffin Oil	4½ gal.
Anhydrous Lanolin	60 oz.

Porcelain Mold Oil

Stearic Acid	24
Ozokerite	1
Paraffin Wax	3
Heavy Mineral Oil	82

Cup Greases

Pressure

Fat	114 parts
Quicklime	16 parts
Petroleum Red Oil	870 parts

preferably 500 visc. at 100° F. or over

No. 1

Fat	123 parts
Quicklime	17 parts
Petroleum Pale Oil	855 parts

100 Visc. at 100° F.

No. 2

Fat	140 parts
Quicklime	19 parts
Petroleum Pale Oil	840 parts
100 Visc. at 100° F.	

No. 3

Fat	157 parts
Quicklime	22 parts
Petroleum Pale Oil	820 parts
100 Visc. at 100° F.	

No. 5

Fat	205 parts
Quicklime	34 parts
Petroleum Oil	760 parts

The weighed fat is placed in a steam jacketed kettle equipped with a paddle type agitator and a small portion of the Petroleum Oil, about half the volume of the fat, is added. Next the lime is hydrated and mixed with sufficient water to form a thin paste. The lime is added to the material in the kettle and the whole is cooked for several hours with continuous agitation. When a small portion of the soap on cooling is firm and brittle the remainder of the Petroleum Oil is added slowly to avoid chilling. The agitation is continued until a uniform grease without lumps is formed.

Graphite Cup Grease

1. Graphite Cup Grease.

	Per cent by weight
Cup Grease No. 2	93.00
Medium Ground Graphite	2.00
American Talc	5.00

2. Graphite Lubricant.

Cup Grease No. 2	86.29
Steam Refined Cylinder Stock	6.80
Powdered Plumbago (Graphite)	6.91

3. Marine Graphite Grease.

Cup Grease No. 2	92.00
Fine Ground Graphite	8.00

4. Special Graphite Grease.

Hard Tallow	10.00
Dark Petrolatum	80.00
Fine Graphite	10.00

Graphite Grease

Ceresin	70
Tallow	70

Heat together to 80° C. and work in

Graphite	30

Driving Journal Grease

Tallow	40 parts
Sodium Hydroxide	7 parts
Steam Refined Cylinder Oil	45 parts
Water	10 parts

Locomotive Rod Cup Grease

Tallow	35 parts
Sodium Hydroxide	6.5 parts
Steam Refined Cylinder Oil	50 parts
Water	10 parts

Thread Grease

Lanolin (dry)	1 lb.
Vaseline	2 oz.

Melt No. 1 and No. 2 and add 3 oz. camphor.

Cordage Grease

Degras	30
Kerosene (Heavy)	60
Caustic Soda (36° Bé.)	10

Warm together and stir until uniform.

Increasing Viscosity of Oils

The viscosity of animal, vegetable or mineral oils is increased by dissolving therein 7–10% Ethyl Cellulose.

Insoluble Oil Lubricant for Wool

Lard Oil, No. 1 Quality	10–20 lb.
Pale Paraffine (debloomed type) Oil	80–90 lb.

Mix cold with stirring, then heat until blended and add some type of artificial odor compound.

Bicycle Chain Lubricant

12 kg. of rosin oil, 25.0 kg. of mineral oil, 1.0 kg. of 10° Bé. caustic potash, 4.5 kg. of slaked lime, 35.0 kg. of flake graphite and 22.5 kg. of mineral oil. The rosin oil and first portion of mineral oil are mixed and emulsified in the alk. soln. The lime and graphite are ground with the second portion of mineral oil, and well mixed with the emulsion.

Dry Powdered Lubricant

Zinc Stearate	50
Talc	50

This is of advantage on machinery in

mills where white goods are handled as this lubricant will not discolor goods.

Gun Lubricant

White Petrolatum	150
Bone Oil (acid free)	50

Lubricant for Dies and Plates
(for moulded clay products)

No. 1.—Thoroughly mix, with both ingredients lukewarm, one part of Saponified Red Oil and five parts of kerosene.

No. 2.—Melt ten pounds of Double Pressed Saponified Stearic Acid to just above the melting point and add ninety pounds of kerosene with brisk agitation to obtain a thorough mixture.

Open Gear Lubricant

A home-made mixture of ½ lb. white lead, ½ gal. cylinder oil and ½ lb. flake graphite makes an especially efficient lubricant for open gears. This mixture adheres well to the gears and can be painted on with a brush as required at intervals of about five hours. Cup grease may be substituted for the oil and the graphite may be omitted. Omission of the graphite is not advisable in warm weather.

Solid Lubricant

1. Rosin	9
2. Machine Oil	82
3. Caustic Soda (40° Bé.)	9

Melt 1 and 2 together and heat to 100° C. and run 3 into it slowly with stirring and raise temperature to 110–120° C.

Stainless Steel Lubricant
Lubricant for Drawing and Forming Stainless Steel

Heavy Drawing Compound	1 gal.
Hot Water	1 gal.
Lithopone	2 lb.
Flowers of Sulfur	½ to 1 lb.
Cresylic Acid	1 oz.

Upper Cylinder Lubricant Tablets

Naphthalene	1
Colloidal Carbon Black	¼
Mineral Oil	1–4

Heat together and stir until uniform.

Valve Lubricant

Unaffected by gas and high temperatures.

1. Barium Stearate	50
2. Mineral Oil	40
3. Talc	10

Heat 1 and 2 together with slow mixing at 120–150° C. until dissolved; work in 3.

Pine Oil, Solidified

Trihydroxyethylamine Linoleate	1
Pine Oil	10
Water	8

Olive Oil, Bleaching

Dark oils are treated with a 12% solution of tannic acid. From 1 to 4% of the acid is necessary, according to the color of the oil, and very thorough mixing of the oil and the solution of the acid is required. A 5% solution of citric acid also gives good results.

Bleach for Animal Fats

Bleach for use with animal fats and oils is to use from 1½ lb. to 4 lb. Manganate of Soda or Permanganate Salts and from 2½ lb. to 6 lb. of Sulfuric Acid to each 100 lb. of fat.

Dissolve required quantity of Manganate of Soda or Permanganate Salts in from 20 to 25 times its own weight of boiling water. Dilute required quantity Sulfuric Acid with 10 times its own weight of water. Liquefy fat thoroughly at as low temperature as possible and then add slowly and with vigorous agitation the Manganate or Permanganate solution, continue agitation actively for 15 to 30 minutes, then add, also with vigorous agitation the dilute Sulfuric Acid and continue stirring for 15 minutes. Then steam is to be turned on and an active boil kept up until all brown stain disappears, which should be from 30 to 60 minutes from time boiling commences. Then settle and draw off spent solution and wash oil with water.

If using Manganate of Soda care must be taken not to add bottoms or undissolved portion. Permanganate Salts cost a little more but are more readily soluble.

Palm Oil, Decoloring

The oil is heated to 90° in the presence of 0.01% Cobalt Resinate and air is blown through it for two hours.

Sulfonated Castor Oil
(Turkey Red Oil)

The apparatus for the manufacture of these oils consists of a lead-lined tank or sulfonator, of about two and a half times the volume of the oil to be treated, containing a lead coil, with agitator, and fitted with a bottom valve. The coil is for cooling the mix by circulating cold brine or well water through it, depending on the size of the installation, prevailing temperature, etc. In warm climates an ice machine is necessary to provide continuous refrigeration. A second lead-lined tank for storage or discharge of spent liquors, and an acid tank or carboy of sulfuric acid are also needed.

Castor oil, which is about the easiest oil to sulfonate, is charged into the sulfonator, and 20% of its weight of concentrated sulfuric acid is then run into the oil in a thin stream, and the mixture agitated. This operation should take about one hour. The temperature should not be permitted to rise above 100° F. Agitation is continued for half an hour longer in order to complete the reaction. The mixture is then allowed to rest for about 24 hours.

Water is next added to the mass, equal to the original volume of oil taken, being stirred to a homogeneous mass, and left standing for another 24 hours. By that time the batch will have separated into two layers, the oil on top and an aqueous layer below. The latter is run off by means of the valve. The oil is then given a wash with brine consisting of 1 lb. to 1½ lb. of salt to the gallon. This should be thoroughly mixed with the oil, allowed to stand over night and run off. The object of this wash is to eliminate as much free sulfuric acid as possible. At this point manufacturers differ in the mode of procedure, some neutralizing with *aqua* ammonia, while others finish off with a 24° Baumé caustic soda solution, which is added slowly, stirring constantly until a clear transparent oil results.

Petroleum Proof Valve Lubricant

Citric Acid, Anhydrous	64 g.
Tetraethylene Glycol	97 g.

Heat at 180-185° C. for 90 minutes; cool. Do not overheat or an infusible product will form.

◆

Rubber Mold Lubricant

Cocoa soapstock, a material containing a large percentage of coconut oil saponified with alkalies to give a pure hard soap, makes a suitable product for lubricating molds to prevent sticking of the vulcanized stock. If properly made, without traces of sodium silicate, it will not cause caking on the molds. The recommended quantity is 8 to 12 lb. to a 55 gal. drum of water. The soap is dissolved in water by cooking, either by open steam or external heat of some kind. For easy spraying the solution is kept warm by steam or a small electric heating unit can be applied at the spray nozzle to prevent clogging.

◆

Cutting Oil

Formula No. 1

a.	Mineral Oil (Spindle Oil)	80 g.
	"Tall-Oil," Refined	20 g.
b.	Caustic Potash (40° Bé.)	6 g.
c.	Methylhexalin	1-2 g.

Saponify a with b, clear with c.

No. 2

Paraffin Oil (28 to 30° Bé.)	250 g.
Rosin	22 g.
Oleic Acid	22 g.
Caustic Soda	3 g.
Water	10 g.
Alcohol	7 g.

◆

Greaseless Lubricating Pencil

Useful for lubricating hinges of automobile doors, etc., as it will not run off and produce stains or accumulate dust.

Beeswax	80 g.
Diglycol Stearate	20 g.
Graphite Powder	100-200 g.

Melt together and stir until just cold enough to pour. Pour into molds and allow to set.

◆

Dynamo Brush Lubricant

Ceresin	20 g.
Tallow, Acid Free	10 g.
Wool Fat, Neutral	10 g.
Castor Oil	10 g.
Vaseline Oil	50 g.

Melt together and add enough organic solvent (Heavy Benzoline, Naphtha or Tetralin).

CHAPTER X

MATERIALS OF CONSTRUCTION

UNDER materials of construction there is included concrete, cement, stucco, tile, terrazzo paving materials, plaster board, wood, metals and alloys and their treatments. Formulae for coloring, acid-proofing, water-proofing and fire-proofing are indicated.

Metals and Alloys

Metals and their numerous alloys are of extreme interest because of their universal use. Much space is given to the coloring and protection of metals and alloys which is of great importance from an aesthetic as well as from a utilitarian standpoint.

Glass, Ceramics, Enamels

The manufacture of glass, ceramics and enamels requires rather specialized equipment so the data given in this chapter in most cases will only be of academic interest.

CONCRETE OR MORTAR

Quantities of Cement, Fine Aggregate and Coarse Aggregate Required for One Cubic Yard of Compact Mortar or Concrete

How to Figure Quantities

Mixtures				Quantities of Materials			
Cement	F. A. (Sand)	C. A. (Gravel or Stone)	Cement in Sacks	Fine Aggregate		Coarse Aggregate	
				Cu. Ft.	Cu. Yd.	Cu. Ft.	Cu. Yd.
1	1.5	...	15.5	23.2	0.86	....	....
1	2.0	...	12.8	25.6	0.95	....	....
1	2.5	...	11.0	27.5	1.02	....	....
1	3.0	...	9.6	28.8	1.07	....	....
1	1.5	3	7.6	11.4	0.42	22.8	0.85
1	2.0	2	8.3	16.6	0.61	16.6	0.61
1	2.0	3	7.0	14.0	0.52	21.0	0.78
1	2.0	4	6.0	12.0	0.44	24.0	0.89
1	2.5	3.5	5.9	14.7	0.54	20.6	0.76
1	2.5	4	5.6	14.0	0.52	22.4	0.83
1	2.5	5	5.0	12.5	0.46	25.0	0.92
1	3.0	5	4.6	13.8	0.51	23.0	0.85

1 sack cement = 1 cu. ft.; 4 sacks = 1 bbl. Based on tables in "Concrete, Plain and Reinforced," by Taylor and Thompson.

Materials Required for 100 Sq. Ft. of Surface for Varying Thicknesses of Concrete or Mortar

C. = Cement in Sacks.
F. A. = Fine Aggregate (Sand) in Cu. Ft.
C. A. = Coarse Aggregate (Pebbles or Broken Stone) in Cu. Ft.
Quantities may vary 10 per cent either way depending upon character of aggregate used. No allowance made in table for waste.

Proportion	1-1½			1-2			1-2½			1-3		
Thickness in inches	C.	F. A.	C. A.	C.	F. A.	C. A.	C.	F. A.	C. A.	C.	F. A.	C. A.
⅜	1.8	2.7		1.5	3.0		1.3	3.2		1.1	3.4	
½	2.4	3.6		2.0	4.0		1.7	4.3		1.5	4.4	
¾	3.6	5.4		3.0	6.0		2.5	6.3		2.2	6.8	
1	4.8	7.2		4.0	7.9		3.4	8.4		3.0	8.9	
1¼	6.0	9.0		4.9	9.9		4.2	10.5		3.7	11.1	
1½	7.2	10.8		5.9	11.9		5.1	12.7		4.4	13.3	
1¾	8.4	12.6		6.9	13.9		5.9	14.7		5.2	15.7	
2	9.6	14.4		7.9	15.8		6.8	16.9		5.9	17.7	

	1-2-2			1-2-3			1-2½-3½			1-3-5		
	C.	F. A.	C. A.	C.	F. A.	C. A.	C.	F. A.	C. A.	C.	F. A.	C. A.
3	7.7	15.4	15.4	6.5	13.0	19.3	5.5	13.6	19.1	4.3	12.8	21.3
4	10.2	20.4	20.4	8.6	17.2	25.8	7.3	18.1	25.4	5.7	17.0	28.4
5	12.8	25.6	25.6	10.8	21.6	32.2	9.1	22.6	31.8	7.1	21.3	35.5
6	15.4	30.7	30.7	12.9	25.8	38.6	10.9	27.2	38.2	8.5	25.6	42.6
8	20.6	41.0	41.0	17.2	34.4	51.6	14.6	36.4	51.0	11.4	34.1	57.0
10	25.6	51.2	51.2	21.5	43.2	64.4	18.2	45.3	63.5	14.2	42.5	71.0
12	30.7	61.4	61.4	25.8	51.6	77.2	21.8	54.5	76.3	17.0	51.1	85.1

Coloring Concrete

Table of Colors to be Used in Concrete Floor Finish

Amounts of pigments given in table are approximate only. Test samples should be made up to determine exact quantities required for the desired color and shade.

Color Desired	Commercial Names of Colors for Use in Cement	Light Shade	Medium Shade
Grays, Blue-black and black	Germantown Lampblack* or	½	1
	Carbon Black* or	½	1
	Black Oxide of Manganese* or	1	2
	Mineral black	1	2
Blue	Ultramarine blue	5	9
Brownish red to dull brick red	Red oxide of iron	5	9
Bright red to vermilion	Mineral turkey red	5	9
Red sandstone to purplish red	Indian red	5	9
Brown to reddish-brown	Metallic brown (oxide)	5	9
Buff, colonial tint and yellow	Yellow ochre or	5	9
	Yellow oxide	2	4
	Green chromium oxide or	5	9
	Greenish blue ultramarine	6	

*Only first quality lampblack should be used. Carbon black is of light weight and requires very thorough mixing. Black oxide or mineral black is probably most advantageous for general use. For black use 11 pounds of oxide for each bag of cement.

Dustproofing Concrete Floors

"Concrete Special" silicate of soda is recommended for this purpose. It is a syrupy solution. Technically, it is 42.25° to 42.75° Baumé, with a ratio of sodium oxide to silica of 1: 3.25. It is diluted as noted below, and applied to the surface of the concrete after it has set. After the concrete is in place, it is desirable to wait at least two weeks before applying the silicate, and four weeks is still better. Also the silicate treatment may be satisfactorily applied to clean concrete at any later time; it is especially good on old concrete.

The diluted "Concrete Special" silicate soaks into the concrete, and a chemical reaction takes place which hardens the surface and makes it more dense.

Method of Application

In ordinary cases it will be found satisfactory to dilute each gallon of the silicate with four gallons of water. The resulting five gallons may be expected to cover 1000 square feet of floor surface, one coat. However, the porosity of floors varies greatly and the above statement is given as an approximate value for estimating purposes.

The floor space should be prepared for the treatment by cleaning free from grease, spots, plaster, etc., and then thoroughly scrubbed with clear water. To get the best penetration the floor should be thoroughly dry, especially before the first application, and if practical it is well to let it dry for several days before the first scrubbing. . . . The solution may be applied with a mop or hair broom and should be continuously brushed over the surface for several minutes to obtain an even penetration. An interval of twenty-four hours should be allowed for the treatment to harden, after which the surface is scrubbed with clear water and allowed to dry for the second application. Three applications made in this manner will usually suffice, but if the floor does not appear to be saturated by the third application a fourth should be applied.

Acid Resistant Concrete

The same treatment with silicate of soda that is recommended for dust proofing is remarkably serviceable in rendering concrete resistant to acid. It works by filling the pores of the concrete with a material that is acid-proof. Concrete itself is rapidly attacked by acids, but when thus protected by an acid-proof filler, it has considerable acid-resistance. For example, a block of concrete was prepared with the silicate treatment applied to one end and not to the other. Concentrated hydrochloric acid was poured over the block. The acid ate rapidly into the untreated end leaving it friable and sandy. The treated end was only slightly affected.

Along this line, therefore, the silicate treatment has frequently done good service where old floors had to be used. The treatment is useful also for protection against dilute acids, and against organic acids. In some cases repeated silicating, perhaps once a year, may be desirable.

Concrete Efflorescence, Removal of

Where efflorescence occurs, it may be dissolved by a dilute solution of muriatic acid (1 part of concentrated acid to 10 parts of water). In using this treatment the surface of the concrete is wetted before applying the acid and is thoroughly washed after the acid treatment.

The length of time required for the acid solution to dissolve efflorescence will depend upon the amount of the latter. In most cases, the acid can be washed off within three or four minutes. It is best not to leave the acid solution on longer than four minutes, for it may etch the colored concrete. If some deposit still remains after the first application, a second can be made. The acid solution should be brushed on smoothly, using the least amount possible for each application.

Efflorescence also can be removed with a solution of equal parts of paraffin oil and benzine rubbed vigorously into the surface when the concrete is dry. This treatment also improves the wearing qualities of the surface by filling the pores and bringing out the color more uniformly. It is frequently applied to concrete surfaces for these reasons only.

Concrete, High Early Strength

Increasing the time of mixing will increase early strength. For concrete cured at normal temperatures, increasing the mixing time from 1 minute to 2 minutes will add about 100 pounds per square inch to the strength at three days. About 200 pounds per square inch are added by increasing the mixing time from 1 to 5 minutes.

Concrete that is to attain high early strength should be kept damp at a temperature of 70° F. or above, beginning soon after it is placed. Concrete cured below 70° F. hardens more slowly and it is not likely to have high strength at an early age.

The admixtures commonly used to increase the rate at which concrete hardens are calcium chloride and calcium oxychloride. These materials may be used within certain limits to hasten hardening and to increase early strengths of concrete.* The quantities of admixtures should not exceed from 2 to 4% of calcium chloride or 7 to 10% of calcium oxychloride by weight of the cement. The calcium chloride is dissolved in the mixing water before adding it to the other materials in the mixer. Most contractors make up a solution of known concentration, adding the desired amount to each batch. Thus, if it is desired to use 2 pounds of calcium chloride per sack of cement a solution containing 1 pound per quart can be made, 2 quarts of the solution being added to the mixture for each sack of cement in the batch. It is important to remember that this solution is to be regarded as part of the mixing water.

Acid Wash for Concrete Surfaces

Aluminum Chloride (Commercial)	1 lb.
Water	10 lbs.

To be flushed over concrete surface and washed off with clean water.

Concrete Floor Hardeners

The fluosilicates of zinc and magnesium, when dissolved in water, have been used with fair success for hardening defective concrete finish. In making up the solutions, ½ pound of the fluosilicate should be dissolved in 1 gallon of water for the first application and 2 pounds to each gallon for subsequent applications. The concrete floor must be clean and free from plaster, oil, paint or other foreign substances, otherwise the solutions will not penetrate sufficiently to react. For the same reason the surface must be absolutely dry. After the floor has dried, the second application may be made. About 3 or 4 hours are generally required for absorption, reaction and drying. In this treatment, with the average floor, one gallon of the liquid will cover approximately 130 square feet. Care should be taken to mop the floor shortly after drying to

* There is evidence to show that calcium chloride and similar compounds do not react in the same manner with all brands of portland cement. Trial batches of the brand of cement and the brand of accelerator proposed to be used should be made up and rate of hardening at the specified temperature noted before proceeding with their use in important work.

remove incrusted salts, otherwise white stains may be formed.

Concrete and Cement Waterproofer

A quantity of naphtha is heated to a temperature of approximately 80° C. and aluminum stearate in the ratio of 2 to 10 parts by weight of stearate to 100 parts of naphtha is added to the hot naphtha. The two materials are then agitated until a complete solution of the stearate in the naphtha is effected. A quantity of anhydrous acetic acid, equivalent to 0.3% to 1.5% by weight of the solution, is then added and the resulting mixture is thoroughly agitated. The product thus obtained is a clear solution having a specific viscosity Engler at 0° F. of 15 to 45 seconds per 100 cc. which can be stored without fear of gelling occurring at ordinary atmospheric temperatures and which may be applied to the substance to be waterproofed by means of a brush, spray or other device, and good penetration be obtained.

Integral Waterproofing for Concrete
Aluminum or Calcium Stearate

About ¼ to ½ lb. to the bag of cement.

Cement Waterproofing (Integral)

Dissolve in gauging water about ½ gal. Ammonium Stearate 26% to every bag of cement.

Dampproofing (Concrete, etc.)

Paraffin Wax	1 lb.
China Wood Oil	¼ gal.
Bodied Linseed Oil (3 Hour heat)	½ gal.
Varnolene	¼ gal.
Benzol	1 gal.
Yield	2⅛ gal.

Heat slightly to dissolve wax.

Masonry and Wall, Waterproofing

Tallow	10
Linseed Oil Bodied	5
Paraffin	1
Naphtha	32
Drier Liquid	0.13

Hydraulic Cement

Portland Cement	90 lb.
Aluminum	2 lb.
Ferro Silicon	8 lb.

Concrete

Recommended Proportions of Water to Cement and Suggested Trial Mixes

Kinds of Work	Add U. S. Gals. of Water to Each Sack Batch if Sand is			Suggested Mixture for Trial Batch			Materials per Cu. Yd. of Concrete*		
	Very Wet	Wet	Damp	Cement Sacks	Aggregates: Fine Cu. Ft.	Aggregates: Coarse Cu. Ft.	Cement Sacks	Aggregates: Fine Cu. Ft.	Aggregates: Coarse Cu. Ft.
5-Gallon Paste for Concrete Subjected to Severe Wear, Weather or Weak Acid and Alkali Solutions									
Colored or plain topping for heavy wearing surfaces as in industrial plants and all other two-course work such as pavements, walks, tennis courts, residence floors, etc.	4¼	Average Sand 4½	4¾	1	1	1½	10	12	15
					Maximum size aggregate ⅜"				
One-course industrial, creamery and dairy plant floors and all other concrete in contact with weak acid or alkali solutions.	3¾	4	4½	1	1¾	2	8	14	16
					Maximum size aggregate ¾"				
6-Gallon Paste for Concrete to be Watertight or Subjected to Moderate Wear and Weather									
Watertight floors such as industrial plant, basement, dairy barn, etc. Watertight foundations. Concrete subjected to moderate wear or frost action such as driveways, walks, tennis courts, etc. All watertight concrete for swimming and wading pools, septic tanks, storage tanks, etc. All base course work such as floors, walks, drives, etc. All reinforced concrete structural beams, columns, slabs, residence floors, etc.	4½	Average Sand 5	5½	1	2¼	3	6¼	14	19
					Maximum size aggregate 1½"				
7-Gallon Paste for Concrete Not Subjected to Wear, Weather or Water									
Foundation walls, footings, mass concrete, etc., not subjected to weather, water pressure or other exposure.	4¾	Aver. Sand 5½	6¼	1	2¾	4	5	14	20
					Maximum size aggregate 1½"				

* Quantities are estimated on wet aggregates using suggested trial mixes and medium consistencies—quantities will vary according to the grading of aggregate and the workability desired.

It may be necessary to use a richer paste than is shown in the table because the concrete may be subjected to more severe conditions than are usual for a structure of that type. For example, a swimming pool ordinarily is made with a 6-gallon paste. However, the pool may be built in a place where soil water is strongly alkaline in which case a 5-gallon paste is required.

Aluminum Sulfate Treatment

This treatment consists in one or more applications of solutions of aluminum sulfate to the clean, dry surface. The solution is made up in a wooden barrel or stoneware vessel and the water should be acidulated with not more than one teaspoonful of commercial sulfuric acid for each gallon of water. The sulfate does not readily dissolve and requires occasional stirring for a few days until the solution is complete. About 2½ pounds of the powdered sulfate will be required for each gallon of water and one gallon of the solution should cover about 100 square feet of floor surface.

Recommended Mixtures for Several Classes of Construction

Intended primarily for use on small jobs

Kind of Work	Gallons of Water to Add to Each One Sack Batch			Trial Mixture for First Batch			Maximum Aggregate Size
	Dry Sand and Pebbles	Moist Sand and Pebbles	Wet Sand and Pebbles	Cement	Sand	Pebbles	
				Sacks	Cu. Ft.	Cu. Ft.	Ins.
Foundation walls which need not be watertight, mass concrete for footings, retaining walls, garden walls, etc.	7½	6	5	1	3	5	2
Watertight basement walls and pits, walls above grounds, dams, lawn rollers, hand tamper, shoe scrape, hot beds, cold frames, storage and cyclone cellar walls, etc.	6½	5	4¼	1	2½	3½	1½
Water storage tanks, well curbs and platforms, cisterns, septic tanks, watertight floors, sidewalks, stepping stone and flagstone walks, driveways, porch floors, basement floors, garden and lawn pools, steps, corner posts, gate posts, piers, columns, chimney caps, concrete for tree surgery, etc.	5½	4¼	3¾	1	2	3	1
Fence posts, clothes line posts, etc., flower boxes and pots, grape arbor posts, mail box posts, benches, bird baths, sundials, pedestals and other garden furniture, work of very thin sections.	4½	3¾	3½	1	2	2	¾

For the first treatment the solution may be diluted with twice its volume of water. Twenty-four hours after this application the stronger solution may be used, and twenty-four hours should elapse between subsequent applications.

Sodium Silicate Treatment

When sodium silicate is used, it is applied in a 20% solution in two or more coats twenty-four hours apart. Ordinarily the sodium silicate requires considerable time to dry before the floor can be used. Commercial sodium silicate varies in strength from 30 to 40% solution. It is quite viscous and requires thinning with water before it will penetrate the floor. It has been found satisfactory to dilute each gallon of the silicate with three gallons of water. Each gallon of the resulting solution will cover approximately 200 square feet of floor surface. The floor should be thoroughly cleaned of all foreign matter, and should be dry before the first application of the silicate solution.

Zinc Sulfate Treatment

This treatment consists of the application of about 16% solution of zinc sulfate made acid with a teaspoonful of commercial sulfuric acid to every gallon. The mixture is applied in two coats, the second coat being applied four hours after the first. The surface should be scrubbed with hot water and mopped dry just before the application of the second coat. This treatment gives the floor a darker appearance.

Cement Accelerator

Commercial Calcium Chloride   4 lb.
Water   96 lb.

The above to be used as gauging water for concrete.

Cement Floor Hardener

Magnesium Fluosilicate   1 lb.
Water   15 lb.

The above to be flushed over a cement

surface. Wash with clean water to remove soluble salts.

Cement Patches

In patching or resurfacing concrete "Concrete Special" silicate of soda can be used to insure a good bond between the old and new cement.

To refill a hole it should be chipped out clean and somewhat under-cut. The fresh surface should then be painted with "Concrete Special" silicate full strength. Neat cement should then be dusted over the surface and worked in with a broom or stiff brush. The new concrete can then be applied in the usual manner.

For resurfacing, the concrete should be roughened with a pick, all loose particles removed and the floor wet thoroughly with water over night. Immediately before the new surface is applied the old one should be washed with a freshly prepared mixture of 10 pounds of neat cement with one quart of "Concrete Special" in fourteen quarts of water. This mixture should be brushed in well and followed at once with the surface layer.

Cement Preservative

Chinawood Fatty Acids	10 lb.
Paraffin Wax	10 lb.
Kerosene	40 gal.

Cement, Resistant to Calcium Chloride Solutions

Aluminum Oxide	40
Lime	40
Iron Oxide or Silicon Dioxide	15
Calcium Chloride	1

Cement Coated Wire

To increase the holding power of fastening devices made from wire, the latter is supported as a coil on a rotating mandrel dipped into one of the following mixtures.

1. Chinawood Oil	30
Ester Gum	20
Naphtha	50
2. Rosin	15
Calcium Hydroxide	0.9
Lead Oxide (PbO)	0.3
Manganese Dioxide	0.2
Chinawood Oil	33.6
Naphtha	50.0

Coloring Cement Gray

Carbon Black Paste	8
Cement or Plaster	100
Water	sufficient

The Carbon Black Paste is dispersed in the water by rapid stirring.

If a darker color is desired the percentage of Carbon Black Paste is increased.

Brickwork, Painting

Use any good quality outside paint. The first coat should seal the pores of the brick; for this the paint is thinned with turpentine and boiled linseed oil, and many painters also add varnish. The second coat is not thinned so much, and for the third the paint is used as it comes in the can.

Removal of Paint from Stone Surfaces

Paper pulp (old newspapers, cement sacks or stock pulp) is prepared by shredding in water by means of a steam jet. Excess water is drained off, 10–15% washing soda is added to the pulp, followed by sufficient fireclay (or lime) to render the mass plastic. Apply as a poultice to the surface to be treated; allow to remain 24 hours.

The poultice can usually be stripped off easily at the end of the above period. In obstinate cases, repeat treatment.

Last traces of pigment are removed by scrubbing with a bristle brush with clear water.

Removal of Pitch, Asphalt, Etc., from Stone Surfaces

Soak one or two thicknesses of blotting paper with carbon bisulfide. Lay over stain and apply a heated flat iron or similar heat retaining body. Remove iron when cool. The bituminous material will be found to be largely or wholly absorbed by the blotting paper. Repeat treatment in case of only partial removal.

Note: As carbon bisulfide is inflammable, the above treatment should not be attempted in the vicinity of sources of ignition.

Bituminous Composition

(for roads, floors, tennis-courts, etc.)

Sand	75–86
Bitumen	11–15
Fire Clay	3–16

Tile and Floor Composition

Asphalt Emulsion	1.75
Cement	1
Crushed Rock	5

Structural Tile

Calcium Carbonate (Marble	
Dust, Fine)	15 parts
Powdered Glass	4 parts
Magnesium Oxide (Heavy)	8 parts
Magnesium Chloride Solution	
(Sp. Gr. 1.19 @ 25° C.)	13 parts

Mix powders and make a thick paste with the solution of magnesium chloride. Pour into paraffined molds on a hard shiny surface. Let stand till dry.

Industrial Flooring Composition

Alpha Gypsum	10–77
Asphalt	4–36
Sand or Gravel	0–86

Terrazzo Floor Finish

1. Base Slab

The surface of the base slab shall be struck off reasonably true at a level not less than 2¼ inches below the required finish grade.

2. Aggregates

No fine aggregate or sand shall be used in the terrazzo finish. The coarse aggregate shall be (insert here the kind and color of marble chips desired). The coarse aggregate shall be graded in three sizes: ⅛ inch, ¼ inch and ½ inch.

3. Mixtures

The mortar base for the terrazzo finish shall be mixed in the proportions of one part of portland cement to 3 parts of clean, coarse sand, mixed with not more than 6 gallons of water per sack of portland cement.

The terrazzo mixture shall be one part of portland cement and 3 parts of stone chips.

Not more than 4 gallons of mixing water, including the moisture in the aggregate, shall be used for each sack of portland cement in the mixture.

4. Consistency

The terrazzo concrete shall be of the driest consistency possible to work with a sawing motion of the strike-off board or straight-edge. Changes in consistency shall be obtained by adjusting the proportions of aggregate and cement. In no case shall the specified amount of mixing water be exceeded.

5. Placing

Before placing the mortar base and the terrazzo finish, the surface of the structural concrete slab shall be covered with a uniform layer of fine sand ¼ inch thick, and covered with an approved tar paper.

The mortar base shall be at least 1¼ inches thick and shall be screeded to an even surface ¾ of an inch below the finished floor level.

Metal dividing strips about 1½ inches wide, at least 20 gauge, shall be inserted in the mortar or supported on the slab to conform to the designs specified by the architect. The top of the strips shall be at least 1/32 of an inch above the finished level of the floor.

When in the opinion of the engineer the mortar base has hardened sufficiently to withstand rolling, the terrazzo mixture shall be placed to the level of the tops of the dividing strips.

6. Finishing

After striking off to the finished level, the concrete topping shall be rolled lengthwise and crosswise so as to secure thorough compaction of the stone chips and cement paste. Additional stone chips of the larger size shall be spread over the topping during rolling until 85% of the finished surface shall be composed of stone. Immediately after rolling, the surface shall be floated and troweled once. No attempt shall be made to remove trowel marks.

After the terrazzo concrete has hardened enough to prevent dislodgments of aggregate particles, it shall be ground down with an approved type of grinding machine shod with free, rapid cutting carborundum stones to expose the coarse aggregate. The floor shall be kept wet during the grinding process. All material ground off shall be removed by squeegeeing and flushing with water.

Air holes, pits and other blemishes shall then be filled with a thin grout composed of neat cement paste. This grout shall be spread over the surface and worked into the pits. After all patch fillers have hardened for seven days the floor surface shall receive a second or final grinding to remove the film of cement paste and to give the floor a polish. It shall then be thoroughly washed and all surplus material removed.

7. Curing and Protection

All freshly placed concrete shall be protected from the elements and from all defacements due to building operations. The contractor shall provide and use when necessary, tarpaulins to cover completely or enclose all freshly finished concrete.

If at any time during the progress of the work the temperature is, or in the opinion of the engineer will, within twenty-four (24) hours, drop to 40° F., the water and aggregate shall be heated and precautions taken to protect the work from freezing for at least three (3) days.

As soon as the concrete has hardened to prevent damage thereby, it shall be covered with at least one (1) inch of wet sand, or other covering satisfactory to the engineer, and shall be kept continually wet by sprinkling with water for at least seven (7) days.

8. Cleaning

After removing all loose material, the finish shall be scrubbed with warm water and soft soap, and mopped dry.

Acid-Proofing Creamery Floors

Paraffin (150° F.)	4
Turpentine	1
Toluol	16

Warm and stir until uniform. Pour into cans and allow to "set." Spread on floor and allow to penetrate for 24 hours. At the end of this time the residual layer should be driven into the concrete by heat. A free flame should not be used due to fire hazards; hot irons will be found safe and effective in forcing the paraffin into the pores and capillaries of the finish for some distance below the surface.

After either treatment, the floor should be given a good waxing with any standard floor wax suited for this purpose. As the wax film is worn away through use, it is replaced by a fresh coating with the use of a polishing machine. Neither of these methods of acid-proofing creamery floors will change the color of the finish appreciably.

Board, Plaster or Wall

Portland Cement	67
Ground Stone	109
Shredded Sugarcane Fiber	24

Patching Plaster

Plaster of Paris	32
Dextrin	4
Pumice Powder	4

Wood, Metal Coating

Wood, stone, textiles, paper, etc., are coated with the following which is first melted, cooled, ground and taken up with water.

Metal (Powder)	40– 70
Paraffin Wax	60– 90
Graphite (Powd.)	60– 90
Precipitated Chalk	100–150
Sodium Silicate	180–220
Casein	40– 70

Wood Strengthener

A solution to help retain nails in wood is made as follows:

Rosin	1 lb.
Benzol	1 gal.

Nails are dipped in this solution, withdrawn, allowed to dry and are then ready for use.

Fireproofing of Wood

The use of sodium acetate for fireproofing wood has been known for a long time, and a solution of 15% strength has been found the most suitable concentration. Better results are obtained if the sodium acetate is reinforced with a small quantity of disodium phosphate. For flame proofing planks a solution containing 228 grams sodium acetate crystals and 33 grams disodium phosphate crystals per litre should be used. The planks are given three coatings with this solution, time being left between each application to allow the liquor to soak in. For efficient working the application of about 70 grams anhydrous sodium acetate per square metre of wood surface is necessary. The depth of penetration depends on the thickness and nature of the wood. In the case of air dried pine boards of 17 mm. thickness a total penetration of 15 mm. was found, the boards being coated on both sides. If the wood has been well dried out it is advisable to give a preliminary treatment with water.

MAKING FUSIBLE ALLOYS

When making fusible alloys, melt the lead and bismuth together. When molten, add the tin with stirring. When the tin has been molten into the mix, adjust the temperature of the mix to about 300° C., and using the cadmium sticks in tongs as stirrers, work in the necessary cadmium. Cadmium burns easily in air, hence the temperature must be watched, and if it rise much above 300° C. this may happen.

Good metal can often be recovered from the dross formed in making fusible alloys by working the dross with the ladle or a stick against the side of the kettle.

Lipowitz Metal

Cadmium	3
Tin	4
Bismuth	15
Lead	8

Melt above together and add

Mercury	2

previously heated to 220° C.

Melting point of above is 143° F.

Rose Alloy

Bismuth	2
Lead	1
Tin	1

Melting point 200° F.

Electrical Fuse Alloy

Tin	94
Lead	344
Bismuth	500

Melting point is 168° F.

METAL COLORING

The coloring of metals depends to a great extent upon the skill of the operator as well as on the different chemicals and methods used. The brushing and relieving operations must be done by one familiar with these operations to produce uniform results. For the brushing operation fine crimped nickel silver or brass wire wheels are used and operated at 800 R.P.M., either wet or dry.

Tampico or muslin buff wheels are used for relieving operations. They are generally used with water and fine pumice and operated at 800 R.P.M.

The use of the sand blast is essential also in producing various shades or colors, as some very beautiful effects may be produced by the proper use of the sand blast machine, both before and after the coloring operation.

The colors produced by chemical means are oxides or sulfides, or a combination of both.

Black Finish for Aluminum

Water	1 gal.
Caustic Soda	1 lb.
Common Salt	4 oz.

Heat the water in an iron or earthenware vessel, and dissolve the caustic soda. Stir well, and add the salt. Keep at about 200° F. and place the aluminum article in for about fifteen minutes. Rinse thoroughly, and immerse in second bath made up as follows:

Hydrochloric Acid	1 gal.
Iron Sulfate	1 lb.
White Arsenic	1 lb.
Water	1 gal.

Dip the aluminum in this bath for a few seconds only. Rinse well in hot water.

Black Finish for Tin

First clean tin thoroughly from grease by soaking in boiling caustic potash solution. Rinse and transfer immediately to bath made up of:

Hot Water	1 gal.
Antimony Chloride	6 oz.
Copper Chloride	12 oz.

Keep in until desired color is obtained, then rinse in hot water.

Bright Tin Finish for Screws

Use the following tin solution to produce a tin deposit on your work:

Aluminum Sulfate	2 oz.
Cream Tartar	2 oz
Tin Crystals	½ oz.
Water	1 gal.

Use a zinc container for the solution; place the screws in the pan and boil for 45 minutes. A new solution is necessary for each batch of work. If the deposit is not bright enough, tumble the screws in an oblique tumbling barrel, using clean hardwood sawdust.

Silver Finishes

The silver finishes are sulfide finishes, and the chemicals used are either sodium, potassium, calcium, or ammonium sulfide. The potassium salt produces the hardest black and the ammonium salt the softest. Either salt is used in the proportion of ½ to 1 oz. per gallon of water, and used hot. To produce a black color the finish is obtained by either wet or dry scratch brushing, and the relief or gray finishes with the use of a rag or tampico wheel with fine pumice and water.

◆

Coloring Copper

There are many formulae for the coloring of copper or copper plated work, and the color will depend upon the chemicals used, the temperature and the length of time the work is left in the coloring solution.

The work should be perfectly clean and free from any grease or finger marks.

Brown on Copper

1

Potassium Chlorate	1 oz.
Copper Sulfate	4 oz.
Water	1 gal.

Use hot, scratch brush wet. If color is uneven, repeat coloring operation and scratch brush dry.

A darker or more red color is produced in this solution:

2

Copper Sulfate	4 oz.
Nickel Sulfate	2 oz.
Potassium Chlorate	1 oz.
Water	1 gal.

Finishing operations are the same as above.

Various shades of bronze from a chocolate color to a black can be produced in a solution made of:

3

Potassium Sulfide	½ to 1 oz.
Water	1 gal.

For the light shades use cold and a short time of immersion. For darker, use hot, with longer immersion.

Various colors are produced in any of the following solutions used either hot or cold.

4

Yellow Barium Sulfide	1 oz.
Water	1 gal.

5

Yellow Barium Sulfide	1 oz.
Calcium Sulfide	½ oz. (fl.)
Water	1 gal.

6

Antimony Sulfide (orange)	½ to 1 oz.
Caustic Soda	1 to 2 oz.
Water	1 gal.

7

Copper Sulfate	12 oz.
Acetic Acid	4 oz.
Caustic Soda	4 oz.
Water	1 gal.

8

Copper Sulfate	4 oz.
Copper Acetate	2 oz.
Potassium Chloride	6 oz.
Water	1 gal.

9

Copper Sulfate	8 oz.
Potassium Permanganate	1 oz.
Water	1 gal.

◆

Verde Antique Finish on Copper

Copper Nitrate	16 oz.
Acetic Acid	4 oz.
Water	1 gal.

Best applied hot and sparingly to previously moistened surface.

◆

Brass, Refinishing Corroded

Saturate vinegar with salt and clean brass with this until all corrosion is removed. Polish with any good metal polish; wash; dry; wash with benzene to remove oil and grease; finish with spar varnish or lacquer.

◆

Green Finish on Brass

Brass articles are colored various shades of green by any of the following baths. When dry they should be lacquered to preserve the coating.

1

Hyposulfite of Soda	8 oz.
Acetate of Lead	2-6 oz.
or Nickel Sulfate	2-6 oz.
or Irin Nitrate	2-6 oz.
or Iron Chloride	2-6 oz.
Water	1 gal.

Use hot.

2

Sodium Bisulfite	4 oz.
Lead Acetate	1½ oz.
Water	1 gal.

Use hot and dip repeatedly.

3

Copper Sulfate	2 oz.
Iron Sulfate	2 oz.
Ammonium Carbonate	2 oz.
Water	1 gal.

Brass, Black Pickling of

Copper Carbonate	750
Ammonia Hydroxide	150

Immersion from 3 to 8 minutes is indicated.

Black Finish on Brass

Solution No. 1.

Yellow brass may be colored blue black by immersion in a solution of water saturated with copper acetate to which ammonium carbonate has been added.

or

Solution No. 2.

Immerse in a solution of ammonium hydroxide which has been saturated with copper carbonate.

or

Solution No. 3

Immerse in

White Arsenic	12 oz.
Yellow Antimony Sulfide	¼ oz.
Water	1 gal.

or

Immerse in

Hyposulfite Soda	8 oz.
Acetate of Lead	4 oz.
Water	1 gal.

These solutions, except the one made up with copper carbonate, should be used hot. Immerse the work until proper color appears. The work should be finished with a coat of lacquer to prevent tarnishing.

Statuary Finish on Naval Bronze

To produce statuary finishes on naval bronze base the following solutions may be used: for light bronze, Potassium Chlorate 1 oz. and Copper Sulfate 4 oz. per gal. water; for dark bronze Potassium Chlorate 1 oz., Nickel Sulfate 2 oz. and Copper Sulfate 4 oz. per gal. water; for dark to blue-black finish, Potassium Sulfide or Ammonium Sulfide ¼–1 oz. per gal. water.

Steel, Blue-Black Finish

A. Place object in molten Sodium Nitrate (700–800° F.) for 2–3 minutes. Remove and allow to cool somewhat; wash in hot water; dry and oil with mineral or linseed oil.

or

B. Place in following solution for 15 minutes.

Copper Sulfate	½ oz.
Iron Chloride	1 lb.
Hydrochloric Acid	4 oz.
Nitric Acid	½ oz.
Water	1 gal.

Then allow to dry for several hours; place in above solution again for 15 min.; remove and dry for 10 hr. Place in boiling water for ½ hr.; dry and scratch brush very lightly. Oil with mineral or linseed oil and wipe dry.

Gun-Metal Finish

After the work has been polished and cleaned, it is placed in the following solution for ten to fifteen minutes:

Ferric Chloride	2 oz.
Mercury Nitrate	2 oz.
Muriatic Acid	2 oz.
Alcohol	8 oz.
Water	8 oz.

After immersing the work in this solution it should be hung up to dry for 10 to 12 hours. Repeat the immersion and drying operation, then brush lightly with a fine crimped steel wire wheel. Finally, oil with paraffin or linseed oil, and remove excess oil with a soft cloth.

Coloring Brass Red

Electroplate in following solution at 110–120° F. at current density of 6 amp./sq. ft. using cast bronze or electrolytic copper anodes.

Copper Cyanide	3 oz.
Zinc Cyanide	½ oz.
Sodium Cyanide	4½ oz.
Sodium Carbonate	1 oz.
Rochelle Salts	2 oz.
Water	1 gal.

By adjustment of current and temperature any shade between copper and yellow brass may be produced. A sufficiently thick coating is needed so that it may stand an acid dip.

Coloring Iron

Etching (*"browning," "bluing,"* etc.)—Solutions of chemical reagents are applied to the steel with a cloth or sponge; the steel is allowed to oxidize for some hours while drying; the rust is then scraped off, leaving a thin adherent coat of oxide. The process is repeated a number of times, depending on the depth of color desired. The surface is then oiled. The following is a representative list of combinations of reagents that have been used for producing the respective colors:

Color, and Reagent for Producing

	Parts by Weight
Black:	
First formula—	
Bismuth chloride	20
Mercuric chloride	40
Copper chloride	20
Hydrochloric acid	120
Alcohol	100
Water	1000
Second formula—	
Copper nitrate solution (10 per cent)	700
Alcohol	300
Third formula—	
Mercuric chloride	50
Ammonium chloride	50
Water	1000
Brown:	
First formula—	
Alcohol	45
Iron chloride solution	45
Mercuric chloride	45
Sweet spirits of niter (ethyl nitrite + alcohol)	45
Copper sulfate	30
Nitric acid	22
Water	1000
Second formula—	
Nitric acid	70
Alcohol	140
Copper sulfate	280
Iron filings	10
Water	1000
Blue:	
Iron chloride	400
Antimony chloride	400
Gallic acid	200
Water	1000
Bronze:	
Manganese nitrate solution (10 per cent)	700
Alcohol	300

Niter bath.—The cleaned steel is heated in fused sodium nitrate or potassium nitrate or a mixture of the two, often with the addition of manganese dioxide. The color acquired by the steel depends on the temperature of the bath, as well as its composition. Other fused oxidizing baths can probably be used also.

Temper colors.—The "temper colors" seen on steel when it is heated between 220° and 320° C. are due to a thin layer of oxide. Such a layer of oxide is often applied as a protecting coating, the blue color being the one usually used. The steel is heated in free air and the various colors will be produced at the following temperatures:

Temper Color	° F.
Pale yellow	418
Straw	446
Brown	491
Purple	536
Pale blue	572
Dark blue	599

The color depends somewhat on the duration of the heating and to a lesser extent on the nature of the steel.

Rustproofing Iron

The article is cleaned by sand-blasting or pickling in acid and plated with a thin layer of Zinc from a bath contg. Sodium Cyanide 4, Zinc Ferrocyanide 5, Caustic Soda 4 oz., and a small amt. of Mercury per gal., zinc anodes contg. 0.5% of Mercury and a c. d. of 25 amp./sq. ft. being used at 5 v. After being washed well, the plated articles are dipped in a soln. contg. Nickel Chloride 4, Ammonium Chloride 6, Sodium Sulfocyanide 2, and Zinc Chloride 0.5 oz. per gal. The black deposit thus obtained may be coated with lacquer or given an oil finish in the usual way.

Rustproofing Small Iron Parts

The articles are immersed in an aqueous soln contg Ferrous Chloride 2% together with 2% of Mercuric Chloride and are then withdrawn and dried in a warm atmosphere. They are then heated to about 100° and subjected to a humidity of 80% and then immediately immersed in boiling water to fix the resulting Iron oxides adhering to the surfaces.

Steel Parts, Preventing Corrosion of

Steel parts exposed to corrosive fumes are coated with

Lanolin	10
Naphtha	20

Rust Prevention

To give temporary protection from rusting, metal articles are coated with a 50% solution of lanolin in naphtha.

Rust Remover

Orthophosphoric Acid	35%
Water	30%
Ethyl Methyl Ketone	10%
Monoethylether of Ethylen-Glycol	25%

Rust Remover

100 parts of stannic chloride are dissolved in 1,000 parts of water. This solution is added to one containing 2 parts of tartaric acid dissolved in 1,000 parts of water and 2,000 parts of water are added. The solution is applied by means

of a brush, after removing grease, and is allowed to remain on for a few moments when the article is rubbed clean, first with a moist cloth and then with a dry cloth, and, if necessary, repolished in the usual way.

Coloring Die Cast Zinc

Zinc weathers to a soft gray. To obtain other effects artificial coloring is necessary. This may be accomplished by electrodeposition or simple immersion (chemical coloring). Since the compounds of zinc are chiefly white, the process of coloring zinc necessitates the production on the zinc surface of a colored compound of some other metal. The compounds of copper are the most useful. By treating zinc with various copper solutions several colors may be obtained. All shades of black and brown are produced by small changes in the procedure, such as time of dip, concentration, etc.

An adherent bright black can be readily produced by electrodeposition in the following bath:

Nickel Ammonium Sulfate (per gal.)	8 oz.
Zinc Sulfate	1 oz.
Sodium Sulfo-Cyanate	2 oz.

A fairly adherent black capable of being brushed to remove the coloring in the high lights, results from a 5-second dip in the following solution:

| Sodium Hydroxide (per gal.) | 4 oz. |
| White Antimony Trioxide | ½ oz. |

Use at 158° to 167° F.

A similar result may be obtained by means of a 30-minute dip in the following solution:

Single Nickel Salts (per gal.)	10 oz.
Sodium Sulfate	15 oz.
Ammonium Chloride	1¾ oz.
Boric Acid	2 oz.

Black, brown, gray, gold, bronze, etc., may be produced in a large range of shades. Oiling with a light oil, or in some cases the use of a coat of clear lacquer will improve the luster and permanence of the deposit.

Colors produced by chemical means are reasonably permanent when used indoors. When exposed to outdoor atmospheres a relatively short life may be expected.

Antique Gold Finish

Gold Cyanide	½ oz.
Silver Cyanide	¼ dwt.
Sodium Cyanide	6 oz.
Sodium Carbonate	2 oz.
Water	1 gal

A very small quantity of lead dissolved in caustic soda is added to this solution. In preparing the lead solution dissolve 1 ounce of lead carbonate and 4 ounces of caustic soda in 1 quart of water, and add 20 to 30 drops to each gallon of solution.

Operate solution at 110° F., with 4 to 5 volts. Use 18 karat green gold anodes. Agitation of the work is essential to produce the antique finish. After the smut is produced relieve on a small rag wheel, using bicarbonate of soda moistened with water. The work is lacquered to protect the finish.

White Gold

An alloy which possesses many of the physical properties of Platinum including some degree of resistance to acids is prepared by alloying a primary alloy with a large proportion of Gold. For a soft (hard, in parentheses) 18-karat white Gold the primary alloy contains Gold, 37 (37.4), Nickel 38.1 (44.5), Copper 16.4 (5.0), Zinc 7.1 (11.1), and Manganese 1.4 (2)%. This alloy is best prepared from granulated metals, and approx. 25% of the alloy is melted with 75% of Gold in the second stage.

Diminishing Corrosion of Aluminum

Aluminum or its alloys are protected against corrosion by chlorine or bromine water by the addition of 0.5 and 5% of sodium silicate respectively.

Etching Aluminum Reflectors

| Water at 45° C. | 950 c.c. |
| Hydrofluoric Acid (48%) | 50 c.c. |

Rotate reflector every 30 seconds
Pour off and wash with running water.
Introduce 50–50 Nitric acid to remove black film.
Pour off and rinse with water.
Swab gently with soft cloth or cotton to remove last thin film of deposit.

Core Binder
(for aluminum castings)

| Sharp Sand | 45 lb. |
| Molding Sand | 45 lb. |

Rosin Powd.	2 lb.
Flour	1 lb.

or

Sharp Sand	71 lb
Molding Sand	25 lb.
Rosin Powd.	4 lb.

Spray with molasses water and bake at 325° F. Remove from oven and coat with soapstone. Return to oven to dry.

Core Oil

1. Tung Oil	10 gal.
2. Linseed Oil	20 gal.
3. Mineral Oil	20 gal.
4. Varnish "foots"	5 gal.
5. Benzine	5 gal.
6 Rosin	200 lb.
7. Lime Slaked	6 lb.
8. Litharge	7 lb.
9. Manganese Dioxide	3 lb.

Melt 1 and 6, stir in 7, 8 and 9.

Heat to 500° F. for 20 minutes. Add 2 a little at a time and keep at 400° F. for 20 minutes. Raise temperature to 480° F. and keep there for two hours. Cool to 300° F. and add with stirring 3, 4 and 5.

Metal Annealing Bath

Sodium Chloride	30 lb.
Potassium Sulfate	44 lb.
Potassium Carbonate	21 lb.
Borax	5 lb.

Carbonizing Steel

The steel blanks are tumbled, burred and tumble finished previous to carbonizing and are then placed in the revolving drum of the carbonizing machine and ¾ pint of carbonia oil with ½ bushel of Burnt Bone added. The drum is closed securely, gas turned on and heated to 700–750 degrees F. for 3 hours. The heat is turned off and the drum allowed to run for 2 hours to cool off. The contents are removed and sifted and tumbled in ½ bushel of No. 2 Granulated cork and 2 pints of japan oil for 5 minutes; then dried and cleaned by tumbling in ½ bushel of sawdust for 5 minutes to put on a high polish.

Bake at 120 degrees F. for 8 to 10 hours to harden oil.

Magnetic Chromium Steel, Heat Treatment of

The best magnetic properties of a steel contg. 1.3% Carbon and 2.1% Chromium

are obtained by quenching from 850° in oil. The steel should not be held too long between 750° and 850°, as a change takes place in the double carbide. Incorrect heat treatment can be remedied by holding at 950–1000° for 1 hr., cooling in air, and then hardening.

Gum for Parting Punch from Die

Beeswax	1 lb.
Rosin	½ lb.
Venice Turpentine	¼ lb.

Battery Terminals, Prevention of Corrosion

Slaked Lime	7
Sodium Bicarbonate	2
Borax	1
Rezinel No 2	

sufficient to make a paste

Magnesium and Its Alloys, Prevention of Corrosion by Water

1% Potassium Dichromate is dissolved in the water used.

Glass Etching or Frosting Compound

Formula 1.

Hot Water	19.0
Ammonium Bifluoride	69.5
Sodium Fluoride	2.5
30% Hydrofluoric Acid	9.0

Formula 2.

Hot Water	18
Ammonium Bifluoride	40
Sodium Fluoride	10
Molasses	20
60% Hydrofluoric Acid	12

These formulas frost well at slightly above room temperature, and weigh approximately 12 pounds per gallon. Mix in order given in a lead lined vessel. The mixture should be kept somewhat warmed until the ingredients are almost all dissolved, and must be thoroly stirred before use. A copper wire basket may be used to dip glassware. Bottles should be fitted with rubber stoppers, and must not be frosted on the inside, as this makes them liable to break easily. Immerse glass into solution for one minute. Remove, drain ten seconds, and wash off at once with hot water. Re-immerse one minute, remove and wash off with hot water as before. Dry. If the frost is not sufficiently opaque, make the first immersion for two minutes, instead of one minute. Use goggles when mixing solution. Adequate ventilation must be maintained as the vapors of Hydrofluoric Acid are exceedingly dangerous. The excess Ammonium Bifluoride provides a reserve etching capacity After being in use for some time, the solution may be further fortified by the addition of Hydrofluoric Acid.

Etch Resist

In etching glass it is necessary at times to block off portions which one desires to keep unetched. A solution for this purpose is composed of the following:

Asphaltum	12.5%
Beeswax	4.5%
Ceresine Wax	58 %
Stearic Acid	25 %

Marking Glass

40° Bé. Sodium Silicate can be used as a marking ink on glass. It adheres well after drying. After a few weeks, the dried silicate is washed off, the glass will be found etched. If desired, colored pigments may be added to the silicate to make it show up better.

Glass Etching Ink

(All ingredients by weight)

Hot Water	12%
Ammonium Bifluoride	15%
Oxalic Acid	8%
Ammonium Sulfate	10%
Glycerin	40%
Barium Sulfate	15%

Molasses may be substituted for the Glycerin, and Talc for the Barium Sulfate. If the ink does not readily adhere to the glass, add an additional very slight amount of water to cut the viscosity. It is not advisable to add free Hydrofluoric Acid, as this causes the ink to run and to blur. The addition of about two per cent of Sodium Fluoride sometimes improves the quality of the ink. The glass should be slightly warmed before writing upon it. Allow the ink to act for about two minutes, then wash off thoroly with hot water and dry. Good legible writing should be obtained easily in not more than 30 seconds when the glass is warm. Use an ordinary steel pen. Wash ink from pen when through. Keep in hard rubber or lead bottles.

Ink for Writing on Glass

Pale Shellac	2 oz.
Venice Turpentine	1 oz.
Sandarac	¼ oz.
Oil of Turpentine	3 fl. oz.

Dissolve by gently heating and then add one of the following pigments.

Black—Lamp Black	½ oz.
Blue—Ultramarine	½ oz.
Green—Brunswick Green	½ oz.
Red—Vermilion	½ oz.

Waterproof Ink for Glass

Shellac Bleached	10
Venice Turpentine	4
Rosin Oil	1
Turpentine	15
Indigo Powder	5

"Horak" Glass

"Horak" glass, made in Czechoslovakia, is said to possess great elasticity, and to be resistant to sudden changes of temperature. The composition is:

	Per cent
Sand	60–70
Boric Acid	15–30
Potassium Carbonate	1–2
Sodium Carbonate	3–6
Zirconia	1–3
Titanium Dioxide	1–3

Glass, Resistant

Silicon Dioxide	70
Boron Oxide	16–20
Litharge	10
Iron Oxide	5

This glass is resistant to high temperatures, quick temperature changes and is easily worked.

Glass, Ruby

The following is added to the basic glass batch

Selenium	2 %
Cadmium Sulfide	1 %
Arsenic Trioxide	1 %
Carbon	0.5%

Belgian Plate Glass

White Sand	50 Kg
Sulfate	17 Kg
Calcite	40 Kg
Coal Dust	1 Kg
Arsenious Acid	230 g

Bohemian Plate Glass—I

Quartz	50.0 Kg
Potash	20.0 Kg
Calcite	8.5 Kg
Arsenic	100.0 g

English Plate Glass

White Sand	50 Kg
Sulfate	14 Kg
Calcite	18 Kg
Coal Dust	520 g
Arsenious Acid	500 g

French Plate Glass

White Sand	50 Kg
Chalk	17 Kg
Sulfate	19 Kg
Coal Dust	500 g
Arsenious Acid	510 g

German Plate Glass

White Sand	50 Kg
Sulfate	17 Kg
Soda	3 Kg
Calcite	18 Kg
Coal Dust	1 Kg
Arsenious Acid	500 g

Glaze, Acid Resisting

Lead Oxide	0.8
Sodium Oxide	0.1
Iron Oxide	0.1
Silicon Oxide	1.5
Boron Oxide	0.4

Glazes, Alkali-free Lime

Satisfactory bright glazes having a maturing temp. of cones 11 to 13 were produced. A good cone 13 bright glaze was produced with 100 limestone, 26 kaolin, 245 calcined kaolin and 396 sand. With mat glazes it was found necessary to use at least 3 mols. of Silicon Dioxide to prevent crazing. A good cone 11 mat was produced with limestone 100, kaolin 26, calcined kaolin 112 and sand 96. These glazes are especially resistant to abrasion and chem. action and therefore are recommended for chem. porcelain, cooking utensils, insulators and tech. stoneware. A good magnesia-lime, alkali-free glaze was produced with calcined magnesite 19, limestone 78, kaolin 26, calcined kaolin 45 and sand 144.

Vitreous Enamel

240 grams borax, 410 grams potash feldspar, 30 grams saltpetre, 120 grams sodium carbonate, 30 grams calcium spar, and 170 grams quartz are fused together to produce 1,000 grams of lump enamel. This is crushed, ground with 60 grams of tinting substance and about 20 grams of zirconia opacifier. The latter should contain about 1 gram of salt of unstable acid, for example, sodium nitrate or formate.

Enameling Steel

The preparation of the steel for enameling consists in giving it such treatment as is necessary to leave a clean surface, free from any foreign matter that will injure the enamel when applied and burned. The treatment required depends upon the nature and size of the piece of ware and the kind of foreign matter that is to be removed. The sand blast is used in cleaning large ware and such as can not be easily cleaned by pickling. When the sand blast is used, no other treatment is required, since grease, rust, and any other foreign matter is readily removed by it. This is the most effective method of cleaning steel and one that gives an excellent surface for enameling. For small pieces it is much more expensive than pickling, and it is economical only in making large pieces or special shapes of comparatively high value.

Treatment Preliminary to Pickling

Nearly all light steel ware is cleaned by the pickling process. The preliminary treatment before the ware is placed in the pickling acid varies. Grease and carbonaceous matter must be removed from the ware before placing in the pickling solution, and three general methods are in use for doing this; scaling, washing in caustic alkali solutions, or the use of proprietary cleaning compounds.

Scaling.—Scaling or heating the ware to redness is the method most generally employed. During the process of shaping the ware from the sheet of steel it invariably collects grease from machinery and workmen's hands, and one method of removing such carbonaceous matter is to burn it off. Especially is this the case when handling large numbers of small pieces. To do this, the ware should be carefully stacked on grates in such a manner as to admit free access of air to all parts of every piece of ware. Care must be taken to prevent flat surfaces from coming into contact with each other, and space must be provided between the different pieces of ware to admit sufficient air to completely oxidize all carbonaceous matter present. It must be remembered that the heat treatment forms an iron scale which must subsequently be removed by acid, and consequently the time and temperature should not be carried beyond that necessary to burn off the oil.

Removing Grease with Caustic Soda.—Caustic soda or potash may be used for removing fatty materials, especially if they are present in small amounts. In this process the steel article is immersed in a boiling solution of caustic soda or potash and allowed to remain for a few minutes. It is then taken out and washed free from alkali in clear water. This precaution is necessary because the adhering alkali solution would rapidly neutralize the pickling acid into which the steel is next placed for the removal of rust and other deleterious impurities.

Pickling

After the oil and carbonaceous matter have been removed from the surface, it is necessary to remove all rust and oxide of iron. The pickling solution used is one of either sulfuric or hydrochloric acid.

1. Mixing the Raw Materials

General practice in mixing the raw materials consists in weighing the batch, which generally approximates 500 pounds, into a box and then turning the mixture over a few times with a hoe or shovel. In the case of colored enamels it is considered mixed when the coloring oxide is uniformly distributed, imparting a uniform gray color to the batch. In white enamels the practice is to turn the mixture a certain number of times, which is considered to be sufficient. Here is one of the places where enamelers can improve their practice and raise the standard of their ware by doing away with slipshod methods and resorting to more thorough, exact, and economical methods. Rotating drums and other forms of mixing machines give much more satisfactory results.

Every enameler, and even the unedu-

cated laborer who has worked around the smelter, has observed that the enamel smelts more quickly when most thoroughly mixed. This is simply the practical application of the well-known scientific principle that the speed of chemical reactions is directly proportional to the area of surface of contact between the reacting substances. If a fire brick were crushed to a powder and mixed into the batch it would go into solution in the melt and disappear with the other ingredients of the batch, while that same brick when laid in the wall of the smelter will stand for months without being eaten away. This same principle applies to all the refractory ingredients of the batch. A large piece of flint stone will go through a melt and come out with only the sharp edges eaten off. The length of time required for smelting the enamel depends directly upon the fineness of the raw material, especially flint and feldspar, and upon the thoroughness with which they are mixed. It follows, then, that better mixing of the raw materials means less labor, less fuel, less time of smelting, and less wear and tear on the smelter.

It is not only from an economic standpoint that thorough mixing is advisable. The quality of the white enamels is inversely proportional to the length of time spent in producing a thorough melt. Long smelting results in a considerable reduction in opacity. Fine grinding and thorough mixing insures a uniform fusion product in the shortest possible time and hence minimum solution of opacifying agents and minimum reduction in opacity.

2. Melting

In the smelter the enamel mixture is melted and fined until no lumps of unfused or undissolved material can be detected in a string of the glass drawn from the melt. The melting process begins with the fusion of the least refractory ingredients or fluxes—borax, soda ash, etc.—at relatively low temperatures. The liquid attacks the more refractory substances both by solution and by chemical reaction. The formation of eutectics between the raw materials and the compounds resulting from chemical reaction facilitates the melting process.

If the smelting process is continued for a sufficient length of time a perfectly homogeneous glass in which all constituents would be in equilibrium would result. Such a condition is not obtainable, especially in white enamels. The melting should proceed only to the point where a stable borosilicate glass is formed, in which the opacifying agents, fluorides, tin oxide, and antimony compounds are carried in suspension. Longer smelting results in a considerable solution of these materials, as well as decomposition of the fluorides and consequent reduction in the opacity of the enamel. No opacity is obtained from tin or antimony oxides after they are once taken into solution. Quick smelting is therefore to be desired, and this again calls attention to the value of fine grinding and thorough mixture of the raw materials.

3. Tempering Enamel Slips

In preparing enamel slips for application to the ware the frit is ground wet and contains 5 to 10 per cent (by weight) of plastic clay. To increase the viscosity of the slip and aid in holding the enamel in suspension, a flocculating agent is added. In white or cover enamels magnesium sulfate is generally used for this purpose. In ground coats borax is almost universally employed, since nearly all other salts which have a similar effect on the slip are likely to cause rusting of the steel during the drying of the ware.

1. Fine grinding makes the frit more easy to float, but enamelers dare not grind too finely, because of difficulty in getting a uniform coating on the ware. Ground coat enamels especially must be coarse, not finer than 100 mesh, and, better, 80 mesh.

2. Lead enamels would, of course, be more difficult to float than lighter ones, but lead is seldom used in enamels for sheet iron. However, all frits are relatively high in specific gravity as compared with clays and therefore settle more readily.

3. Settling is easily prevented by making the slip thick, approaching a paste, but in order to apply them by dipping or spraying, slips must be sufficiently fluid to flow. With such a consistency heavier substances will settle unless a floating agent is used.

4. Viscosity has been described as the friction between two liquids flowing in contact with each other, or between a liquid and a solid moving in it; in other words, resistance to flow. The efficiency of a floating medium in preventing the settling of heavier particles, therefore, depends upon its viscosity or resistance to the motion of particles passing through it. The floating medium in the case of enamels is not to be considered as the water, but as the clay substance in suspension in water.

High viscosity is also required in enamel slips to prevent them from flowing down the sides and into the corners of the ware after dipping. A steel body, being nonabsorbent, offers a different problem from that of a porous body dipped in a glaze slip. The absorption by the porous body prevents the flowing of the glaze, but the enamel slip must stay in place by virtue of its viscosity, although it is possible that surface tension also plays an important rôle here.

5. It is evident that a sufficient amount of the floating medium to prevent settling can readily be added, but other considerations limit the amount of clay which can be used with any glaze or enamel, about 10 per cent being the maximum permissible in the latter. The efficiency of the clay as a floating agent is therefore highly important, especially in enamels where the frit is of higher specific gravity and more coarsely ground than in glazes or engobes, and where the amount of clay used is necessarily small.

1. Application of the Enamel

There is no more vitally important operation in the entire process of enameling than the application of the first coat of enamel. A piece of ware which has passed through the operations of forming and cleaning has acquired considerable value to the manufacturer on account of the labor expended upon it. In the application of the ground coat it is possible to enhance this value or to destroy it, or, still worse, to so treat it that it will pass through the succeeding operations and still be worthless as a finished piece of ware. Given a good ground coat, properly applied and burned, the finishing of the ware is simple. The very best ground coat improperly applied or burned can give only a poor piece of ware, regardless of what its previous cost or future treatment may be. Every possible precaution should therefore be taken to insure a suitable coating on the steel.

Four different methods are used for applying the enamel to the steel—slushing, draining, spraying, and dusting. The choice of method depends upon the size and shape of the ware and the nature of the enamel. The chief factor to be considered in the application of the enamel is to obtain a coating of uniform and sufficient thickness on the surface of the ware. If a thin and uniform coating is not obtained, the enamel will burn off the portions where it is too thin and will not be sufficiently burned where it is thick. Either of these defects will cause the finished ware to be defective. The method best suited to produce this result, with due consideration to the cost of the operation, is the one generally used.

Slushing.—By far the greatest proportion of enameled ware is slushed, especially in the case of all light wares and such as can be easily shaken to distribute the enamel uniformly. The operation consists in dipping the piece of ware into the enamel slip, removing it and shaking it in such a way as to leave a thin and uniform coating over the entire surface of the metal. There are two factors of vital importance in securing proper results by this method —the consistency of the enamel slip and the skill of the operator. The consistency of slip for slushing is such as is termed "short"; that is, it has a high viscosity and will not run down or drain off from vertical surfaces after dipping.

To the novice it would seem a simple matter to dip a piece of steel into a tub of slip, shake off a little, and obtain a nicely coated piece of ware. As a matter of fact, considerable practice is required to acquire skill sufficient to slush even simple shapes uniformly, while extensive training and a very high degree of skill is required in the handling of complicated shapes.

Draining.—This method is frequently applied to perfectly flat ware, such as signs, and to simple shapes. The piece of ware is dipped in the slip and is then set on edge to allow the excess to run off at the bottom. The consistency of the slip, which is very different from that used in slushing, is the principal factor in the success of this operation. In this case the viscosity is much lower, so that the slip will flow down the vertical surface, but at the same time its consistency must be such that it will

form a good coating and adhere to the ware after the excess drains off. It must also be sufficiently viscous to keep the enamel in suspension and not allow it to settle onto the bottom of the tank.

Spraying.—For applying enamel to complex shapes and heavy ware, spraying is frequently resorted to. It is too expensive to use on the ordinary grades of ware, but for special shapes with many corners and sharp angles, or any piece of ware which can not be slushed uniformly, spraying is the best method of coating. It is wasteful of material and requires skill to obtain good results, but if proper care is used any piece of ware can be very uniformly coated by spraying. The piece may be placed on a whirling rack and turned while the spray is being applied.

The consistency of the enamel is highly important again in this case. The enamel must be ground sufficiently fine to prevent stopping the nozzle of the sprayer, but for best results it must not be too finely ground. Its viscosity must be high to prevent flowing. Since the distribution of the slip over the surface is accomplished in this case by the movement of the spray and not by shaking the piece, it is possible to work with a higher viscosity than in slushing.

Dusting.—This method of application is very common in cast-iron work, but in steelwork it is used only on heavy wares, such as condensers for chemical works, etc. It has a decided advantage in the production of acid-resisting wares, because no raw materials are added to the frit, whereas when any of the other methods of application are used, it is necessary to add some raw clay and soluble salts to the frit in order to get a slip of the proper consistency. These raw materials are invariably decidedly injurious to the enamel, especially where resistance to chemical corrosion is desired. While an enamel is a glassy coating, it is far from being a solid glass; and the more raw material added in grinding the frit the further is the finished enamel removed from this condition, since these raw materials are only to a very slight extent combined with the frit during the brief burning operation. Because of this fact the dusting method is decidedly the best to use for making enameled ware to resist chemical corrosion.

In carrying out this process, the ground coat, as well as cover coat, is frequently dusted on. The metal is wiped with a wet sponge or cloth, and the powder dusted on while the metal is still wet. Sometimes an adhesive agent is added to prevent the enamel from falling off when dry.

The methods used for cover enamels are the same as those used for ground coats. The quality of workmanship in applying cover coats is far less important than in applying ground coats. If a piece of ware is perfectly coated with the ground coat, the cover coat may be quite imperfectly applied and still give good results. Of course there are limits to this, and the more uniformly the enamel is applied the better it will be. It should be said, further, that best results are always obtained with thin enamels. Barring the properties of whiteness and opacity, the excellence of enamels is inversely proportional to their thickness. This is especially true of the ability of the ware to withstand bending and abrasion. In view of these facts the aim should always be to keep the enamel as thin as possible, while at the same time obtaining the desired opacity and color.

2. Drying

Ground-coat enamels should be dried as rapidly as possible to prevent rusting of the steel. This will be controlled to some extent by the flocculating agents used in the slip, but rapid drying is the best practice in any case. If an alkaline flocculating agent is used for tempering the ground coat, it can be dried in the open air without serious rusting; but if chlorides or sulfates are used, rusting is almost sure to result even with rapid drying. This rust may or may not be visible after the ware is dry, but it is quite sure to make its appearance, when the ground coat is burned, in the form of spots where the iron oxide has reacted with the enamel to such an extent as to form a spot-like iron scale. When these spots are formed, it is practically impossible to cover them with cover enamel. They will show in the finished ware either as dark spots or as pits in the surface. While proper drying of the ground coat can not entirely prevent this trouble in an improperly tempered enamel, it will always reduce the trouble, and when the ware is not dried rapidly the

trouble is likely to come even in the best tempered enamel.

The rate of drying of cover enamels is of less importance than in drying ground coats. However, rapid drying is here again desirable. One of the chief reasons for this, especially in white enamel, is the fact that dirt in the form of factory dust sticks to the ware while wet, and therefore rapid drying of the white enamel makes for pure white ware. Another point in favor of rapid drying of finished ware is the need of space for storing the ware. After the enamel is dry the ware can be handled and stored in much less space than when wet, and in making some classes of wares, such as cooking utensils, the problem of finding room for storing sufficient ware to keep the furnaces going is sometimes troublesome. There are two common defects caused by improper drying. Water streaking, caused by moisture from drying ware condensing on the cold surface of wet ware and running down vertical surfaces in streaks, can be avoided by proper circulation of air in the drier. When ware is dried too rapidly the enamel will crawl. This is caused by the formation of shrinkage cracks due to driving off the moisture from the clay too rapidly. These cracks do not show in the dry ware, but when it is burned the enamel crawls and collects in beads. This defect will be caused when a piece of wet ware is set on a hot piece of metal or when the drying is very sudden. The same defect may result from rough handling of the dry ware, a sudden sharp blow breaking the bond between the dry enamel and steel, which results in crawling.

3. Burning Enamels

General Description.—Muffle furnaces are almost invariably used for burning light wares and especially white ware. For burning heavy steel wares open furnaces are used.

The ware is set on pointed projections from iron grates, which should be kept sharp so that the least possible part of the grates comes in contact with the enamel. Only pieces of approximately the same size and weight should be burned together, since only a few minutes are allowed for burning a fork of light steel ware, and if there is much difference in the size of the ware it will heat up to the temperature of the furnace at different rates. As a result of this the lighter ware will be sufficiently fired before larger pieces have acquired the desired temperature, and some of the ware will be sure to be imperfectly fired.

In setting the ware on the grates preparatory to firing, care should be taken to see that ample space is left between all surfaces. Heavy parts like handles on dishpans and ears on kettles should be removed as far as possible from all other surfaces. The reason for this is not only to permit these heavy parts to heat up as rapidly as possible but also to prevent them from absorbing radiated heat from parts near *them, thereby retarding the rate at which these parts are heated.

It frequently happens that there will be a small area on a piece of ware underburned while the piece as a whole is properly burned. Investigation of the cause of this will reveal the fact that this underburned spot was in close proximity to some heavy piece of metal or other surface which absorbed the heat while the main body of the piece of ware was free to heat up rapidly. A good burner will strike the happy medium and leave his ware in the furnace long enough to fire the heavy parts properly but not long enough to burn off the light parts. The nature of the enamel influences very materially the burner's ability to properly burn light and heavy parts, but he can greatly facilitate matters by using proper care in setting his ware on the grates.

The temperatures used for burning enamels differ widely, depending upon the enamel and the ware. General practice is to burn the ground coat at much higher temperature than the finishing coats. This is not due to the fact that the ground coat necessarily has a higher softening temperature than the finishing coats, but rather to the fact that it has been found that the general excellence of the ware is improved by this procedure.

Crucibles, Refractory

Flake Graphite	21
Crushed Silicon Carbide	45
Flint	11
Borax	5
Tar	18

Low Temperature Glaze for Art Ware and Enameled Brick

White Lead	35 lb.
Feldspar	17 lb.
Flint	20 lb.
Whiting	8 lb.
China Clay	8 lb.
Colemanite	12 lb.
Tin Oxide	5 lb.
Matte Glaze—Cone 06 to Cone 02:	
White Lead	490 lb.
Whiting	138 lb.
Cornwall Stone	114 lb.
China Clay	210 lb.
Feldspar	98 lb.
Flint	60 lb.

For light green use 2 to 3% copper oxide; for light brown 2% manganese dioxide; for blue 1% cobalt oxide; for yellow 2% sodium uranate; for yellow-brown ½ to 2% Crocus Martis.

Cresylic Wood Impregnation Bath

Cresylic Acid	100 lb.
Red Oil (Double Pressed)	100 lb.

Caustic Soda Solution, 32° Bé. 20 lb.
Manipulation: Add caustic soda solution to red oil at 50° C., add cresylic acid slowly with constant agitation and cool rapidly.

Arsenic Cement Coating for Wood Piling

Sand	12 lb.
Cement	3 lb.
Arsenic, White	1 oz.

Mix dry and add water before use. Then apply to piling by air gun.

Stone Waterproofing

An economical treatment that is very durable may be made by dissolving from 6 to 12 oz. of a high-melting-point paraffin to the gallon of solvent, such as mineral spirits, naphtha, gasoline, etc. This usually gives high waterproofing values on materials of medium to coarse textures. For fine-pore structures it will be desirable to add from 3 to 6 oz. of china wood oil to the gallon of gasoline.

CHAPTER XI

PAPER

PAPER is made to specifications to suit it for its ultimate use. Thus a wrapping paper is made from different raw materials than a fine writing paper. It is likewise finished differently. The different types of paper and methods of manufacture are given below.

Types of Paper Stock
Groundwood

A flour of wood produced by grinding barked logs against stone. The process is purely mechanical.

Sulfite

Prepared by cooking wood chips at 70 to 80 lb. pressure 15 to 18 hours with a solution of sufurous acid which has been passed through a tower of lime or dolomite. The final solution varies greatly but a total sulfur dioxide content, 4.5%, 3.5% free and the rest combined is considered good practice.

Sulfate (or Kraft)

Prepared by cooking wood chips at 120 to 140 lb. pressure about 8 hours with a solution of sodium hydroxide and sodium sulphide. The solution may have a formula approximating sodium carbonate, 11, sodium hydroxide 90, sodium sulphide 25 gm. per liter.

Soda

Prepared by cooking wood chips at 110 to 120 lb. pressure about 8 hours with a 6–8% sodium hydroxide solution.

Jute

Prepared by cooking cut burlap sacks (old bags) at normal or increased pressures with mild alkali such as 1–5% sodium hydroxide or 5–10% calcium hydroxide from 4 to 18 hours, washing and beating the product to pulp.

Rope
(Hemp or Manilla)

Prepared by cooking rope (old rope) as outlined for jute.

Note: There is more variation in method for production for the last two pulps than in the others. For instance there is one secret process which produces an excellent product, bleached, washed and ready for the beater continuously. All other methods are intermittent. The complete cycle is less than forty minutes. No other cycle is less than seven hours.

Principal Types of Paper

All papers are formed on a screen catching the suspended fibers and passing through the water. The resulting mat is dried by squeezing through felts and heating on hot cylinders.

Book

Chiefly prepared from sulphite and soda pulp.

News

About eighty per cent ground wood.

Wrapping

Sulfite, Sulfate, Jute, Rope, or mixtures.

Writing

May be old rag, but usually sulfite or sulfite and soda.

Minor Types
Waxed

A paper that has been run through paraffine.

Parchment

A paper that has been treated with concentrated sulfuric acid.

Glassine

A heavily beaten, unloaded paper. Supercalendered.

Grease Proof

Prepared as above, but not supercalendered.

Cellophane

Not technically a paper. A film of regenerated cellulose, cellulose nitrate or acetate.

Basic Weight

Paper is sold by basic weight. Official basic weight is the weight of 500 sheets, 25 by 40 inches. Trade custom basic weights vary. To convert from official to trade figures the following factors are useful.

Trade Name	Trade Size (inches)	Factor
Book	25 × 38	0.950
News	24 × 36	0.864
Wrapping	24 × 36	0.864
Writing	17 × 22	0.374

To Waterproof Paper

Waterproofing is best accomplished by parchmentizing paper but this treatment leaves a surface that is too irregular to make a good writing surface. One part of any of the following to six parts of water are supposed to give a good waterproofed paper. Glue, gelatin, shellac or aluminum acetate. Excellent results are obtained by using one part of borax, five parts of shellac and ten parts of water. The mixture is brought nearly to the boil, but not boiled and kept hot until all the shellac has passed into solution. The paper may be dipped into the solution, or it may be applied with a wide brush. The surface is a satisfactory vehicle for ink or water color.

To Parchmentize Paper

Prepare a fifty per cent solution of sulfuric acid. Pass a water-leaf (unloaded) paper through the solution being careful that no air bubbles prevent even contact with acid. Each part of the paper should remain in contact with acid for about 5 seconds. Promptly plunge the paper into a large quantity of cold water. Then wash with a running stream of water from the faucet or a wash bottle with a wide-mouthed tip. Next wash with a weak solution of ammonia to remove the last trace of acid and finally wash with water to remove any ammonia. An excellent parchment-like effect is acquired by thick papers. However, there is an art in this and only experience can guide the operator in the length of time the paper should be in contact with the acid. If a longer time is required stronger acid may be used.

To Fireproof Paper

Prepare a solution as follows:

Ammonium Sulfate	8	gm.
Boric Acid	3	
Borax	1.7	
Water	100	cc.

The solution should be heated to 122° F. and kept at this temperature. The paper is dipped in the solution and hung to dry. Wrinkles can be prevented by drying in a press, or the paper may be subsequently ironed.

To Remove Creases from Paper

Creases may be removed from even fine engravings if a little care is exercised. Place the sheet smoothed as far as possible by hand on a clean sheet of paper on top of a well-covered ironing board or similar surface. Cover with another clean sheet. Finally dampen a third sheet, place on top of the others and press with a moderately warm iron.

Temporary Tracing Paper

It is sometimes necessary to make a tracing on a regular sheet of writing or bond paper. Temporary translucence may be created by sponging the paper with benzine. As soon as the benzine evaporates the paper reverts to its normal condition. The last trace of odor can be removed with a draft of warm air. While still translucent the paper will take either pen or ink drawing without difficulty. The use of benzine provides a quick accurate method for tracing graphs.

PAPER COATINGS
Casein Glue

Casein	100	lb.
Water	50	gal.
Borax	17	lb.
Ammonia 26°	1	qt.

The casein is preferably soaked a few hours in the water, the borax dissolved in a little hot water—added, and the whole cooked to 160° F. till no undissolved particles of casein remain. Then the ammonia is added and the glue cooled.

Wax Emulsion for Paper

Carnauba Wax	50 lb.
Water	50 gal.
Soap	12 lb.

The soap is dissolved in the water and brought to boiling. The wax is added and boiling continued until all is emulsified. The emulsion is preferably stirred continuously until cold. The soap may be any good grade of washing soap free from rosin.

Yellow Coating

Clay	50 lb.
Blanc Fixe Pulp (70% dry)	50 lb.
Chrome Yellow Pulp (50% dry)	125 lb.
Talc	12 lb.
Casein Glue	11 gal.
Carnauba Wax Emulsion	4 gal.

Blue Coating

Prussian Blue Pulp (30% dry)	100 lb.
Violet Lake Pulp (35% dry)	75 lb.
Maroon Lake Pulp (35% dry)	75 lb.
Casein Glue	8 gal.
Carnauba Wax Emulsion	3 gal.
Talc	4 lb.

Pearl Coating

Clay	50 lb.	
Blanc Fixe Pulp	50 lb.	
Italian Talc	4 lb.	Pulped together
Ultramarine Blue	5 lb.	
Water	4 gal.	
Casein Glue	12 gal.	
Carnauba Wax Emulsion	4 gal.	

Red Coating

Red Pulp (40% dry)	200 lb.
Talc Italian	4 lb.
Casein Glue	12 gal.
Carnauba Wax Emulsion	6 gal.

White Coating

Clay	300 lb.
Water	20 gal.

Italian Talc	18 lb.
Casein Glue	25 gal.
Carnauba Wax Emulsion	12 gal.

Coating—Special for High Finish— White

Water	65 gal.
Soda Ash	3 lb.
Ammonia	4 gills
Satin White	440 lb.
Clay	650 lb.

Mix thoroughly and add the following solution

Water		50 gal.
Casein		100 lb.
Soda Ash	10 lb.	Dissolved in
Tri Sodium Phosphate	7 lb.	3 gal. of
Borax	5 lb.	hot water
Ammonia		6 gills

Coating—White—Soft Sized

Water	165 gal.
Clay	1300 lb.

Stir 15 minutes in a rapid dissolver and add

Dry Casein	140 lb.

Stir 15 minutes and add

Dry Borax	18 lb.

Stir 5 minutes and add

Ammonia	4 qt.

Heat to 140° F. and stir till casein is dissolved and cool to room temperature. Strain before using.

If hard sized coating is desired, increase the amount of casein until the desired degree of sizing is obtained.

Coating—Friction Finish—Yellow

Casein	200 lb.
Borax	12 lb.
Ammonia	5 qt.
Water to make	150 gal.

Water	43 gal.
Talc	23 lb.
Clay	200 lb.
Blanc Fixe Pulp	390 lb.
Medium Yellow Pulp	18 lb.
Carnauba Wax Emulsion	16 gal.
Casein as above	32 gal.

"Imitation Parchment" Coating

1.

Sodium Silicate	30 gm.
Sodium Sulforicinoleate	20 gm.

Heat together on water bath and add 30 cc. boiling water.

Dip paper into this and draw out immediately. This gives a parchment like effect to the paper.

Keep the mix boiling for five minutes and dip second piece of paper into it. This gives a translucent paper.

2.

A small amount of Tricresyl Phosphate is added to a thin alcohol solution of bleached shellac. Paper dipped in this solution and dried will resemble parchment, except that it will be very resistant to moisture.

◆

"Glassine" Paper

Paper is coated with or dipped in the following:

Copal	100
Alcohol	300
Castor Oil	8–12

◆

Shellac Solution for Paper Waterproofing

In a wooden tank, fitted with steam injector place

Water	25	gal.
Orange Shellac	150	lb.
Ammonium Hydroxide	6½	gal.

Allow to stand overnight and then turn on steam until dissolved. Bring volume to 100 gal. with cold water. Two coatings of this solution are given to the paper.

◆

Waterproofing Solution for Blueprints

To waterproof blueprints and give them a sheen and greater legibility, rub them with a soft cloth that has been dampened with a solution of rosin, 50 grains, paraffin, 100 grains, and turpentine, 1 oz.

◆

Waterproofing Paper and Fibreboard

The following composition and method of application will render uncalendered paper, fibreboard and similar porous material waterproof and proof against the passage or penetration of water.

Paraffin (M.P. about 130° F.)	22.5%
Trihydroxyethylamine Stearate	3.0%
Water	74.5%

The paraffin or a wax is melted and the stearate added to same. The water is then heated to nearly boiling and then vigorously agitated with a suitable mechanical stirring device, while the above mixture of melted wax and emulsifier is slowly added. This mixture is cooled while it is stirred.

The paper or fibreboard is coated on the side which is to be in contact with water. This is then quickly heated to the melting point of the wax, which then coalesces into a continuous film that does not soak into the paper which is preferentially wetted by the water. This method works most effectively on paper pulp moulded containers and possesses the advantages of being much cheaper than dipping in melted paraffin as only about a tenth as much paraffin is needed. In addition, the outside of the container is not greasy, and can be printed upon after treatment which is not the case when treated with melted wax.

◆

Oil and Greaseproofing Paper and Fibreboard

This solution applied by brush, spray, or dipping will leave a thin film which is impervious to oils and greases. Applied to paper or fibre containers, it will enable them to retain oils and greases. All the following ingredients are by weight:

Starch (preferably Cassava)	6.6%
Caustic Soda (76% Na_2O)	0.1%
Glycerol	2.0%
Sugar	0.6%
Water	90.5%
Sodium Salicylate	0.2%

The caustic soda is dissolved in the water and the starch is then made into a thick paste by adding a portion of this solution. This paste is then added to the water. This mixture is placed in a water jacket and heated to about 85° C. until all the starch granules have broken and the temperature maintained for about half hour longer. The other substances are then added and thoroughly mixed and the composition is ready for application. A smaller water content may be used if applied hot and a thicker coating will result. Two coats will result in a very considerable resistance to oil penetration.

◆

Glaze, Paper

100 parts Carnauba Wax are melted together at 120–130° C. with 25 parts curd soap, while stirring well. 900 parts boiling water are then added while stir-

ring well, very slowly at first and then more rapidly, the whole being boiled up, and stirred until cold.

Glaze for Paper, Wood or Metal

Casein	100	lb.
Borax	7–15	lb.
Trisodium Phosphate	7–15	lb.
Hexamethylene Tetramine	0.5–8	lb.
Castor Oil	1– 5	oz.
Clovel	1	oz.

Mimeograph Paper

The substance used for the coating consists of a mixture of hydrocarbons of the fatty series plus ozokerite, oleine, and palmitine.

The carrier for the coating is a light cellulose paper weighing about 12 gm. per square meter. This is placed on a metal plate, heated to 100° C. The coating is melted and painted on the surface with a soft sponge. The operation is done on the reverse side to the one on which the tracing is to be made. The molten coating penetrates the pores of the cellulose by dialysis and it thus becomes incorporated in a uniform manner which, when it comes into contact with the hot plate gives perfect glazing to that side of the sheet.

Formula for coating.

Tricosane	1250	parts
Ozokerite	55	parts
Oleine	32.5	parts
Palmitine	12.5	parts

RESISTANCE OF WRAPPING MATERIALS TO THE PASSAGE OF WATER VAPOUR

Materials Examined	Loss, in Grammes per Square Metre, in 24 Hours
Waxed paper	Down to 10
Waxed paper, after severe creasing	90 to 100
Coated viscose film	16 to 20
Viscose film	150 to 190
Coated glassine paper	100 to 150
Glassine paper	280
Vegetable parchment	185 to 320
Kraft papers	200 to 250
M.G. sulphite papers	Up to 480

"Safety" Paper

Paper treated to prevent fraudulent alteration and useful for checks, drafts, etc., is made by incorporating in it or coating it with a 10% water solution of a leuco indophenol and drying it. It is then passed through a bath containing 5 lb. of Manganous Sulfate per 20 gallons of water.

Paper Softener

Paper dipped in a 10% water solution of glycerin and dried will thereafter be very soft and cloth-like.

Transparent Wrapping Material
(Similar to "Cellophane")

Ethyl Cellulose or Benzyl Cellulose dissolved in Ethyl Acetate and spread on a glass plate to dry will produce a perfectly transparent sheet with a high gloss. A small quantity of Tricresyl Phosphate or Dibutyl Phthalate will increase the flexibility of the same. This material may be colored as desired by the addition to the solution of Benzyl soluble dyes. The dyes are dissolved in Benzyl and added to the solution.

Soap, Rosin Size

Into a suitable boiler or heater an amount for instance 100 kilogrs. of rosin is placed and as much water, then a mixture of carbonated and bicarbonated alkalis is added in a quantity necessary for saturating say 88% of the rosin put in operation. If the bicarbonate is employed in about the proportion of half the carbonate, then approximately 11 kilogrs. of carbonate of soda and 5 kilogrs. of bicarbonate of soda will be required.

The boiler is heated by steam for example and when cooking is considered sufficient, water and a volatile alkali (ammonia) are added, the amount of alkali being sufficient to saturate the 12 kilogrs. of rosin which have not been affected by the carbonated alkali. For this second phase of saponification by means of ammonia liquid it is necessary to employ about 4 kilograms of aqueous ammonia solution having a density of 0.930 (which would contain about 18% of pure ammonia) when the quantity of hydrated rosin to saponify is 12 kilogrs. that is to say, the proportion of ammonia liquid is ⅓ to ⅔ hydrated rosin. The heating by steam is continued so as to bring the mixture up to boiling point for some minutes, at the end of which time the product is finished.

CHAPTER XII

PHOTOGRAPHY

THE ACTION OF PHOTOGRAPHIC SOLUTIONS

The sensitive photographic emulsion is composed of a number of light-sensitive crystals of certain silver salts (silver halides) incorporated in a layer of gelatin. When light falls upon the sensitive salts, we know that a definite reaction takes place in the emulsion; although this change in the silver compound is not visible, its occurrence is proven by the fact that a developing solution will act selectively on an emulsion which has been affected by light, reducing the affected silver halide crystals to metallic silver without changing the unaffected crystals.

Developing Solutions

A developing solution is composed of a number of chemicals, each of which is called upon to exert a certain action.

The reducing agents (so called because they "reduce" silver halide salts to metallic silver) most commonly used are metol, hydrochinon, pyro, glycin and amidol.

The other ingredients in the developing solution are used to control the action of the reducing agent. Most reducing agents in water solution are neutral or slightly acid and in this condition do not readily attack reducible silver salts. To correct this condition an alkaline sodium salt such as sodium carbonate must be added in order to facilitate the desired action. The addition of carbonate to the developer solution also has the effect of softening the gelatin, thus permitting the reducing agent quickly to penetrate the gelatin and act upon all of the sensitive crystals which have been affected by the light. This quick penetration by the reducing agent makes for more rapid development and increased contrast in the resulting image.

Strong caustic alkalines such as sodium or potassium hydroxide, may in theory be substituted for sodium carbonate; but these chemicals are seldom used in prac-

tice because of their great strength, which makes the developing agent so active as to affect the silver salts throughout the entire emulsion whether acted upon by light or not, causing a heavy chemical fog. The addition of a very strong alkali also softens the gelatin excessively, causing frilling and blistering with even a short immersion in the developer.

To restrain the action of the alkali and to limit the action of the developer to light-affected crystals alone, a small amount of potassium bromide is incorporated in the solution, making it possible to use a larger concentration of sodium carbonate, thus increasing the speed of development without causing any corresponding increase in the fog.

Summing up: the active agents in the developing solution are (1) a reducing agent, (2) an alkali to facilitate the action of the agent, and (3) a bromide to control the action of the alkali. However, a solution containing only the chemicals which have been enumerated will quickly form undesirable oxidation products, because all reducing agents have a strong affinity for oxygen. To prevent premature oxidation of the developer, which would cause stain and fog, a small amount of sodium sulfite is added. Sodium sulfite has a stronger attraction for oxygen than does a reducing agent, and by its own absorption of oxygen, delays oxidation of the reducing agent. The sulfite, then, acts only as a preservative in the solution and has nothing to do with the reduction of the silver salts.

Pyro, a developing agent in wide use, gives added printing contrast through the stain caused by the pyro oxidation. This stain is in proportion to the density; and though the density of the silver image may not appear heavy to the eye, yet the pyro stain will give a very high effective printing contrast. Pyro oxidizes quickly, consequently bisulfite and metabisulfite of sodium or potassium, because they are acid in character, must be used with a pyro developer. Pyro is more readily oxidized in an alkaline solution than other reducing agents so that the acidity of the bisulfite more effectively prevents oxidation of pyro than does alkaline sodium sulfite.

Perhaps the most common developers used are those combining metol and hydrochinon. Metol rapidly builds up density in the early stages of development, while hydrochinon acts more slowly. Under prolonged development the hydrochinon builds up a greater density than the metol.

It should be noted in this connection that when using a metol-hydrochinon developer, temperature control is extremely important. This is vitally necessary, because of the fact that while metol reacts to an inconsequential degree to temperature changes, hydrochinon becomes practically inactive below sixty degrees and is extremely active at temperatures higher than seventy degrees.

Thus, if a developing solution combining these two chemicals is used below the normal operating temperature, the image produced is almost entirely a metol image, full of detail but lacking in contrast; whereas if the temperature is too high, the hydrochinon acts quicker and blocks up the higher densities.

Amidol is commonly used by photographers who are susceptible to metol poisoning and who consequently find it impossible to use the regular MQ developers. The amidol developing formula is extremely simple to mix, as it requires (in addition to the reducing agent) only the sulfite preservative and the potassium bromide restrainer.

Glycin is used at times in connection with hydrochinon or it may be used alone for tank development because of its excellent keeping properties. It is commonly included in paper formulas for production of olive or warm black tones.

Though contrast is primarily a function of the emulsion, yet the time of development has its effect upon contrast, increasing it as the time is prolonged. However, too long immersion in the developer will cause an increase of chemical fog; there is always a certain amount of fog present in a photographic emulsion, and though this fog is restrained to a great extent by the potassium bromide, yet over-prolonged development causes it to become noticeable and at times objec-

tionable. As the fog increases, it becomes proportionately greater in the lower densities than in the higher; therefore when extreme development times are given, the growth of fog will cause a decrease in the effective contrast.

Another factor which influences contrast is the agitation of the developer. If the developer is agitated, a higher contrast will be built up; because as the emulsion is developed, bromide is liberated from the emulsion, which, if allowed to remain in contact with the emulsion, exerts a restraining effect.

It is advisable to keep the temperature of the developer around 68° F., especially if an MQ developer is used.

Fixing Solutions

It is common practice today to use an acid-alum fixing bath. Though the cost of mixing such a bath will be somewhat higher than that of a plain bath, the acid-alum is much more efficient as regards keeping qualities. The importance of mixing ingredients according to the order given in the formula holds as true for the acid fixing bath as it does for the developer solution. It is the custom at times for photographers to add acid directly to the hypo in the mixing of the acid hypo bath; this addition of free acid to the hypo decomposes the latter and forms sulfur, which renders the bath inactive. When the hardening solution is mixed, it should be mixed separately from the hypo, and only added after the latter has cooled to normal temperature. The sulfite should be mixed first, and then the acid. If the alum is added to the sulfite before the acid, a white precipitate (aluminum sulfite) will form, thus removing some of the necessary ingredients from active participation in the fixing process.

For the hypo bath, as well as the developing solution, it is important to hold the temperature at 68° F. If the fixing bath becomes too warm, the acid will exert a stronger action and will decompose the hypo into free sulfur.

General Remarks

In all photographic formulas, the ingredients should be mixed in their tabulated order. If the developer agent is mixed first, and then to this is added the carbonate and after that the sulfite, the developing agent will rapidly oxidize. The addition of sulfite directly after the developing agent prevents oxidation of the latter, which might result in stain on the emulsion.

In the mixing of developing solutions care must be taken that each chemical is completely dissolved before the next is added. The temperature of the water may be about 125° F., while the developing agents and sulfite are being dissolved; but the solution should be cooled somewhat before the sodium carbonate is added.

Accuracy is important in weighing, especially when only a small quantity of solution is being mixed; approximating weights is a practice through which lack of uniformity in results often arises, and should be avoided.

In the following formulas the quantities of sodium sulfite and sodium carbonate are given for the anhydrous form of the sulfite and the monohydrated sodium carbonate. If other forms of these chemicals are to be used, due allowance must be made for difference in strength; for example, if it is desired that sodium sulfite crystals be used, the quantities shown for anhydrous sulfite must be doubled. If anhydrous sodium carbonate is to be used, the amounts given must be decreased 17%. If sodium crystals are to be used in place of monohydrated sodium carbonate the amounts shown must be increased to 2¾ ounces for each ounce specified in the formula.

The development times given in this booklet are based upon a developer temperature of 68° F.

Developers for Negative Emulsions

The following developers may be used with negative emulsions:

MQ Developer

TANK

Metol	170 gr.	12.0 gm.
Sodium Sulfite	12 oz.	360.0 gm.
Sodium Bisulfite	¼ oz.	7.5 gm.
Hydrochinon	¾ oz.	22.5 gm.
Sodium Carbonate	2 oz.	60.0 gm.
Potassium Bromide	100 gr.	7.0 gm.
Water to	3½ gal.	14,000.0 cc.

Development time about 12 minutes.

TRAY

Metol	15 gr.	.9 gm.
Sodium Sulfite	400 gr.	25.7 gm.
Sodium Bisulfite	8 gr.	.5 gm.
Hydrochinon	25 gr.	1.6 gm.
Sodium Carbonate	65 gr.	4.3 gm.
Potassium Bromide	8 gr.	.5 gm.
Water to	32 oz.	1,000.0 cc.

Development time 5–7 minutes.

For a softer working MQ formula which is very satisfactory for portraiture, use the following:

TANK

Metol	285 gr.	18.5 gm.
Sodium Sulfite	9 oz.	255.2 gm.
Sodium Bisulfite	82 gr.	5.3 gm.
Hydrochinon	100 gr.	6.5 gm.
Sodium Carbonate	1½ oz.	42.5 gm.
Potassium Bromide	75 gr.	4.9 gm.
Water to	3½ gal.	12,500.0 cc.

Development time 10–15 minutes.

TRAY

Metol	76 gr.	5.0 gm.
Sodium Sulfite	1¾ oz.	50.0 gm.
Sodium Bisulfite	16 gr.	1.0 gm.
Hydrochinon	20 gr.	1.3 gm.
Sodium Carbonate	130 gr.	8.5 gm.
Potassium Bromide	16 gr.	1.0 gm.
Water to	32 oz.	1,000.0 cc.

Development time 5–7 minutes.

Pyro Tank Formula

Solution "A"

Sodium Bisulfite or Potassium Metabisulfite	140 gr.	9.8 gm.
Pyro	2 oz.	60.0 gm.
Potassium Bromide	16 gr.	1.1 gm.
Water to	32 oz.	1,000.0 cc.

Solution "B"

Sodium Sulfite	3½ oz.	105.0 gm.
Water to	32 oz.	1,000.0 cc.

Solution "C"

Sodium Carbonate	2½ oz.	75.0 gm.
Water to	32 oz.	1,000.0 cc.

For tank use, take 5½ oz. of solutions A, B and C, and add water to make 1 gallon. Developing time about 12 minutes.

For use in tray, take 1 oz. each of A, B and C, and add 7 oz. of water. Development time about 6 minutes.

Pyro Soda Developer

Solution "A"

Pyro	135 gr.	7.0 gm.
Potassium Metabisulfite	20 gr.	1.3 gm.
Sodium Sulfite	1 oz.	28.4 gm.
Potassium Bromide	10 gr.	.7 gm.
Water to	20 oz.	500.0 cc.

Solution "B"

Sodium Carbonate	1 oz.	28.4 gm.
Water to	20 oz.	500.0 cc.

Normal Exposure—use equal amounts of A and B.

Over Exposure—use more of A.

Under Exposure—use more of B.

Pyro-Metol

Solution "A"

Sodium Bisulfite	¼ oz.	7.5 gm.
Metol	¼ oz.	7.5 gm.
Pyro	1 oz.	30.0 gm.
Potassium Bromide	60 gr.	4.2 gm.
Water to	32 oz.	1,000.0 cc.

Solution "B"

Sodium Sulfite	5 oz.	150.0 gm.
Water to	32 oz.	1,000.0 cc.

Solution "C"

Sodium Carbonate	2½ oz.	75.0 gm.
Water to	32 oz.	1,000.0 cc.

For tank use take 8 oz. each of solutions A, B and C and add water to make 1 gal.

For tray use take 2 oz. each of solutions A, B and C and add water to make 16 oz.

Pyro-Soda

(Non-staining Formula)

Make up two solutions according to the following formulas:

Solution "A"

Sodium Sulfite	4 oz.	200.0 gm.
Potassium Metabisulfite	1 oz.	50.0 gm.
Pyro	1 oz.	50.0 gm.
Water to	60 oz.	3,000.0 cc.

Solution "B"

Sodium Carbonate	4½ oz.	225.0 gm.
Water to	60 oz.	3,000.0 cc.

Mix A, 1 part; B, 1 part; water, 2 parts.

In making the "A" solution the sulfite and metabisulfite should be mixed together dry, and put together into hot water. When they are dissolved, the solution should preferably be brought to the boil and boiled for about a minute. After the solution has cooled, the pyro is added. The boiling greatly improves the keeping qualities of the solution.

This developer will produce negatives free from pyro stain, and 4 to 6 minutes development at normal temperature with full exposure will yield negatives which are well suited to enlarging. The advantages of the developer are its cleanliness and the extraordinary keeping qualities of the "A" solution, which must be made up as directed above.

When stronger negatives are required the developer can be made up by taking equal parts of "A," "B" and of water; or equal parts of "A" and "B" alone can be used, this giving a developer containing about 4 grains of pyro to the ounce.

For underexposures and normal exposures, use

Metol	27 gr.	1.9 gm.
Sodium Sulfite	1⅛ oz.	35.0 gm.
Adurol	20 gr.	1.4 gm.
Hydrochinon	88 gr.	6.2 gm.
Potassium Carbonate	380 gr.	27.0 gm.
Potassium Bromide	25 gr.	1.7 gm.
Water to	32 oz.	1,000.0 cc.

For tray use take 1 part of stock solution and 2 parts of water. Develop 3–4 minutes.

For tank development use 1 part of stock to 8 parts of water. Develop 12–15 minutes.

Amidol

As there is no alkali present in this developer there is no softening of the gelatin, and the tendency to frill at high temperature is reduced to a minimum; so that development of plates, films and papers can be safely conducted at temperatures up to 85° F. This formula will give excellent results as a negative developer. On gaslight papers and bromide papers it will produce beautiful blue-black tones; and the developer will keep well for two days, although to insure best results it should be freshly prepared each time.

Sodium Sulfite	½ oz.	15.0 gm.
Potassium Bromide	7 gr.	.5 gm.
Amidol	60 gr.	4.0 gm.
Water to	20 oz.	600.0 cc.

A Soft Working Metol Pyro Formula

Solution "A"

Metol	30 gr.	3.8 gm.
Potassium Metabisulfite	100 gr.	13.0 gm.
Pyro	100 gr.	13.0 gm.
Potassium Bromide	10 gr.	1.5 gm.
Sodium Sulfite	¾ oz.	42.5 gm.
Water to	20 oz.	1,000.0 cc.

Solution "B"

Sodium Carbonate	2 oz.	113.0 gm.
Water to	20 oz.	1,000.0 cc.

Normal Exposure—1 part solution "A" and 1 part solution "B."

Over Exposure—2 parts solution "A" and 1 part solution "B."

Under Exposure—1 part solution "A" and 2 parts solution "B."

Press Plate Developer

For general press photography, the following formula may be used where negatives of higher than normal contrast are desired.

Metol	210 gr.	13.0 gm.
Sodium Sulfite	6 oz.	170.0 gm.
Hydrochinon	325 gr.	20.0 gm.
Sodium Carbonate	2¾ oz.	78.0 gm.
Potassium Bromide	100 gr.	5.0 gm.
Water to	1 gal.	6,000.0 cc.

Use 1 part stock to 1 part water. Development time 4–6 minutes.

Developers for Fine Graininess

At the present time many photographers are primarily interested in securing negatives with a minimum of graininess to make possible very extreme enlargements. Where negatives of particularly fine graininess are desired, the following formulas are suggested:

Metol	15 gr.	1.0 gm.
Sodium Sulfite	1⅛ oz.	32.0 gm.
Glycin	8 gr.	.5 gm.
Hydrochinon	8 gr.	.5 gm.
Sodium Carbonate	1 oz.	28.0 gm.
Potassium Bromide	23 gr.	1.5 gm.
Citric Acid	15 gr.	1.0 gm.
Water to	40 oz.	1,000.0 cc.

If exposure has been correct, the film will be properly developed in 10–12 minutes.

If a still finer graininess is desired, the

use of the following formula is recommended.

Metol	18 gr.	2.0 gm.
Sodium Sulfite	2 oz.	100.0 gm.
Hydrochinon	26 gr.	3.0 gm.
Resorcine (metadiozyd benzolum)	18 gr.	2.0 gm.
Borax	18 gr.	2.0 gm.
Water to	20 oz.	1,000.0 cc.

Time of development 8 minutes.

The preparation of this developer must be extremely accurate, and we therefore advise proceeding as follows:

Solution 1. First the metol is dissolved in 4 oz. of water at 120° F. Separately 200 grains sodium sulfite are dissolved in 4 oz. water, to which the hydrochinon and resorcine are added. This last solution is then added to the metol solution.

Solution 2. The remnant of the sodium sulfite and the borax are dissolved in 7 oz. water at 158° F. When entirely cold, solution 2 must be poured slowly and carefully into solution 1, which must be kept in movement. Then water is to be added to 20 oz.

The following simple formula may also be used to obtain fine graininess.

Metol	115 gr.	7.5 gm.
Sodium Sulfite	1¼ oz.	35.0 gm.
Water to	20 oz.	500.0 gm.

Add 15 minims of a 10% solution of potassium bromide to every 2 oz. of developing bath. Time of development 16 minutes.

DEVELOPERS FOR GRAPHIC EMULSIONS

For Extreme Contrast

Solution "A"

Sodium Bisulfite	¾ oz.	25.0 gm.
Hydrochinon	¾ oz.	25.0 gm.
Potassium Bromide	¾ oz.	25.0 gm.
Water to	40 oz.	1,000.0 cc.

Solution "B"

Sodium Hydrate (Caustic Soda)	1¾ oz.	50.0 gm.
Water to	40 oz.	1,000.0 cc.

Use equal parts of solutions "A" and "B." These stock solutions keep well separately but when mixed should be used within 1 hour. Development time should be less than 4 minutes.

After developing plates in above developer it is imperative that plates should be well rinsed in water before placing in fixing bath. This will prevent the yellow stains which are produced when using caustic developers. Care should be taken to guard hands and clothes from this caustic developer.

For Line and Halftone Work

Metol	50 gr.	.9 gm.
Sodium Sulfite	8 oz.	62.8 gm.
Hydrochinon	2 oz.	15.7 gm.

Sodium Carbonate	3 oz.	23.5 gm.
Potassium Bromide	120 gr.	2.1 gm.
Water to	1 gal.	1,000.0 cc.

Time of development will be 4–5 minutes.

Use above solution full strength. Developer keeps very well and gives remarkably clear results.

For Normal Gradation

Metol	½ oz.	3.9 gm.
Sodium Sulfite	7 oz.	55.0 gm.
Hydrochinon	1 oz.	7.9 gm.
Sodium Carbonate	5 oz.	39.2 gm.
Potassium Bromide	100 gr.	1.8 gm.
Water to	1 gal.	1,000.0 cc.

Use 1 to 3 for Normal results. Time of development about 3 minutes.

Use 1 to 4 for Soft results. Time of development about 4 minutes.

For Normal Negatives or Positives

Metol	60 gr.	1.0 gm.
Sodium Sulfite	4 oz.	31.4 gm.
Hydrochinon	½ oz.	3.9 gm.
Sodium Carbonate	3½ oz.	27.4 gm.
Potassium Bromide	75 gr.	1.3 gm.
Citric Acid (Crystals)	75 gr.	1.3 gm.
Water to	1 gal.	1,000.0 cc.

Time of development about 3 to 4 minutes. When diluted with equal part of water, 4 to 5 minutes.

For Normal Negatives or Positives

Metol	80 gr.	1.5 gm.
Sodium Sulfite	6 oz.	47.1 gm.
Hydrochinon	1 oz.	7.9 gm.
Sodium Carbonate	6 oz.	47.1 gm.
Potassium Bromide	80 gr.	1.5 gm.
Water to	1 gal.	1,000.0 cc.

Use 1 to 1 for snappy negatives or positives. Time of development, 2 to 3 minutes.

Use 1 to 2 for normal negatives or positives. Time of development, 3 to 4 minutes.

Use 1 to 3 for soft negatives or positives. Time of development, 4 to 5 minutes.

For Soft Gradation

Metol	2 oz.	15.7 gm.
Sodium Sulfite	6 oz.	47.1 gm.
Hydrochinon	¼ oz.	2.0 gm.
Sodium Carbonate	3 oz.	23.5 gm.
Potassium Bromide	80 gr.	1.5 gm.
Water to	1 gal.	1,000.0 cc.

Use 1 oz. stock to 4 oz. water. Time of development will be about 4 minutes.

For Commercial Use

Metol	30 gr.	1.8 gm.
Sodium Sulfite	1⅛ oz.	32.0 gm.
Hydrochinon	150 gr.	9.7 gm.
Potassium Carbonate	¾ oz.	21.0 gm.
Potassium Bromide	35 gr.	2.3 gm.
Water to	32 oz.	1,000.0 cc.

Development time 4 to 6 minutes.

LANTERN SLIDE DEVELOPERS

For Slides from Normal (average) Negatives

Metol	25 gr.	1.5 gm.
Sodium Sulfite	2 oz.	57.0 gm.
Hydrochinon	½ oz.	14.0 gm.
Sodium Carbonate	2 oz.	57.0 gm.
Potassium Bromide	15 gr.	1.0 gm.
Water to	40 oz.	1,000.0 cc.

For Slides from Strong (contrasty) Negatives

Metol	25 gr.	1.6 gm.
Sodium Sulfite	175 gr.	11.3 gm.
Hydrochinon	9 gr.	.6 gm.
Sodium Carbonate	1 oz.	29.0 gm.
Potassium Bromide	9 gr.	.6 gm.
Water to	20 oz.	500.0 cc.

For Slides from Weak (flat) Negatives

Metol	15 gr.	1.0 gm.
Sodium Sulphite	½ oz.	14.0 gm.
Hydrochinon	55 gr.	3.6 gm.
Sodium Carbonate	270 gr.	17.5 gm.
Potassium Bromide	9 gr.	.6 gm.
Water to	20 oz.	500.0 cc.

DEVELOPER FOR RADIOGRAPHIC FILM

Metol	¼ oz.	1.0 gm.
Sodium Sulfite	9½ oz.	71.7 gm.
Potassium Metabisulfite	235 gr.	4.0 gm.
Hydrochinon	1 oz.	7.6 gm.
Sodium Carbonate	4¾ oz.	36.0 gm.
Potassium Bromide	235 gr.	4.0 gm.
Water to	1 gal.	1,000.0 cc.

Development time 5 minutes.

DEVELOPERS FOR CINÉ FILM
For Soft Negatives

Metol	25 gr.	2.8 gm.
Sodium Sulfite	125 gr.	14.0 gm.
Hydrochinon	20 gr.	2.3 gm.
Sodium Carbonate	92 gr.	10.7 gm.
Potassium Bromide	13 gr.	1.5 gm.
Water to	20 oz.	1,000.0 cc.

Development time about 7 minutes.

For Negatives of High Contrast

Metol	9 gr.	1.0 gm.
Sodium Sulfite	260 gr.	30.0 gm.
Glycin	4½ gr.	.5 gm.
Hydrochinon	18 gr.	2.0 gm.
Sodium Carbonate	½ oz.	25.0 gm.
Potassium Bromide	13 gr.	1.5 gm.
Citric Acid	9 gr.	1.0 gm.
Water to	20 oz.	1,000.0 cc.

Development time 6–8 minutes.

For Soft Positives

Metol	2 oz.	.5 gm.
Sodium Sulfite	12¼ lb.	50.0 gm.
Hydrochinon	1¼ lb.	5.0 gm.
Sodium Carbonate	7½ lb.	30.0 gm.
Potassium Bromide	6 oz.	1.5 gm.
Citric Acid	4 oz.	1.0 gm.
Water to	30 gal.	1,000.0 cc.

Development time about 5 minutes.

For Titles and other Contrast Work

Hydrochinon	2½ oz.	10.0 gm.
Sodium Sulfite	6¼ lb.	25.0 gm.
Sodium Carbonate	25 lb.	100.0 gm.
Potassium Bromide	12 oz.	3.0 gm.
Water to	30 gal.	1,000.0 cc.

Development time about 5 minutes.

DEVELOPERS FOR PAPERS

The following formula is a standard MQ formula which may be used with all papers.

Metol	50 gr.	3.3 gm.
Sodium Sulfite	1½ oz.	42.5 gm.
Hydrochinon	150 gr.	9.7 gm.
Sodium Carbonate	2½ oz.	71.0 gm.
Potassium Bromide	15 gr.	1.0 gm.
Water to	32 oz.	1,000.0 cc.

Dilute 1–2 or 1–2 for use with Gevaert Printex, Novarex, Novages and Novaflex. For use with Novabrom, dilute 1–4.

As this is extensively used in amateur finishing, where larger quantities are used than those given above, the following amounts are given for preparing a greater volume of solution.

Quantity Formula

Metol	½ oz.	5 oz.
Sodium Sulfite	6 oz.	3¾ lb.
Hydrochinon	1½ oz.	15 oz.
Sodium Carbonate	10 oz.	6¼ lb.
Potassium Bromide	60 gr.	1½ oz.
Water to	1 gal.	10 gal.

For use dilute with equal part of water.

For Soft Effects with Novagas

Metol	35 gr.	2.3 gm.
Sodium Sulfite	¾ oz.	21.3 gm.
Hydrochinon	45 gr.	3.0 gm.
Sodium Carbonate	¾ oz.	21.3 gm.
Potassium Bromide	15 gr.	1.0 gm.
Water to	32 oz.	1,000.0 cc.

For use, dilute with equal part of water.

For More Contrast with Novages

Metol	10 gr.	.7 gm.
Sodium Sulfite	¾ oz.	23.4 gm.
Hydrochinon	135 gr.	9.7 gm.
Sodium Carbonate	1⅛ oz.	35.1 gm.
Potassium Bromide	10 gr.	.7 gm.
Water to	32 oz.	1,000.0 cc.

For Vigorous Effects with Novabrom

Metol	10 gr.	.7 gm.
Sodium Sulfite	165 gr.	10.7 gm.
Hydrochinon	35 gr.	2.3 gm.
Sodium Carbonate	½ oz.	14.2 gm.
Potassium Bromide	10 gr.	.7 gm.
Water to	20 oz.	500.0 cc.

For Softer Effects with Novabrom

Metol	50 gr.	3.3 gm.
Sodium Sulfite	275 gr.	17.8 gm.
Hydrochinon	15 gr.	1.0 gm.
Potassium Carbonate	185 gr.	12.0 gm.
Potassium Bromide	10 gr.	.7 gm.
Water to	20 oz.	500.0 cc.

A Special Developer for Novaflex

Metol	11 gr.	.7 gm.
Sodium Sulfite	350 gr.	23.0 gm.
Hydrochinon	45 gr.	3.0 gm.
Sodium Carbonate	265 gr.	17.0 gm.
Water to	32 oz.	1,000.0 cc.

Add enough potassium bromide to keep whites clean.

Developer for Production of very Soft Prints or High Key Effects

Metol	30 gr.	2.0 gm.
Sodium sulfite	1 oz.	28.5 gm.
Sodium Carbonate	1½ oz.	42.5 gm.
Potassium Bromide	5 gr.	.3 gm.
Water to	20 oz.	500.0 cc.

For use, dilute with 4 parts of water.

------Note------

With some formulas, especially in cold weather, it may be found difficult to keep the developing agent in solution; this can be remedied (without affecting the developing properties of the solution), by the addition of one ounce of alcohol to each ten ounces of solution.

◆

Warm Tone Developers for Artex Paper

With these formulas, variation of tone may be obtained by changing the bromide concentration, time of development and temperature of developer. When these factors are controlled, the tone of the prints will be constant through a long run. The longer the exposure and the shorter the development, the warmer the tone; and vice versa.

Metol	40 gr.	1.3 gm.
Sodium Sulfite	300 gr.	9.7 gm.
Hydrochinon	90 gr.	3.0 gm.
Sodium Carbonate	½ oz.	7.0 gm.
Potassium Bromide	45 gr.	1.5 gm.
Water to	64 oz.	1,000.0 cc.

Sodium Sulfite	1½ oz.	35.5 gm.
Hydrochinon	150 gr.	10.0 gm.
Glycin	¼ oz.	7.0 gm.
Sodium Carbonate	2¾ oz.	78.0 gm.
Potassium Bromide	35 gr.	2.3 gm.
Water to	32 oz.	1,000.0 cc.

For use dilute with 3 parts of water.

Sodium Sulfite	2 oz.	56.8 gm.
Hydrochinon	70 gr.	4.6 gm.
Glycin	35 gr.	2.4 gm.
Sodium Carbonate	1¼ oz.	35.0 gm.
Hypo	18 gr.	1.2 gm.
Potassium Bromide	18 gr.	1.2 gm.
Water to	32 oz.	1,000.0 cc.

Sodium Sulfite	1 oz.	28.4 gm.
Glycin	½ oz.	14.2 gm.
Potassium Carbonate	2⅝ oz.	74.4 gm.
Potassium Bromide	¼ oz.	7.0 gm.
Water to	32 oz.	1,000.0 cc.

◆

Acid Short Stop Bath

The use of an acid bath between the developer and the fixing bath is recommended, to neutralize the alkali of the developer. This prevents further development action, tending to eliminate yellow stains on the print and lengthen the life of the hypo.

Acetic Acid 28%	1½ oz.	50.0 cc.
Water to	32 oz.	1,000.0 cc.

◆

FIXING BATHS

The plain hypo fixing bath does not keep well, there being no acid to neutralize the alkali carried over from the developer. This objection is overcome when the bath is used in conjunction with an acid short stop bath. A simple acidified fixing bath is the following:

Hypo	5 oz.	250.0 gm.
Potassium Metabisulfite	½ oz.	25.0 gm.
Water to	20 oz.	1,000.0 cc.

Probably the most frequently used acid fixing bath is the following formula, which contains both an acidifier and a hardening agent. It is especially recommended for use in warm weather to prevent possible frilling of the emulsion.

Hypo	8 oz.	250.0 gm.
Water to	32 oz.	1,000.0 cc.

To this add two ounces of the following solution:

Sodium Sulfite	2 oz.	60.0 gm.
Acetic Acid 28%	6 oz.	180.0 cc.
Potassium Alum (Powdered)	2 oz.	60.0 gm.
Water to	10 oz.	320.0 cc.

The above stock solution hardener should be mixed in hot water and allowed to cool before adding to the hypo.

◆

Fixing Bath for Ronix Paper

Hypo	4 oz.	125.0 gm.
Water to	32 oz.	1,000.0 cc.

◆

Chrome Alum Fixing Bath for Plates and Films

Solution "A"

Hypo	2 lb.	907.0 gm.
Sodium Sulfite	3 oz.	85.0 gm.
Water to	96 oz.	3,000.0 cc.

Solution "B"

Potassium Chrome Alum	2 oz.	57.0 gm.
Sulfuric Acid C. P.	¼ oz.	7.0 gm.
Water to	32 oz.	1,000.0 cc.

For use, add "B" to "A" slowly—stir vigorously.

◆

TONERS
Hypo Alum Toning Bath

This method is used by studios making sepia prints on a commercial scale. The tones obtained range from a true sepia to a warm purplish brown.

Hypo	4¼ oz.	117.0 gm.
Hot water to	35 oz.	1,000.0 cc.

Dissolve and add

Alum	1¼ oz.	32.0 gm.

Allow the above solution to cool and add the following solution:

Silver Nitrate	8 gr.	.5 gm.
Common Table Salt	8 gr.	.5 gm.
Water to	2¼ oz.	70.0 cc.

The above mixture or solution should not be filtered. It works better as it becomes older and may be strengthened from time to time with the addition of a small quantity of fresh solution. The best results are obtained by keeping the solution hot, or as warm as the emulsion will stand—from 90° to 115° F. In this solution, prints will tone in from 30 to 60 minutes. When the desired tone is obtained, remove the prints from the toning bath and wipe with tufts of cotton dipped in water of approximately the same temperature as the toning bath, to eliminate any sediment or surface deposit on the prints; then wash for at least one-half hour in running water.

Note: Keep prints separated while toning. When removing prints from the hot toning bath, do not put them immediately into cold water, but step down the temperature of the water through several changes. If the toning bath becomes badly discolored, a new bath should be used. The temperatures of the toning bath and the length of time in which the print is immersed therein will determine to a great extent the tone of the finished print. The colder the toning bath, the slower the toning and the colder the tones; the hotter the toning bath, the more rapidly it works and the warmer the tones. Satisfactory results will be obtained at temperatures varying from 90° to 115° F.

Sulfide Re-Developer

After prints are thoroughly fixed and washed, they are bleached in a bleach bath made as follows:

Potassium Bromide	115 gr.	7.5 gm.
Potassium Ferricyanide	300 gr.	19.5 gm.
Water to	20 oz.	500.0 cc.

After which they are rinsed for not more than one minute in water. Longer washing may result in impure tones in the finished prints. Throw away the sulfide bath after a day's use—stale sulfide solution is the most frequent cause of bad tones or the failure of prints to darken in the sulfide bath. Redevelop or tone in a sulfide bath made as follows:

Sodium Sulfide	4 oz.	115.0 gm.
(NOT SULFITE)		
Water to	20 oz.	500.0 cc.

For use, take three parts of stock solution to 20 parts of water. Prints are placed in the sulfide bath, where they should darken to the full brown or sepia in a very short time, after which they are washed in running water for half an hour. Prints by the sulfide process of redevelopment are permanent, providing solutions are correctly made from pure chemicals and the instructions given above are carefully followed.

Note: If prints fail to bleach out to a faint brown color in the bleach bath within three minutes, it indicates that the bleach bath is becoming exhausted. Remedy: Make a new bath.

Liver of Sulfur Toner

The liver of sulfur toning bath is useful where only a few prints are to be made, or where it is found that the hypo-alum and re-developer are too troublesome.

Liver of Sulfur	30 gr.	2.0 gm.
Water to	40 oz.	1,000.0 cc.

Use at a temperature of 80° F.

The tones obtained in this bath are quite similar to those obtained with the hot hypo alum bath.

Toning Prints in Color

Before using toning baths it is essential that the prints be well washed in running water. It will be found that a given formula will not produce an identical tone on all papers; as a rule the commercial type bromide and chloride papers respond better to toning than do the more specialized emulsions. The majority of toning formulas exert a bleaching action in addition to creating the change of color, and for that reason prints to be toned should be made considerably deeper than normal. Although the following formulas have been found satisfactory for general use, there are many other interesting toners which will give successful results; and if the formulas given do not produce the exact tone desired, experiments may be made with other solutions.

Blue Toner

Solution "A"

Ferric Ammonium		
Citrate	85 gr.	5.6 gm.
Water to	2 oz.	56.7 cc.

Solution "B"

Potassium Ferricyanide	85 gr.	5.6 gm.
Water to	2 oz.	56.7 cc.

Solution "C"

Acetic Acid 28%	8 oz.	227.3 cc.
Water to	12 oz.	341.0 cc.

When ready to use, mix the three solutions together. Soak prints until the image becomes bright blue. After toning, wash in clear water until the high lights are clean.

Red Toner

Solution "A"

Potassium Citrate	1½ oz.	100.0 gm.
Water to	16 oz.	500.0 cc.

Solution "B"

Copper Sulfate	115 gr.	7.5 gm.
Water to	8 oz.	250.0 cc.

Solution "C"

Potassium Ferricyanide	100 gr.	6.5 gm.
Water to	8 oz.	250.0 cc.

Mix solution "B" into solution "A," then slowly add solution "C," stirring well. Remove prints from toning bath when desired tone is obtained and wash thoroughly.

Green Toner

Solution "A"

Oxalic Acid	120 gr.	7.8 gm.
Ferric Chloride	16 gr.	1.0 gm.
Ferric Oxalate	16 gr.	1.0 gm.
Water to	10 oz.	285.0 cc.

Solution "B"

Potassium Ferricyanide	32 gr.	2.0 gm.
Water to	10 oz.	285.0 cc.

Solution "C"

Hydrochloric Acid	1 oz.	28.4 cc.
Vanadium Chloride	32 gr.	2.0 gm.
Water to	10 oz.	285.0 cc.

In solution "C" the acid is first added to the water; the solution is then heated almost to the boiling point before the vanadium is added.

Mix solution "B" into "A," then introduce solution "C," stirring well.

Tone in the mixed solution until the prints are deep blue. Remove and place in wash water until tone changes to green.

If there is any yellowish stain in the whites it may be removed by immersion in the following solution:

Ammonium Sulfocyanide	25 gr.	1.6 gm.
Water to	10 oz.	285.0 cc.

Toning Lantern Slides

In toning latern slides the same procedure is followed as that used when toning papers, and it is just as important that the slides be well washed before and after toning. Many of the toning formulas employed with papers will give good results with slides, but the following are the two most commonly used in commercial lantern slide toning.

Blue Toner—Lantern Slides

Bichromate of Potash	4 oz.	.1 gm.
Ferric Alum	80 gr.	1.3 gm.
Oxalic Acid	180 gr.	3.0 gm.
Red Prussiate Potash	60 gr.	1.0 gm.

Powdered Alum	300 gr.	4.8 gm.
Muriatic Acid	100 mm.	1.4 cc.
Water to	1 gal.	1,000.0 cc.

Brown Toner—Lantern Slides

Uranium Nitrate	155 gr.	2.4 gm.
Oxalate Potash	150 gr.	2.5 gm.
Red Prussiate Potash	55 gr.	1.0 gm.
Ammonium Alum	340 gr.	5.5 gm.
Muriatic Acid	320 mm.	4.8 cc.
Water to	1 gal.	1,000.0 cc.

INTENSIFIERS

The following intensifier is used when it is desired to obtain the maximum increase of density and contrast.

First bleach in:

Bichloride of Mercury	120 gr.	8.0 gm.
Potassium Bromide	120 gr.	8.0 gm.
Water to	16 oz.	500.0 cc.

After a short rinse re-develop in

Sodium Bisulfite	¾ oz.	21.0 gm.
Hydrochinon	¾ oz.	21.0 gm.
Potassium Bromide	¾ oz.	21.0 gm.
Water to	32 oz.	1,000.0 cc.

If it is desired to soften the contrast after development, clear the re-developed image in the following solutions:

Ammonia	1 oz.	100.0 cc.
Water to	10 oz.	1,000.0 cc.

If an increase in contrast is desired, clear in this solution instead of the ammonia bath.

Hypo	1 oz.	100.0 gm.
Water to	10 oz.	1,000.0 cc.

For moderate intensification with proportionate increase of density and contrast, use the following:

Solution "A"

Potassium Bichromate	1 oz.	60.0 gm.
Water to	16 oz.	1,000.0 cc.

Solution "B"

Hydrochloric Acid	1 oz.	100.0 cc.
Water to	10 oz.	1,000.0 cc.

Use 10 parts of "A" to 1 part of "B' and 40 parts of water.

Keep the negative in this solution until bleached and then wash until the stain disappears. Then re-develop in a staining developer, such as the following:

Pyro	55 gr.	6.3 gm.
Sodium Carbonate	½ oz.	25.0 gm.
Potassium Bromide	10 gr.	1.0 gm.
Water to	32 oz.	1,000.0 cc.

Probably the most efficient intensifier is the Red Mercuric Iodide Formula which follows:

Red Mercuric Iodide Intensifier

Sodium Sulfite	750 gr.	100.0 gm.
Red Iodide of Mercury	75 gr.	10.0 gm.
Water to	16 oz.	1,000.0 cc.

If the red iodide of mercury is not available, mercury perchloride (corrosive sublimate) can be substituted—using the quantity given for the red iodide and adding double the quantity of potassium iodide.

This is fast working, and intensifying operations can be repeated several times on the same negative until utmost intensification is the result. Solution will keep well in a tightly corked bottle, and should be well shaken before use. Negatives must be well washed before using this intensifier, as the presence of hypo will not permit its proper action.

◆

REDUCERS
For Reducing the Contrast

This formula reduces the heavier densities to a greater degree than any of the other densities.

Ammonium Persulfate	750 gr.	50.0 gm.
Sodium Sulfite (Dry)	75 gr.	5.0 gm.
Sulfuric Acid	75 min.	4.2 cc.
Water to	16 oz.	500.0 cc.

For use, dilute with 9 parts of water. Agitate the solution. Wash rapidly, then place negative in 10% hypo solution acidified with a little potassium metabisulfite.

A plate reduced in this formula may later be intensified or reduced as desired.

◆

For Increasing the Contrast

Commonly known as "Farmer's Reducer," this is perhaps the most widely known of all photographic reducers. It is simple in composition and is extremely useful where it is desired to build up contrast, especially in the case of process emulsions.

Solution "A"

Hypo	1 oz.	28.5 gm.
Water to	16 oz.	500.0 cc.

Solution "B"

Potassium Ferricyanide	2 oz.	57.0 gm.
Water to	16 oz.	500.0 cc.

Use equal parts of "A" and "B"; after negative is reduced, rinse, and then place in hypo for 10 minutes.

◆

For Reducing the Density Proportionately

This formula affects all of the densities in proportion to their original values, increasing the transmission of light proportionately through all parts of the negative. It is a desirable formula to use where the negative is of proper contrast but is too heavy to print satisfactorily with a normal exposure.

Solution "A"

Potassium Permanganate	2 gr.	.1 gm.
Concentrated Sul-		

phuric Acid	12 min.	1.0 cc.
Water to	16 oz.	500.0 cc.

Solution "B"

Ammonium Sulpho-cyanide	190 gr.	12.3 gm.
Water to	16 oz.	500.0 cc.

Use equal parts of solutions "A" and "B."

The time in the reduction bath varies from one to three miutes, according to the effect desired. After sufficient reduction has been obtained, bathe in a 1% solution of potassium metabisulfite. After a short immersion in this bath, wash well.

◆

Cleaner for Photographic Trays

Potassium Bichromate	4 oz.	120.0 gm.
Sulphuric Acid	3 oz.	100.0 cc.
Water to	32 oz.	1,000.0 cc.

After cleaning with this solution, it is recommended that the tray be thoroughly washed and then wiped out with cotton or a clean rag.

◆

Removing Developer Stain from Hands

Potassium Permanganate	½ oz.	15.0 gm.
Water to	32 oz.	1,000.0 cc.

Place the hands in this solution until they take on a dark purple stain, then rinse the hands in water and place in the following solution:

Potassium Metabisulfite	8 oz.	250.0 gm.
Water to	32 oz.	1,000.0 cc.

Occasionally the photographer has a negative which is badly stained, due to the use of an old pyro developer or improper fixing. The easiest way in which the stain can be removed is to bleach the film and re-develop in an MQ formula.

◆

Photographic Bleach Solution

Solution "A"

Potassium Permanganate	75 gr.	5.0 gm.
Water to	32 oz.	1,000.0 cc.

Solution "B"

Sodium Chloride (Table Salt)	2½ oz.	70.0 gm.
Sulphuric Acid (Concentrated)	½ oz.	2.0 cc.
Water to	32 oz.	1,000.0 cc.

Use equal parts of "A" and "B."

After bleaching, rinse in a one per cent solution of sodium bisulfite and then re-develop in daylight, using any of the regular MQ formulas.

A simple formula which is particularly effective in the removal of dichroic fog is the thiocarbamid clearing bath.

Thiocarbamid	¾ oz.	20.0 gm.
Citric Acid	150 gr.	10.0 gm.
Water to	32 oz.	1,000.0 cc.

Before using the thiocarbamid, the negatives must be well washed.

◆

Acetate Film Cement

Ethyl Acetate	3½ oz.	100.0 cc.
Acetone	3½ oz.	100.0 cc.
Acetate Base	30 gr.	2.0 gm.
Acetic Acid	1 oz.	30.0 cc.

———

Blue Stains

With some varieties of panchromatic emulsions, a light blue stain will appear. While it is in no way objectionable as regards printing quality, it may be removed by bathing the negative in a weak solution of ammonia or a ten per cent solution of sodium sulfite.

◆

Desensitizing

The use of desensitizers has become quite popular with photographers in all branches of photography. The use of a desensitizer permits the inspection of a negative under a brighter safelight than could normally be employed. Desensitizer in the developer also helps to prevent fog, when the development is forced. If desired, the desensitizer can be used as a separate bath before the negative is placed in the developer.

Care should be taken to see that the bright safelight is not turned on until the development is well under way. Do not turn on the safelight as soon as the negative is placed in the developer. It will be found that the only developers to which the desensitizer cannot be added are the borax developers.

◆

Etching Filler

A filler for etched lines in metal to make them more distinctive has the following formula:

White Beeswax	10 gr.
French Chalk	5 gr.

Melt together.

———

Etching Steel

The following solution is used.

Nitric Acid	32 oz.
Hydrochloric Acid	3 oz.
Denatured Alcohol	16 oz.
Water	96 oz.

ETCH SOLUTIONS FOR LITHOGRAPHIC PLATES

Etches for Zinc Plates

Ammonium Nitrate	3	oz.
Ammonium Phosphate	3½	oz.
Calcium Chloride	¼	oz.
Hydrofluoric Acid	½	oz.
Gum Arabic Solution (Saturated)	80	oz.

◆

Phosphoric Acid	1 part
Gallic Acid	2 parts
Gum Arabic Soln.	8 parts
Water	14 parts

◆

Gum Arabic Solution	32 oz.
Ammonia Water (16%)	3 oz.
Phosphoric Acid	1 oz.
Hydrofluoric Acid	5 or 6 dr.

Pour each of the above ingredients into gum separately and stir continuously.

Keep 24 hours before using.

◆

Etches for Aluminum Plates

(a) Dissolve 2 oz. of Pulverized Ammonium Bichromate in 16 oz. water.

(b) Mix 1 oz. of the soln. resulting from ''A'' with the following:

(1) (20%) Phosphoric Acid	1 oz.
(2) Gum Arabic Soln.	8 oz.
(3) Water	8 oz.

(a) Sodium Phosphate	½ oz.
(b) Sodium Nitrate	½ oz.

Dissolve (a) and (b) in ½ gal. of hot water and add 1 oz. (80%) Phosphoric Acid. Use this etch without gum, spreading it evenly over the Plate, by means of a soft sponge or a brush made of camels or badgers hair.

———

(a) Phosphoric Acid (85%)	1 oz.
(b) Gum Arabic Soln.	32 to 40 oz.

———

(a) 1 gal. of chemically pure Nitric Acid with 7 gals. of water.

(b) Dissolve zinc to the point of saturation in this Nitric Acid solution.

(c) Take 1 oz. of resulting soln. and ½ oz. of gum arabic soln. and mix with a gallon of water.

———

2 oz. Bichromate of Ammonia.
1 pt. Gum Arabic Solution
1 tps. of the following:
2½ oz. Phosphoric Acid (85%)

into

84 oz. Gum Solution

Gum Solution is water saturated with gum arabic and filtered.

◆

Etches for Stone

Nitric Acid added to gum solution until action of acid is plainly visible when it is applied to the stone.

◆

Photo Engravers Collodion

Nitrocellulose (15–20 sec.)	3
Ether	48.5
Alcohol	48.5

Filter and bottle.

◆

Photographic Developing Fixer

Metol	5– 10
Hydroquinone	15– 20
Sodium Sulfite	50– 80
Sodium Carbonate (Anhyd.)	30– 40
Caustic Soda	20– 30
Potassium Bromide	5– 10
Sodium Hyposulfite	250–300
Ammonium Picrate	3– 5
Water	1000

◆

Photographic Film, Reclaiming

Forty kg. of discarded pieces of old film is washed for 15–20 min. in a soln. prepd. by heating to 70° 100 l. of water and 0.7 kg. of sodium hydroxide. The alkali-contg. gelatin and silver are drawn off and the celluloid is further washed with hot water before being used for other purposes. The alkali is returned to the washer for treating another 40 kg. of film. After the alkali has been used on 80 kg. of film it is boiled with steam in a wooden vessel and hydrochloric acid (d. 1.19) is added to complete coagulation of the gelatin. After settling, the supernatant liquid is removed, the ppt. is filtered and then ashed in a muffle at 500°–600°. The dried substance is mixed with sodium carbonate 1 : 3 and heated in a crucible until it is liquid. To eliminate the admixtures, silver is melted with potassium nitrate until the surface is mirror-bright. The pptd. silver sulfide from the fixing soln. is treated in the same way except that it is melted with iron. To regenerate fixing solution silver it is pptd. by sodium sulfide.

Photographic Negatives, Removing Water Spots from

For removing water-spot drying marks on negatives, bleach in the following soln.: potassium dichromate 1 g., water 100 cc., hydrochloric acid 2 cc. and re-develop with an elon-hydroquinone developer.

◆

Photographic Lens Cleaner

Water	3 oz.
Alcohol	1 oz.
Nitric acid	3 drops

After dusting the lens, rub with an old clean cotton cloth dipped in this solution and polish with a dry piece of the same cloth.

◆

Gelatin Film Cleaner

Alcohol	98–99
Diethylamine	2– 1

◆

Gelatin Films, Hardening

Gelatin or other protein layers are rendered very insoluble by treatment with the following:

Formaldehyde	100 cc.
Potassium Carbonate	100 gm.
Water	1000 cc.

◆

Paste Acids

(for etching, cleaning and soldering)

1. Activated Colloidal Clay	6 lb.
2. Water	70 lb.
3. Muriatic Acid	28 lb.

Allow 1 and 2 to stand over-night and then mix until smooth. If necessary use warm water. When cold add 3 slowly and stir until uniform.

◆

Electrotyping

The first step in the production of an electrotype consists in the preparation of an impression or "mold" in wax of the form to be reproduced. The molding wax usually consists of ozokerite to which various substances have been added to produce the desired physical properties. The molten wax is poured upon one side of a metallic plate, consisting of lead, copper, or aluminum. The wax-coated metal is termed a "case." After taking the impression of the form by the use of suitable pressure at a slightly elevated temperature, usually by means of a hydraulic press, the resultant "mold" is "trimmed" and "built up" to produce the desired degree of relief in the finished plate.

The mold is then coated with graphite, applied by a wet or a dry process, or both. After washing out the excess of graphite, the form is either introduced directly into the depositing bath, or, in some cases, is given a preliminary treatment (so-called "oxidizing") with copper sulfate solution and fine iron filings, whereby a thin film of copper is deposited by "immersion" upon the graphite. The baths are usually contained in lead-lined wooden tanks, with copper cross bars, from which the anodes and cathodes are suspended. Electrical connection to the graphited cathode surface is made by means of the suspending hook by either of two methods. In the one known as the "case connections," the hook is in direct contact with the metallic plate of the case, portions of the wax being removed in order to bring the metal and the graphite surface into electrical contact, while the back of the case is insulated with wax. In the method now more generally employed, and known as the "face connection," the hook is in contact with a small copper plate imbedded in the wax near the top of the form and in contact with the graphite surface. In the latter method the metallic plate itself is not in the circuit, and there is less tendency for copper to deposit upon any accidentally exposed portions of it.

After the copper is deposited to the desired thickness (usually 0.006 to 0.010 inch (0.15 to 0.25 mm.)) the case is taken from the bath, and the copper "shell" is loosened by means of hot water. After trimming the edges, the back of the shell is treated with soldering fluid (usually an acidified solution of zinc chloride) and coated with "tin foil" containing about 35 per cent of tin and 65 per cent of lead,

after which it is laid face downward upon a heated pan. After the tin foil is melted upon the back of the shells, molten electrotype metal (usually containing from 3 to 4 per cent each of tin and antimony and from 92 to 94 per cent of lead) is poured over them to the desired depth. The electrotypes thus produced are cleaned, cut, and trimmed to the desired size, "finished" to a plane surface and shaved to the proper thickness. They may be subsequently curved if desired.

In many cases, for the most perfect reproduction of halftone or other work in low relief, molding in thin sheet lead at high pressures is practiced. The lead mold thus produced is cleaned with alcohol to remove grease, and is then treated with a dilute solution of chromic acid or a chromate. This forms a thin film of lead chromate, which prevents the deposited metal shell from adhering too tenaciously. The subsequent steps are similar to those involved when wax molds are used.

For the better classes of work, especially color process halftones, or for plates requiring very severe service, nickel electrotypes (commonly called "steel" or "nickel steel") are frequently employed. In their preparation, a thin layer of nickel (usually about 0.001 inch or 0.025 mm.) is first deposited upon the wax or lead mold, copper is then deposited back of the nickel, and the resultant nickel-copper shell is treated as above. The true "nickel electrotype" thus made should not be confused with a nickel-plated electrotype in which nickel is deposited upon the surface of a finished copper electrotype.

During recent years a thin coating of chromium, usually about 0.0002 inch (0.005 mm.) has been often applied to nickel or copper electrotypes that are to be used for very long runs, for example in the printing of cartons and labels.

Blue for Drawings

Saturate 10 g. of oxalic acid in a little water with ferric hydroxide, filter off excess of ferric hydroxide, add concentrated solutions of 27 g. sodium oxalate and 11.6 g. sodium ferrocyanide, apply the mixture to paper with a brush and dry in a dark room. Develop the prints with dilute hydrochloric acid or sulphuric acid.

Waterproof Coating for Wooden Photographic Trays

Methyl Alcohol	500 cc.
Orange Shellac	100 g.
Rosin	25 g.
Venice Turpentine	25 g.

The ingredients are heated on a water-bath until completely dissolved.

CHAPTER XIII

PLATING

PLATING is resorted to to obtain a coating of a more desirable metal on the surface of another. This may be done to produce an enhanced appearance, protect it against corrosion, give it a harder surface or for other reasons. Good plating is dependent on divers conditions which are developed in this chapter. The preliminary and after treatments are as important as the plating itself.

Plating without Electricity

The following mixed powders are rubbed with a wet cloth onto the surface to be coated.

Nickel

Double Nickel Salts	60%
Magnesium Powder	3%
Powdered Chalk	30%

Tin

Stannous Chloride	15%
Ammonium Sulfate	15%
Magnesium Powder	3%
Powdered Chalk	67%

Zinc

Zinc Dust	45%
Ammonium Sulfate	15%
Magnesium Powder	3%
Powdered Chalk	37%

PREPARATION OF METALS FOR ELECTROPLATING

For the production of impervious adherent metal electrodeposits, the preparation of the articles for plating is of the greatest importance.

A. Polishing and Buffing

No general procedure can be given for all objects due to the large number of factors to be taken into account, such as composition of the object, shape, size, plate and surface finish desired, etc. The directions given here will be of a general nature, with some specific procedures for the common base metals iron and steel, and copper and brass. Treatises on the subject should be consulted for further information on these and other substances.

Naturally the smoothness and polish of the finished plate is greatly influenced by the same properties of the object before plating, particularly if the plate is thin, as is usually the case. Therefore, proper attention must be given to the operations of polishing and buffing the object before plating, and in some cases afterwards. The particular choice of cutting and finishing tools, abrasives, etc., is determined by the metal, the degree of finish on the final surface, etc.

For objects covered by a considerable amount of rust or millscale, sand-blasting or sand-rolling will greatly reduce the labor required for the final polishing. In sand-rolling the objects are rolled in steel barrels with abrasives such as sand, alundum, carborundum or emery mixed with water or oil. Where the number of objects is small a steel wire brush is best for removing coarse scale.

A certain amount of polishing should be used in all cases before plating, whether a high luster is desired or not. This is because the surface will be rendered more uniform, which will improve the quality of finish and corrosion resistance of the final plate. However, the polishing and subsequent treatments must be carefully studied and controlled in order not to weaken the surface layers with subsequent peeling after plating.

Under ordinary circumstances finishing is a two-step operation: "cutting down" to produce a smooth surface and "coloring" to produce a high final luster. It is often possible by proper choice of cloths, abrasives, speed of wheel, etc., to accomplish this with but two wheels, one for each step. However, in some cases more wheels are necessary for hard metals containing deep scratch marks, especially in the cutting-down step. Materials used for the wheels include muslin, flannel, felt, canvas, brushes of various kinds, leather and wood depending upon the nature of the material being polished, the coarseness of the abrasive, the finish desired, the preference of the polisher, etc. The abrasive composition is of much greater importance, since it is the medium doing the actual work. Excessive wheel wear means that the wheel is doing the work rather than the composition, and is due to improper choice or insufficient amount of composition. For efficiency the wheels are run at the maximum allowable speed. In some cases the limit is set by the material of or composition on the wheel, and in others by the material being buffed. Thus in the cutting-down step, where the abrasive is held on by glue a speed higher than 7,500 surface feet per minute will soften the glue and allow it to be torn from its setting on the wheel face. For soft metals on the other hand a speed this high generates enough heat to soften the metal and cause it to flow.

The first or cutting down step (often called simply polishing) is done by wheels faced with abrasive and glue. The abrasives used are either emery or artificial alumina, the latter being usually more desirable for most purposes. The glue should be the best quality hide glue; high viscosity, strength and flexibility being of prime importance. Application of the abrasive composition to the wheel is by rolling the wheel in a warm glue abrasive mixture and allowing to dry. If run at high speeds, polishing wheels should be faced with tallow to prevent burning.

The second or coloring step (often called simply buffing) is done by wheels faced with abrasive and grease. The abrasives used are of all kinds and grades, lime, silica, tripoli, emery, rouge, etc., being used. The melting point of the grease used will depend on the speed, a hard, high melting point grease being selected for buffing at high speeds. The grease should be of the saponifiable variety, because of the easier and quicker removal by alkaline cleaners.

For steel containing mill marks on which a high final luster is desired, the following combinations are suitable.

For very deep mill marks, two canvas wheels faced with glue and abrasive should be used. Suggested abrasive sizes are 120 and 220 mesh. These should be followed by one or two buffing steps on cloth wheels, depending upon the final finish desired.

In cases where the object is not deeply scratched to begin with, the following three-wheel combination offers advantages. One canvas wheel faced with glue and 180 mesh abrasive; one tampico brush wheel faced with fine emery paste; and one cloth wheel faced with chrome or steel rouge. The brush wheel offers the advantages of reaching backgrounds that cannot be reached with the usual polishing wheel, and of not requiring the glue-dressing step needed for the latter.

In going from one wheel to the next, the object should be rotated 90°, so that the new scratch marks are perpendicular to the old ones. The object must be kept on any one wheel until all the scratch marks of the previous step have been eradicated. If this takes an excessively long time, another wheel with an intermediate grade of abrasive should be used.

After polishing, the next step and the one of greatest importance is the cleaning of the article to be plated. The foreign materials likely to be present on metallic surfaces are of two classes: first, grease, dirt and organic substances; and second, oxides, scale, tarnish, and rust.

B. Removal of Grease

Grease of all kinds whether saponifiable or not can be removed by solution in organic solvents. In cases where the objects are heavily coated with grease, a cheap organic solvent such as gasoline, or better a non-inflammable one such as carbon tetrachloride or mixture containing it, should be used. However, this will not give complete cleansing, as the solvent on evaporation will leave a thin film of grease, making another operation such as dipping into fresh solvent necessary. The latter is obviated in a recently designed apparatus, where the articles are suspended in the vapor above a boiling apparatus. The condensing solvent washes them free of grease, and since it is being continually distilled, no second step is necessary. A non-inflammable solvent must be used in this case—trichloroethylene has met with considerable favor recently because it does not hydrolyze as readily as carbon tetrachloride in the presence of moisture.

The common method of removing grease is by emulsification with alkaline solutions, which should be used as hot as possible. The detergents used in these solutions are soap of all kinds, caustic soda and potash, soda ash, trisodium phosphate, sodium metasilicate, sodium cyanide, borax, sodium sesquicarbonate, sodium aluminate, etc., and all kinds of mixtures thereof. Sometimes finely divided insoluble substances such as silica, alumina, etc., are added. These are not fillers but help to clean either by scouring of the surface or by adsorption of the dirt. Each plater, seller of plating supplies, etc., has a particular composition and procedure that he swears by. Since the kind and degree of contamination of metallic surfaces vary considerably in different plating shops, naturally certain particular mixtures used in conjunction with a specific procedure will clean more quickly than others. However, probably any hot alkaline solution will work if given sufficient time. In general either soap with one builder (alkaline salt) or a mixture of two alkaline salts is used. The soap should be of a very soluble variety so as to be quick and free rinsing; fish oil soaps have been found very satisfactory. Soda ash has been used in the past as an alkaline soap builder because of its cheapness. Even today practically all commercial cleaners contain much soda ash. However, it is being gradually replaced by the more efficient detergents trisodium phosphate and sodium metasilicate. These seem to act more quickly not only because of higher alkalinity, but also due to specific emulsifying action. Caustic soda is used in many mixtures; it cleans not only by its emulsifying action, but also by saponifying the fats present on the metal. (Since any alkaline solution will have some saponifying action, the greases used in the manufacturing and polishing operations should be of the saponifiable variety.)

Electrolytic cleaning is frequent practice in plating shops. In this method an electric current is passed through the object, which is made one electrode in a hot alkaline solution. Usually the object is made the cathode, both because of the greater gas evolution (hydrogen) which gives a scouring action, and the higher free alkali concentration giving an increased cleaning action. Furthermore, cathode metals will not dissolve and some reduction of the oxides on the surface may take place. The voltage applied should be sufficient to produce a current density of 10 amp. per sq. ft. (1 amp. per sq. dm.) or greater. Any of the solutions used ordinarily for cleaning may be employed; the alkali or alkaline salt content should be high to give good conductivity. Cleaners containing suspended solids should be avoided, as solids are often occluded to an electrode during electrolysis. Iron bars or the containing tank may be used as anodes.

Special procedures must be used when the objects contain aluminum, zinc, tin or lead. For ordinary cleaning caustic soda or potash must be avoided as these substances will dissolve. In cathodic electrolytic cleaning these will dissolve to some extent in any case whether caustic is added or not, due to the formation of free alkali at the cathode. Sometimes small amounts of the zinc, tin or lead may be redeposited from such cleaners, giving a film which will cause subsequent peeling of the electrodeposit. In such cases the object should be made the anode for short time, either in the same or in a separate bath. An alternative procedure is to use anodic cleaning. The mechanism of anodic cleaning is quite different from that of cathodic. In the latter, as stated above, the action is due to the bubbles of gas and the increased alkali concentration. However, with anodic cleaning the action is largely due to the etching (solution) of the surface. Since the impurities are on the surface only, they will thus drop off. Anodic cleaning is often used for brass and copper. Zinc should not be cleaned anodically as it is attacked so rapidly the surface blackens due to the finely divided metal formed.

A simple cleaning bath base may be made of the following:

Soda Ash

(60 gm. per l.) 8 oz. per gal.

or

Washing Soda

(165 gm. per l.) 22 oz. per gal.

Trisodium Phosphate

(120 gm. per l.) 16 oz. per gal.

or

Sodium Metasilicate

(30 gm. per l.) 4 oz. per gal.

To this should be added 1–2 oz. per gallon of soap and 1–2 oz. per gallon of caustic soda. If used electrolytically, most or all of the soap should be eliminated—0.1 oz. per gallon is sufficient.

For large scale production a double system will be found desirable. The greater part of the grease by solvent dip or by a strong hot soap solution; and then the object put into the electrolytic cleanser. Usually 3–4 minutes of the electrolytic cleaning is sufficient. When clean there should be a continuous film of water left on the object. Rinse thoroughly before proceeding with the pickling.

C. Removal of Oxides and Tarnish

Oxides, scale and tarnish are usually removed by solution in a suitable reagent, the process being usually called pickling. For iron and steel, sulfuric or hydrochloric acid is used; and for copper and brass sulfuric and nitric acids.

If the copper or brass is polished and clean, a short immersion in a "bright dip," composed of 425 ml. conc. sulfuric acid and 75 ml. conc. nitric acid in 500 ml. water is sufficient. For brass with appreciable amounts of oxide scales, a preliminary "scaling dip" in a solution composed of 375 ml. conc. sulfuric acid and 75 ml. conc. nitric acid in 550 ml. water should be used. The brass is dulled by the latter process and should subsequently be immersed in a bright dip.

For large scale treatment of iron and steel, sulfuric acid should be used because of its cheapness. The proper concentration is about 10% by weight (1 part conc. sulfuric acid by volume to 16 of water). For smaller jobs hydrochloric acid is to be preferred because of its more rapid action. The concentration should be 7% by weight (5 parts commercial hydrochloric acid by volume to 32 of water). The time taken will depend naturally upon the amount of scale present and will vary from several minutes to an hour. These acids act not only by actual solution of the oxide, but also by attack of the metal with evolution of gas, which helps detach the scale. For objects with imbedded sand (from castings or sand blasting) hydrofluoric acid should be added to 4% by weight (1 part commercial hydrofluoric acid by volume to 16 of water). This will dissolve the silica.

After pickling thoroughly rinse the object and immerse immediately in the plating bath with the current on. The latter precaution is particularly important for acid plating baths to avoid partial solution of the metal before the current starts to flow. The exposure to the air of the prepared object should be a minimum, because the surface is unusually clean and particularly susceptible to oxidation.

D. Combination Procedures and Special Processes

In many cases some of these cleaning procedures can be combined or shortened. Thus if the metal has been highly buffer, the pickling step can be omitted. The oxides have been removed during buffing, and further oxidation prevented by the grease of the buffing composition. This grease may be removed either by solvent treatment or alkaline cleaning. Often a single solvent dip alone is satisfactory if the object is to be chromium plated, because the strongly oxidizing chromic acid bath will oxidize the traces of grease remaining. However, in some cases unsuccessful adhesion of the deposit occurs with this simplified treatment. This may be due to the presence of absorbed matter which is not removed by the solvent. In such cases the alkaline cleansers may yield better results, or a light scrubbing of the surface with Vienna lime may help.

In preparing highly polished brass for plating, the pickling step may be dispensed with by the addition of sodium cyanide to the alkaline cleansing bath. This will dissolve the traces of oxides and tarnish present. Cyanides should not be used for copper, as a film is formed which is very difficult to wash off.

The pickling step induces the following detrimental factors when used on iron and steel:

(1) Formation of surface carbon preventing adhesion of the plate.

(2) Formation of hydrogen on the surface, which is occluded and adsorbed preventing adherence and causing brittleness. The factors have caused the failure of plates (especially nickel) often in the past. The remedy found in recent years (Madsenell process—patented) is degasification. After pickling the metal to be plated is made the anode on a 12-volt circuit in concentrated sulfuric acid at room temperature. Usually a lead cathode is used. The current starts at about 5 amp. per sq. dm. and subsides over a period of from 30 seconds to 10 minutes to practically zero, when evolution of gas ceases. By this process the occluded and adsorbed gases and embedded oils and greases are removed. Although a passive film of metal is probably formed, this does not seem to be detrimental to the adhesion of the plate.

An alternative method is to use solutions of dichromates or chromic acid; old chromium plating baths serve admirably.

◆

Chromium Plating

1

Chromium is deposited from

Chromic Acid   22% solution
Chromium Sulfate   .5% solution

At. 35° C. onto a graphite rod with a current density of 50 amps. per square decimeter. Only 10 amps. are necessary for a chromium cathode.

2

Chromic Acid   245 gms./liter
Chromium Sulfate   3 gms./liter
Anodes—Two chromium metal rods
Cathodes—Iron sheets
Current density—125 amps. per sq. ft.
Voltage—2-3. Temp. 15° C.
Time—2 hr.

◆

Nickel Plating—Still Tanks

Nickel solution:

Nickel Ammonium Sul-
    fate   8 oz. per gal.
Nickel Sulfate   4 oz. per gal.
Boric Acid   2 oz. per gal.

Ph. value of above solution is kept at 5.8; nickel content, should be 3½ oz. nickel per gal. Tanks used at room temperature. Additions for nickel are made by adding double nickel salt according to analysis shown. Practice is about 5 lb. every ten days. Nickel anodes should be 99 plus, and maximum copper content .30%. Amperage and voltage is limited to type of work, usually about 25 amperes and 6 volts for one hour.

◆

Machine Nickel Plating Solution

Nickel Sulfate   4 oz. per gal.
Nickel Ammonium
    Sulfate   12 oz. per gal.
Magnesium Sulfate   2 oz per gal.
Boric Acid   3 oz. per gal.

◆

Black Nickel Finish

Formula

Nickel Ammonium Sulfate   8 oz.
Sodium Sulfocyanate   2 oz.
Zinc Sulfate   1 oz.
Water   1 gal.

Procedure for Plating

Work is strung on racks.
Hung on mild alkaline solution to remove grease.
Wash in water.
All above work is done in the dip room. The following work is finished in the buff room plating department.
Bright dip work is washed in mild alkaline solution again before going through the following operations.
Buffed parts to be plated are first dipped and brushed with gasoline and dried in sawdust, after which they are dipped and brushed with mild alkaline solution.

Wash in water.
Dip in cyanide solution.
Washed with water.
Plate in black nickel solution.
Wash in cold water.
Wash in hot water.
Bright dipped parts are dried in sawdust. Buffed parts are dried in hot box.

◆

Cadmium Plating

Formula:

Sodium Cyanide   9 oz.
Cadmium Oxide   3 oz.
Sodium Hydroxide   2 oz.
Water   1 gal.

Use at room temperature using 8 to 10 amperes per sq. ft.

Procedure for Plating:

Very greasy work is washed in gasoline and dried in sawdust.
Wash and brush in mild alkaline solution.
Wash in water.
Dip in Muriatic Acid.
Wash in water.
Wash and brush in mild alkaline solution.
Wash in water.
Dip in Cyanide.
Wash in water.
Plate in cadmium solution from 20 minutes to 1½ hours depending on type of work and quantity of cadmium desired.
Wash in cold water.
Wash in hot water.
Dry in sawdust or hot box whichever the type of work requires.
Some work is rubbed with steel wool to brighten the metal finish.

Silver Plating

Silver bath formula:

Silver Cyanide	3½ oz.
Sodium Cyanide	5 oz.
Water	1 gal.

Silver strike formula:

Silver Cyanide	½ oz.
Sodium Cyanide	8 oz.
Water	1 gal.

Procedure for Plating:

Wash and brush in mild alkaline solution.

Wash in water.

Dip in Cyanide solution.

Wash in water.

Flash in silver strike at 6 volts.

Plate in silver bath for 30 minutes at 2 volts.

Wash in cold water.

Wash in hot water.

Dry in hot box.

------◆------

Silver Plating

Formula for silver solution

1. | | |
   |---|---|
   | Silver Cyanide | 3½ oz. |
   | Sodium Cyanide | 5 oz. |
   | Ammonium Chloride | ½ oz. |
   | Water | 1 gal. |

2. | | |
   |---|---|
   | Silver Chloride | 3½ oz. |
   | Sodium Cyanide | 8 oz. |
   | Ammonium Chloride | ½ oz. |
   | Water | 1 gal. |

Either of the two solutions will give good results if operated at a temperature of 75° F. with a cathode current density of 4 or 5 amperes per sq. ft.; ¾ to 1 volt.

Solution 1 is generally used, but No. 2 is whiter.

Silver strike:

Silver Cyanide	½ oz.
Sodium Cyanide	8 oz.
Water	1 gal.

Use steel or carbon anodes; 6 volts.

Blue dip:

Bichloride of Mercury	1 oz.
Sodium Cyanide	6 oz.
Ammonium Chloride	1 oz.
Water	1 gal.

Brightener for silver solution:

Silver Solution	1 qt.
Sodium Cyanide	8 oz.
Carbon Bisulfide	1 oz.
Ether	1 oz.

To prepare the brightener place the carbon bisulfide and ether in a quart bottle and shake thoroughly. Dissolve the cyanide in the silver solution and fill bottle. Shake bottle from time to time until the carbon bisulfide is thoroughly dissolved and then filter.

One ounce of this stock solution should be sufficient for an addition to each 15 gallons of the regular plating solution. Care must be taken to avoid an excess or else the deposit will be rough and patchy. If an excess has been added, remove by raising the temperature of the solution to 140° F.

Silver strip solutions:

1. | | |
   |---|---|
   | Sodium Cyanide | 12 oz. |
   | Caustic Soda | 2 oz. |
   | Water | 1 gal. |

Reverse current with cold rolled steel as cathodes. Voltage 6 to 8. Agitate the work for a cleaner job.

2. | | |
   |---|---|
   | Sulfuric Acid | 5 gal. |
   | Nitric Acid | 1 gal. |

Place crock that contains the strip in a hot water container. If all water is kept from the strip, brass or copper work will be attacked but very slightly.

------◆------

Silver Plating Powder

Silver Nitrate	20
Ammonium Chloride	10
Sodium Bisulfate	40
Water	40
Potassium Carbonate to make a paste	

Keep in dark bottles.

------◆------

Gold Plating

1. Cyanide solution:

Metallic Gold as Fulminate or Cyanide	5 dwt.
Sodium Cyanide	2 oz.
Phosphate Soda	1 oz.
Water	1 gal.

Temperature 130 to 160° F.; 1 volt; 24 kt. gold anodes.

2. Chloride solution:

Gold Chloride	6 oz.
Hydrochloric Acid	10 oz.
Water	1 gal.

Room temperature; 2 to 3 volts.

In preparing the solution dissolve the gold chloride in dilute hydrochloric acid before adding it to the solution. The amount of free hydrochloric acid that the solution contains does not seem to make a great deal of difference in the operation of the bath, but it does have a decided effect upon anode. The greater the amount of free acid the faster the anode dissolves.

This solution is used where heavy deposits of gold are desired. The work is plated in the cyanide bath for a few minutes before placing in the acid bath.

3. Immersion gold solution:

Fulminate of Gold	4 dwt.
Yellow Prussiate Potash	12 oz.
Carbonate Soda	24 oz.
Caustic Soda	¼ oz.
Water	1 gal.

Solution should be boiled in a cast iron tank for an hour and allowed to cool to 180° F. before using.

If color is too light, it may be darkened by adding a very small amount of copper carbonate which has been taken up with yellow prussiate of potash.

4. Salt Water gold:

Yellow Prussiate of Potash	64 oz.
Sodium Phosphate	32 oz.
Sodium Carbonate	16 oz.
Sodium Sulfite	8 oz.
Gold as Fulminate	12 dwt.
Water	4 gal.

Boil for an hour and add to solution as required.

Solution is boiled for one hour, then diluted with water to make four gallons of solution. The solution is placed in a porous pot which is put in a tank that contains a saturated solution of sodium chloride heated to 190° F.

The porous pot is surrounded with a cylinder of zinc which is provided with a rest rod, on which the work to be plated is suspended in the gold solution.

The advantage of this type of solution over the cyanide solution is that a more uniform color may be obtained, although the deposit is not as rapid as with the cyanide solution, unless used with outside current. This is accomplished by connecting the zinc cylinder with the positive lead from the generator and the work rod with the negative lead. The amount of voltage is regulated with the class of work being done. If the work is wired or racked, 1 to 2 volts is sufficient. If basket work is being done, 5 to 6 volts give good results.

The solution is replenished from a stock solution:

Yellow Prussiate of Potash	16 oz.
Sodium Phosphate	8 oz.
Sodium Carbonate	4 oz.
Sodium Sulfite	2 oz.
Gold as Fulminate	1 oz.
Water	1 gal.

Green gold:

Metallic Gold as Fulminate or Cyanide	4 dwt.
Silver Cyanide	¼ dwt.
Sodium Cyanide	2 oz.
Water	1 gal.

Temperature 105° F.; 2 volts; 18 kt. green gold anodes.

Dark or antique green gold solutions are produced by adding to the green gold solution a small quantity of lead carbonate that has been dissolved in caustic soda, and increasing voltage to 5 or 6. Agitation of the work produces best results.

White Gold

White gold and other karat gold solutions are best prepared by running the gold into solution with the porous pot method. This consists of making a cyanide solution of four ounces to a gallon of water which is to be the plating solution. Connect up tank for plating in the usual way. Place anodes on anode rod and on cathode rod suspend a porous pot which contains a fairly strong solution of sodium cyanide, 4 to 6 oz. per gallon. Into the porous pot suspend a sheet of copper, or better still a copper rod formed into a coil, and operate solution until the desired amount of gold has been dissolved from the anode. This can be readily determined by weighing the anode from time to time.

Rose gold solution:

Yellow Prussiate of Potash	4 oz.
Potassium Carbonate	4 oz.
Sodium Cyanide	¼ oz.
Gold as Fulminate	10 dwt.
Water	1 gal.

Temperature 175° F.; 6 volts. If a red color is desired, add small quantity of copper carbonate.

Cheap rose gold finish:

The work which must be brass is placed in the following dip until a smut is produced:

Copper Sulfate	16 oz.
Muriatic Acid	½ gal.
Water	1 gal.

Dissolve the copper sulfate in the water and then add the acid. The work should have a deep red smut which should be lightened somewhat by placing in a saturated salt solution for a few seconds. Plate in the regular fine gold solution, then relieve the high lights with bicarbonate of soda, replate in gold solution for a few seconds, dry and lacquer.

To remove fire scale after soldering on solid and karat gold, the work is pickled in a dip composed of: sulfuric acid 12 ounces, sodium bichromate 4 ounces, water 1 gallon; used hot.

It is then made the anode in the following solution:

Yellow Prussiate of Potash	2 oz.
Sodium Cyanide	8 oz.
Rochelle Salts	2 oz.
Water	1 gal.

Temperature 150° F. to 175° F., 6 volts, and lead cathodes.

Iron Plating

Formula for iron solution:

Ferrous Chloride	40 oz.
Calcium Chloride	20 oz.
Water	1 gal.

Temp. 200° F.; current density 40 to 50 amp. per sq. ft.; 2 to 2½ volts; pH 1.5 to 2. Pure iron anodes.

This bath is used to produce heavy deposits of iron.

For thin deposits of iron use the following:

Dissolve 16 ounces of ammonium chloride in each gallon of water. Connect up tank, same as for plating, using cold rolled iron for anodes. On the cathode rod suspend some old plating racks or other work, and work solution with highest current density obtainable. After four or five hours of working the solution, there will be enough iron dissolved from the anodes and the solution will produce a deposit of iron. Operate solution at 80° F.; 1.5 to 2 amperes per sq. ft.; 1 volt.

Lead Plating

Formula for lead solution:

Lead Carbonate	20 oz.
Hydrofluoric Acid (50%)	32 oz.
Boric Acid	14 oz.
Glue	.025 oz.

To prepare the solution, place the hydrofluoric acid in a lead-lined tank and add the boric acid with constant stirring. When the boric acid is completely dissolved, the solution is allowed to stand until cool when the lead carbonate is added in the form of a paste with water. The solution is allowed to settle when the clear solution is siphoned off and placed in the plating tank. The solution is then diluted to the proper volume with water and the glue added by dissolving the same in warm water. Mechanical agitation of the solution is essential.

A cathode current density of 10 to 20 amperes per sq. ft., 3 to 4 volts, and lead anodes are employed.

For thin deposits of lead, use the following:

Carbonate of Lead	2 oz.
Caustic Soda	6 oz.
Water	1 gal.

Lead anodes. Temperature 175° F.; 3 to 4 volts.

Brass and Bronze Plating

Formula for brass solution:

Copper Cyanide	4 oz.
Zinc Cyanide	1 oz.
Sodium Cyanide	6 oz.
Sodium Carbonate	2 oz.
Water	1 gal.

Temperature 90° F. Cathode current density 2.5 to 3 amperes per sq. ft.; 2 to 3 volts. Use rolled anodes, 80% copper, 20% zinc.

This solution will produce a good yellow deposit. If a green deposit is desired, for instance, such as is used for a flash deposit, in the novelty trade, previous to gold plating, use 1 ounce less of each, copper cyanide and sodium cyanide, and a small quantity of ammonium hydroxide.

As temperature plays a very important part in controlling a uniform deposit, it is advisable to have the tank equipped with a steam coil for proper regulation.

In operating a brass solution, it is well to keep in mind that a high current density tends to produce a deposit that is high in zinc; also, that the addition of ammonia or caustic soda to a brass solution has the same effect.

Bronze solution:

"Bronze plate" (really a high-copper brass deposit) is generally produced in an alkaline solution, one similar to a brass solution, but with a higher copper content.

Copper Cyanide	4 oz.
Zinc Cyanide	½ oz.
Sodium Cyanide	5 oz.
Sodium Carbonate	2 oz.
Rochelle Salts	2 oz.
Water	1 gal.

Temperature 95° F. Cathode current density, 2 to 2.5 amperes per sq. ft.; 2

to 3 volts. Rolled bronze anodes, 90% copper, 10% zinc.

Temperature always plays a very important part in the control of this solution, so the tank should be equipped with a steam coil to keep the temperature constant. When rochelle salts are added to a bronze solution, better anode corrosion is obtained, and therefore, a more uniform deposit.

In replenishing the metal content of a brass or bronze solution, it is not advisable to make a stock from copper cyanide, zinc cyanide and sodium cyanide, as it would be impossible to control the constituents in their proper proportion to produce a uniform color in the deposit. A separate stock solution of the zinc salt and copper salt is recommended. They should be prepared by dissolving equal parts of copper cyanide and sodium cyanide, and zinc cyanide and sodium cyanide in water and placed in separate containers until wanted for use.

It is a known fact that when a zinc salt is added to a brass or bronze solution (and especially the latter), it takes considerable time before a uniform color of the deposit is obtained. This is probably due to the difference in potentials at which the two metals are deposited. It is by the formation of the double cyanides that it is possible to deposit these two metals from the same solution in different proportions.

Remarks on Brass and Bronze Solutions:

Rochelle salts, when added to a brass or bronze solution, have the property of dissolving the oxides that form on the anodes, thereby permitting a more uniform deposit. One to two ounces per gallon is to be recommended.

It should be remembered that the factors that tend to make the zinc predominate in the deposits are a high zinc content, high current density, low free cyanide content, decrease in temperature, and the addition of ammonia or caustic soda to the bath.

When arsenic is added to a brass solution to produce a bright deposit, care should be used to avoid an excess as a light colored deposit will be the result. To prepare the arsenic stock solution, take two pounds of caustic soda and dissolve same into two quarts of cold water. Then add one pound of white arsenic and when all has been dissolved, dilute to one gallon. One ounce of this stock solution is enough to add to each 100 gallons of solution. It is impossible to bright dip

a piece of work that has been plated in a brass solution that contains an excess of arsenic. Arsenic should never be added to a bronze solution; neither should ammonium salts be added.

The free cyanide of a bronze solution is usually less than that of a brass bath. The color desired should be regulated by the proportion of the copper and zinc salts used and the temperature at which the bath is operated.

◆

Bronzes, Restoration of Ancient

This article is made the cathode in 2% caustic soda solution, and a weak current is passed for some hours, a sheet-iron anode being used. In this way the incrustation is reduced again to metallic copper and the outer layers of dirt and loose sponge copper are then readily removed by gentle brushing, this leaving a clean surface which usually shows all the original surface details. Malignant patina is due to the presence of copper oxychloride in the corrosion products; the above electrolytic process effectively eliminates the patina, especially when the malignant salts impregnate the mass of the bronze. Another method which gives satisfactory results is to brush the parts affected with diluted silver sulfate solution, which converts the chlorides into insoluble silver chloride after being dried with blotting-paper, the surface is brushed with barium hydroxide solution, which is allowed to dry, leaving a white powder, which is readily brushed away.

◆

Brass Plating on Steel
(for rubber adhesion)

Copper Cyanide	4 oz.
Zinc Cyanide	1 oz.
Sodium Cyanide	6 oz.
Carbonate of Soda	2 oz.
Water	1 gal.

Temperature 80° F. to 85° F.

Cathode current density, 2.5 to 3 amps. per square foot. Rolled anodes should be used consisting of 80% Copper and 20% Zinc.

The work must be perfectly clean and it is necessary to maintain a regulated temperature and current density.

◆

Copper Plating

There are two types of solutions that are used for the deposition of copper, namely, the acid (sulfate) and the alkaline (cyanide) baths. Their use is dependent upon the class of work to be plated and the finish desired.

The cyanide solution is always used for depositing copper upon the ferrous metals, so as to prevent the deposition of copper by immersion which would be the result of the use of the acid bath on this class of work. There are two formulae for the cyanide solution, either of which will give satisfactory deposits—carbonate or cyanide.

Cyanide copper solutions:

Copper Cyanide	3½	oz.
Sodium Cyanide	4½	oz.
Carbonate of Soda	2	oz.
Hyposulfite of Soda	1/32	oz.
Water	1	gal.
Copper Carbonate	5	oz.
Sodium Cyanide	10	oz.
Hyposulfite of Soda	1/32	oz.
Water	1	gal.

Either solution should be operated at 100° F. to 110° F. Cathode current density 4 to 6 amperes per sq. ft., 1½ to 2 volts. Use rolled copper anodes. The free cyanide content of the bath should not be allowed to rise too high or else gassing will be produced at the cathode causing a blistered deposit. Enough cyanide should be used to keep the anodes fairly clean from the formation of basic copper salts, but not enough to prevent the dark discoloration which is produced by the use of the hyposulfite of soda. This discoloration usually disappears when the current is off for a few hours.

If the cyanide solution is operated at room temperature, a higher free cyanide content is necessary than at 110° F. With a metal content of approximately 2.50 oz. of metallic copper per gallon and operated at room temperature, a free cyanide content of 1 to 1.25 oz. per gallon will produce good results. If operated at 110° F. use a free cyanide content of .50 to .75 oz. per gallon.

Pitted deposits of copper are caused when the carbonate content becomes too high. When this occurs the carbonates may be precipitated from the solution by the addition of barium chloride. The precipitated carbonates are allowed to settle, the solution siphoned off, the carbonates removed from the tank, the solution is then replaced in the tank which is filled with water to proper solution level when the solution is ready for use. It is not advisable to remove all of the carbonates, for without any carbonates a hard deposit will be produced.

Acid copper solution:

Copper Sulfate	28 oz.
Sulfuric Acid	3 to 5 fl. oz.
Water	1 gal.

Temperature 75° F. Cathode current density for still solution 10 to 15 amperes per sq. ft.; ¾ to 1 volt. Agitation of the cathode or of the solution allows the use of higher current density. Use rolled copper anodes.

———◆———

Remarks on Copper Solutions

Bright deposits of copper from the cyanide solution may be obtained by adding to the bath lead carbonate which has been dissolved in a caustic soda solution. Agitation of the cathode is also necessary. The deposit from newly prepared cyanide solutions is usually hard and at times blistered. The addition of one or two ounces per gallon of caustic soda helps to overcome this condition.

Oxidized finishes are hard to produce uniformly from a cyanide solution that contains hyposulfite of soda.

More uniform bronze finishes are produced from an acid copper deposit. An excess of sulfuric acid in the acid solution produces a deposit that is hard and streaky; so will an excessive current density. The higher the sulfuric acid, the greater the conductivity of the bath.

A high acid content is indicated by the formation of copper sulfate crystals, especially when the temperature of the bath is below normal.

Coppering by immersion:

Copper Sulfate	1 to 2 oz.
Sulfuric Acid	½ to 1 oz.
Water	1 gal.

Where only a very thin film of copper is desired, the above solution will give good results. The work is free from grease by the usual cleansing methods and then immersed in the solution just long enough to become coated with copper. Rinse thoroughly in clean cold water and dry in sawdust.

———◆———

Metalizing Non-Metallic Articles

Plastics, bone, etc., are washed with naphtha to remove grease; dried and soaked in 3–4% aqueous quinol; then immersed in a solution of silver nitrate. Silver is deposited which may be polished. Other metals may be then plated thereon.

Zinc Plating

The two types of zinc solutions that are in common use are the acid and alkaline solutions. The acid solution is usually preferred when cost is considered, as it can be made more cheaply, but the throwing power of this solution is lower than that of the cyanide bath.

Formula for acid zinc solution:

Zinc Sulfate	32 oz.
Ammonium Chloride	2 oz.
Sodium Acetate	2 oz.
Water	1 gal.

Temperature 80° F. Cathode current density, 15 to 20 amperes per sq. ft.; 3 to 4 volts.

Formula for cyanide zinc solution:

Zinc Cyanide	4 oz.
Sodium Cyanide	4 oz.
Caustic Soda	3 oz.
Water	1 gal.

Temperature 100° F. Cathode current density 10 to 25 amperes per sq. ft.; 2 to 3 volts.

Use pure zinc anodes in both solutions. Corn sugar may be used in the proportion of one ounce per gallon in either solution to obtain a finer structure of deposit.

Remarks on Zinc Solutions

The throwing power of the acid zinc solution is quite poor. The addition of one ounce of stannous chloride to a 100 gallon solution will improve the throwing power. An excess should be avoided, as it has a tendency to discolor the deposit. The pH is the most important factor to control in the acid solution. A pH of 3.5 to 4.5 using thymol blue as an indicator is about right. This should be maintained by adding the required sulfuric acid.

In the cyanide bath, the free cyanide is the most important factor to control. If the free cyanide is equal to the metal content best results will be had. An excess of free cyanides causes a bright, rough deposit.

Care should be used in drying zinc deposit to prevent stains. A thorough rinsing in clean cold water followed by hot water and hardwood sawdust is good procedure.

Plating Baths

Basic recipes for still solutions have been developed for the guidance of the plater. However, the proportions of the constituents should be changed according to special requirements for individual needs. The following procedure is recommended for making up new solutions or replenishing old baths:

Fill the tank with one-third the amount of water required. Dissolve the Sodium Cyanide in this water, which should be at a temperature of about 50° C. (120° F.). Then add the Metal Cyanide and stir until it is in solution. Finally add the balance of the ingredients and mix in the remaining two-thirds of water.

Spotting, Prevention of Plating

After plating and rinsing, dry in an oven at a temperature of 400 to 450° F. for several hours, then perform the final finishing operations. Still another method that has been used with some success is to rinse the work in a solution of 2 ounces of cream of tartar to the gallon of water, letting it remain in this rinse for 10 to 15 minutes, and then drying it after passing through cold and hot water rinses several times.

Immersion Tin—Caustic Soda Method

This method is used to tin by immersion, small brass or copper articles.

Formula for Immersion Tin:

Caustic Soda	12 oz.
Stannous Chloride	4 oz.
Sodium Chloride	1 oz.
Water	1 gal.

The solution is placed in an iron tank which is heated with a steam coil. The bottom of the tank is covered with moss tin over which is placed an iron wire screen.

The work to be tinned is bright dipped or tumbled clean, placed in brass wire baskets and separated with sheets of perforated tin, placed in the solution at boiling temperature for 15 to 30 minutes, or until completely covered with tin. It is rinsed thoroughly in clean cold water and dried with the aid of hot water and sawdust.

The brightness may be increased somewhat by tumbling for a few minutes in hardwood sawdust.

Moss tin is prepared by melting the tin and pouring same into cold water at a slight elevation.

Electric Cleaner

Mild Alkaline Solution 8 oz. per gal. This solution is used with an E. M. F. of 6 to 12 volts, on work requiring exceptionally clean surface. It can be augmented by addition of stronger detergents but care must be used to prevent staining of colored work. Use at 200° F.

Removing Fire Scale

To remove the fire scale from sterling silver use:

Nitric Acid	2 parts
Water	1 part

Use hot and agitate work.

Remove fire scale by reverse current with:

Sodium Cyanide	8 oz.
Water	1 gal.

Use hot and agitate work. Lead anodes; 4–6 v.

Bright dip:

Sulfuric Acid	2 gal.
Nitric Acid	1 gal.
Water	1 qt.

One ounce of muriatic acid for five gallons of above.

It is necessary to add water only when a new bright dip is made. Dip must be operated cold.

Matt dip:

Sulfuric Acid	1 gal.
Nitric Acid	1 gal.
Zinc Oxide	2 lb.

Operate hot and keep all water and chlorides from dip.

If the matt is coarse, add sulfuric; if too fine, nitric.

Burn Off Dip

If the work has been annealed, the fire scale should be removed in a hot sulfuric acid solution, 1 part acid, 3 parts water, rinsed in water and then placed in what is known as the ''burn off'' dip, made by using 2 parts of sulfuric acid, 1 part of nitric acid, and 5 parts of water.

The work is left in the ''burn off'' dip for five to twenty seconds, then rinsed in water, and bright dipped. If not bright enough, repeat the ''burn off''' and bright dip.

Anodic Treatment of Aluminum

The aluminum or aluminum alloy is made the anode in a chromic or sulphuric acid solution, and 10-100 amperes per square foot is passed through for 10-20 minutes.

Formula No. 1

The chromic acid solution contains 5-15% chromic acid. The current density for this bath varies from 10 amperes per square foot to 100 amperes per square foot. The temperature of this bath is important and should be kept between 90-100° F.

Fumes of chromic acid develop as the process continues. A ventilating system should be in operation at all times as the fumes are injurious.

No. 2

The sulphuric acid method consists of anodizing the aluminum or its alloy in a solution containing 5-60% sulphuric acid by volume. The current density varies from 10 to 25 amperes per square foot.

The temperature control is not as important as in the chromic acid solution.

Sulphuric acid spray is released during the process, and for this reason the bath should have a ventilating system applied to it.

After the work has been removed from the solution, it is essential to wash with water until all traces of sulphuric acid or chromic acid has been removed. For this purpose two rinses in running water for 10 minutes each will suffice.

Non-Electric Nickel Plating Compound

Formula No. 1

Nickel Ammonium Phosphate	5 oz.
Nickel Sulphate	3 oz.
Cream of Tartar	2 oz.
Tin Chloride	2 oz.
Ammonium Chloride	1 oz.
Codium Chloride	1 oz.
Copper Powder	2 oz.
Chalk Powder (Whiting or Precipitated Carbonate)	4-5 oz.
Water	until pasty

CHAPTER XIV

POLISHES, ABRASIVES, ETC.

POLISHES are principally used for restoring the original lustre and finish of a surface. Most polishes are built on an oil or wax base, with or without the addition of water. When the latter is present an emulsifying agent is included so as to form a non-separating emulsion. When the latter is not properly made separation results which necessitates shaking before use (when the products is liquid).

Wax polishes give more durable finishes than oil polishes. But the rubbing necessary for their proper application has limited their use by the general public. Until recently the wax type of polish consisted of various waxes "dissolved" in solvents such as turpentine or naphtha. The most recent development in this field is the "bright-drying" type of water wax.

All polishes have their limitations and no one polish is suitable for all conditions.

Abrasives are composed of hard substances of mineral or synthetic origin fixed on or suspended in various media. The hardest abrasive used is the diamond, followed by carborundum, corundum, quartz, garnet and felspar in order of hardness.

Sand- and emery-paper are made of special sands and garnet of uniform sizes bonded to paper or cloth. Abrasive wheels are moulded from special abrasions bonded with "Bakelite" or other synthetic resins. For special purposes abrasives are suspended in oil or fatty emulsions, greases, etc., which act as lubricants and carriers for the cutting or polishing particles.

Auto Polish

1

Paraffin Oil	5	gal.
Linseed Oil Raw	2	gal.
China Wood Oil	½	gal.
Benzol 90%	1	qt.
Kerosene	1	qt.
Odor to suit.		

Mix oils together. Mix Benzol and Kerosene, then add to oils and stir thoroughly.

2

Fuller's Earth	4	oz.
China Clay	3	oz.
Kerosene	1¼	pt.
Mineral Oil	1¼	pt.
Turkey Red Oil	1	qt.

Ammonia Water (10%)	4	oz.
Water	2½	pt.
Formaldehyde (40%)	4	oz.
Glycerin	½	pt.

3

Carnauba Wax	9	lb.
Beeswax	4	lb.
Ceresin Wax	4	lb.
Naphtha	75	lb.
Stearic Acid	7	lb.
Triethanolamine	2.5	lb.
Water	75	lb.
Abrasive	25 to 60	lb.

Preparation

Add the Triethanolamine and stearic acid to the water, heat to 100° C. and stir to obtain a smooth soap solution.

Then melt the waxes in the naphtha and, when the solution is about 85° to 90° C., add it to the hot soap solution. Stir vigorously until a smooth emulsion is obtained and then slowly until cold. If any separation occurs shortly after the emulsion has cooled, stir vigorously until the emulsion is creamy.

The method of adding the abrasive is dependent upon the type of abrasive used. An oil-absorbing abrasive should be well mixed with the hot oil solution before it is added to the soap solution, but an abrasive that absorbs water is best stirred into the finished emulsion. The latter type, like Bentonite, to the extent of 25 pounds, produces a paste with the above emulsion, while 60 pounds of the former, as Tripoli, makes a liquid polish.

4

A.	Carnauba Wax	30 lb.
	Glyco Wax	20 lb.
	Naphtha or Varnolene	68 lb.
	Turpentine	17 lb.
B.	Water	70 lb.
	Borax	10 lb.

Melt "A" together but do not heat above the boiling point of water. Meanwhile dissolve "B" while heating to a boil.

Run "A" into "B" *slowly* while stirring *vigorously*.

5

A good formula for a cleanser and polisher is:

Yellow Wax	20.0
Commercial Silica, Very Finely Powdered	40.0
Turpentine Substitute	40.0
Soft Soap	1.0
Water	5.0

Melt the wax and incorporate the powder, slowly adding the turpentine substitute, finally stir in the soap, previously dissolved in the water. Some may prefer it to be without the soap, but experience shows it to be worth its slight softening effect in yielding a higher and better polish. The paste may be tinted with ferric oxide.

Another formula is as follows:

6

Kieselguhr (Levigated)	11 parts
Silica (Levigated)	9 parts
Yellow Ochre	1 part
Red Ochre	1/10 part

Kerosene	16 parts
Soft Paraffin	2 parts
Powdered Soap	1 part

The following formula is suitable for polishing fabric bodies:

7

Oleic Acid	80.0
Liquid Paraffin	250.0
Potassium Hydroxide	16.0
Tragacanth	6.0
Water	to 1,000.0

Mix the oleic acid with the paraffin and slowly add the potassium hydroxide, previously dissolved in 200.0 of water. Soak the tragacanth in 500 cc. of water until fully absorbed, then heat to boiling, and when cool stir into the above emulsion.

Once a good surface has been produced by the above it is not an advantage to use too frequently, as frictional powders are bound to show the effect sooner or later if unwisely used. A thin film of wax once deposited on paintwork of the highly polished variety is best kept in condition by a hard wax polish. Beeswax is too soft, and the best for the purpose is Carnauba wax. This, however, is intractable and likely to crumble; it needs rubbing up with the cloth in order to soften it before applying. A modification enabling the polish to be easily applied and which does not modify in any way its polishing and surfacing effect is made as follows:

8

Grey Carnauba Wax	25.0
Japan Wax	5.0
Rosin	5.0
Melt and stir in	
Turpentine Substitute	60.0
Strain and add	
Solution of potash (1%)	5.0

This last addition has been found to give just sufficient saponification to prevent the paste crumbling. The preparation gives a highly polished hard surface, and where dirt and grease are not present its direct application forms a perfect protection of enamelled paintwork which can easily be kept clean with a dry cloth.

9

Carnauba Wax	20
Beeswax	30
Japan Wax	30
Paraffin Wax	60
Turpentine	326

10
Automobile and Floor Polish
(Wax Paste Type)—(Rubbing Type)

Yellow Beeswax	6 lb.
Ceraflux	16 lb.
Carnauba Wax	27 lb.
Montan Wax	8 lb.
Naphtha or Varnolene	89 lb.
Turpentine	10 lb.
Pine Oil	3 lb.

Melt together and pour into cans. Do not disturb until solidified. This makes an excellent auto polish of great durability and luster. Variations can be made to suit individual requirements.

11

Floss Powder	8	parts
Paraffin Oil	8	parts
Methylated Spirits	2	parts
Glycerin	2	parts
Gum Tragacanth	⅛	part
Water	40	parts

Furniture or Auto Polish
1

Light Mineral Oil	1 gal.
Powd. Carnauba Wax	2½ oz.

Heat until wax is dissolved.

2

1. Blendene	10 parts by vol.
Spindle Oil	60 parts by vol.
2. Water	40 parts by vol.

Stir (1) with a high speed mixer. Add (2), stir five minutes. Blendene will give clear soluble oils with mineral oils, depending on grade, from two to six times its volume. The cruder the mineral oil, the higher percentage of oil will mix clear with Blendene. They emulsify readily on stirring in water.

3

Nelgin	8 lb.
Water	126 lb.

Allow the above to soak a few hours, stir and then add the following mixture to it slowly with good stirring.

Light Mineral (Spindle) Oil	26 lb.
Blown Castor Oil	18 lb.
Varnolene or Solvent Naphtha	16 lb.
Lemenone Crude	16 lb.

This polish works exceptionally well on lacquered, painted or varnished metal surfaces.

4

Carnauba Wax	30 lb.
Beeswax	15 lb.

Ceresin Wax	15 lb.
Turpentine	26 lb.
Naphtha	24 lb.
Stearic Acid	8 lb.
Triethanolamine	4 lb.
Water	65 lb.

Liquid Wax
5

Carnauba Wax	10	lb.
Beeswax	4	lb.
Ceresin Wax	4	lb.
Naphtha	80	lb.
Stearic Acid	8	lb.
Triethanolamine	4.5	lb.
Water	200	lb.

Preparation

Melt the waxes and stearic acid and add the triethanolamine. Temperature should be about 90° C. Add the naphtha slowly so that a clear solution is maintained. Using a water or steam jacketed kettle prevents overheating and also caking of the waxes on the sides of the container. Add the boiling water to the naphtha solution and stir vigorously until a good emulsion is obtained and then slowly until the emulsion is cold.

6

(Packages in glass only. No tin cans.)
200 gals.

Turpentine	8 gal.
Naphtha	30 gal.
Spindle Oil (Light)	49 gal.
Acetic Acid 36%	6 gal.
Water	100 gal.
Antimony Chloride	4 gal.
Gum Arabic	10 lb.
Gum Tragacanth	10 lb.
Perfume	1 gal.

Make up with water to 200 gallons and run through colloid mill.

7

Yellow Ceresin	3 lb.
Japan Wax	1 lb.
Beeswax	2 lb.
Linseed Oil Raw	4 gal.
Turpentine	1 gal.
Paraffin Oil 28° gr.	1 gal.
Water	7 gal.
Carbonate of Potash	3 oz.
Soap Chips (Animal Fat Soap)	1 lb.

Mix the above thoroughly.

8

Carnauba Wax Bleached	6
Japan Wax	3½
Paraffin Wax	1½

Turpentine	12
White Curd Soap	3
Rosin Pale	2
Water	30
Clovel	Trace

9

A. Carnauba Wax	60
Turpentine	60
Stearic Acid	2
B. Trihydroxethylamine	
Stearate	12
Water	62

Heat (A) and (B) in separate vessels to 200° F. and run (B) into (A) slowly with vigorous stirring. Stop when homogeneous.

10

Carnauba Wax	6
Paraffin Wax	9
Ceresin	2
Naphtha	43
Turpentine	4
Stearic Acid	1
Trihydroxyethylamine	
Stearate	4.5
Water	130

Procedure—as above.

10

| Pale Paraffin Oil | 3 parts by vol. |
| Benzol | 2 parts by vol. |

This polish is being used by one of the largest furniture houses in America. The benzol softens the surface permitting the oil to leave a thin film on surface.

11

Furniture Gloss Oils

These are essentially emulsions of oil and gum in water. A little glycerin aids the ease of application.

Water	10	parts
Nut Oil	1	part
Mineral Oil	1	part
Acetic Acid	⅛	part
Gum Arabic	11	parts

"Dry Bright" Polish

Carnauba Wax	13.2	lb.
Oleic Acid	1.5	lb.
Triethanolamine	2.1	lb.
Borax	1.0	lb.
Water	108	lb.
Shellac	2.2	lb.
Ammonia (28%)	0.32	lb.

Melt the wax and add the oleic acid. The temperature should not be above 90° C. Using a hot water or steam jacketed kettle maintains a good temperature and prevents wax caking along the sides of the container. Add the triethanolamine slowly, stirring constantly. The solution should be *clear* at this point. Dissolve the borax in about a pint of boiling water and add to the wax solution to obtain a clear jelly-like mass. Stir for about 5 minutes. Add 92 pounds water, previously heated to boiling temperature, slowly with constant stirring. An opaque solution should be obtained. Cool. Add 16 pounds of water to the shellac and then the ammonia and heat until the shellac is in solution. Cool. Add this to the above wax solution and stir well to obtain an even mixture.

Floor Wax, "Rubless"

| Hydromalin | 138 lb. |
| Carnauba Wax No. 2 | 250 lb. |

Heat to 120–140° C. half hour. Cool to 100–105° C.

Add to the above slowly with stirring.

| Water (in 2 or more portions) | 1780 lb. |

Heated to 100° C. Keep as close to 100° C. as possible for 15 minutes.

This formula can stand additional water if a lower cost product is desired. The more water added, however, the lower the gloss will be.

When properly made this gives a "bright-drying" water wax of highest durability. If allowed to dry over-night it becomes water resistant. Its gloss is excellent and ordinary wear improves it. If a higher gloss is desired it may be gotten by light polishing with a soft dry mop.

Floor Polish

| Carnauba Wax | 30 |
| Rosin | 6 |

Heat above to 140° C., cool to 100° C. and add following with vigorous stirring which has been heated to 95–100° C.

Soap Flakes	10
Turpentine	1
Water	270

Wax Floor Polishes

1

Carnauba Wax	15 parts
Paraffin	26 parts
Ceresine	32 parts
Benzine	170–180 parts

Color to suit with any oil soluble color.

2

Carnauba Wax	60 parts
Paraffin	104 parts
Ceresin	128 parts
Turpentine	600 parts
Naphtha	100 parts

Dance Floor Wax

Ceresin	44
Stearic Acid	12
Scale Wax	140
Carnauba Wax	4
Oil Soluble Color	to suit

Prepared Floor Wax

| Carnauba Wax (grade 3) | 5 lb. |
| Yellow Ceresin | 3 lb. |

Melt together, then stir in

| Turpentine | 2 gal. |
| Kerosene | ½ gal. |

Keep warm while mixing. *Caution:* fumes are very inflammable! Pour into tins. Allow to stand undisturbed overnight.

Non-Slippery Rubless Floor Polish

| Carnauba Wax No. 1 or 2 | 500 lb. |
| Hydromalin | 276 lb. |

Heat with stirring for ½ hour to 120–140° C. Cool to 100° C. and add slowly with vigorous mixing

| Water (Boiling) | 3560 lb. |

Stir until uniform; allow to stand overnight and add slowly while stirring

| Sodium Silicate | 80 lb. |

Wood Polish

Carnauba Wax	33 parts
Beeswax	66 parts
Dipentene	75 parts
Turpentine or White Spirit	225 parts
Soap	1 part
Water	10 parts

The soap is dissolved in water (hot) and the waxes are dissolved in the dipentene. When cool the solutions are mixed with vigorous shaking or stirring.

Glass Polish

1

1. Ammonium Linoleate Paste	20
2. Orthodichlor Benzol	100
3. Water	200
4. Infusorial Earth	60

Dissolve (1) in (3) overnight and run (2) in while beating with high speed mixer. Then beat (4) in until uniform.

2

Precipitated Chalk	50
Kieselguhr	20
White Bole	30

Make into a slurry with water for use.

3

Whiting	54
Silica "Smoke"	18
Starch	15
Cream of Tartar	11
Magnesium Oxide	10
Infusorial Earth	2

For use make into a cream with water or benzine.

Leather Polish

Carnauba Wax	11 lb.
Turpentine	16 lb.
Stearic Acid	3 lb.
Oil Soluble Nigrosine	2 lb.
Triethanolamine	1 lb.
Water	66 lb.
Water Soluble Nigrosine	1 lb.

Preparation

Dissolve the water soluble Nigrosine in the water, add the Triethanolamine and stearic acid and heat to boiling. Stir until a smooth soap solution is obtained. In a separate container, melt the carnauba wax in the turpentine and add the oil soluble Nigrosine. When this solution has reached a temperature of 85–90° C., add it to the soap solution. Stir vigorously to obtain a good dispersion of the wax and then stir slowly until the emulsion is cold.

Leather Belt Polish

A polish for unfinished edges of leather belting is composed of the following:

| Water | 1 gal. |
| Gum Tragacanth | 2 oz. |

Bismarck Brown Solution—in amount to obtain desired color.

Leather Dressing

Tallow	70
Petroleum Jelly	3.5
Diglycol Stearate	13
Beeswax	9
Rosin	2
Water	2

Military Leather Paste Polish

Carnauba Wax	18
Candelilla Wax	2
Japan Wax	10
Paraffin Wax	2
Turpentine	20

Linoleum Polish

Carnauba Wax	1 lb.
Paraffin Wax	1 oz.
Yellow Wax	7 oz.
Turpentine	1 gal.

Metal Polish

1

Naphtha	62	lb.
Oleic Acid	1	lb.
Abrasive	7	lb.
Triethanolamine	0.33	lb.
Ammonia (26°)	1	lb.
Water	128	lb.

Preparation

In one container mix together the naphtha and oleic acid to a clear solution. Dissolve the Triethanolamine in water separately, stir in the abrasive, if it is of a clay type, and then add the naphtha solution. Stir the resulting mixture at a high speed until a uniform creamy emulsion results. Then add the ammonia and mix well, but do not agitate as vigorously as before.

2

Tank A

Dissolve thirteen (13) pounds of Oxalic Acid in forty (40) gallons of water. Heat to not more than 80° C. Add twelve (12) pounds of 26° Bé. Ammonia.

Tank B

Mix twenty-five (25) pounds of Red Oil with twenty-five (25) pounds of Denatured Alcohol. Add twelve (12) pounds of 26° Bé. Ammonia, to be warmed slightly to affect saponification.

Add contents of Tank A to Tank B while mixing. This can be done successfully in the cold, also with varying degrees of heat, but the mixture should not be too hot.

While adding Tank A to Tank B, Schultz Silica should be added slowly and the whole mixture stirred gently. The amount of Silica to be added ranges from 100 to 200 pounds to above proportions. 200 pounds are necessary if you desire a thicker and creamier polish. The above proportions produce approximately sixty to sixty-five gallons of polish.

3

Palm Oil	20	lb.
Yellow Petrolatum	8	lb.
Paraffin Wax	4	lb.
Crocus "B"	12½	lb.
Silex Double Ground	12½	lb.
English Rottenstone Powd.	6	lb.
Bright Red Iron Oxide Powd.	2	lb.
Oxalic Acid	10	oz.
Clovel	8	oz.

Melt the first three items and when clear, while heat is on, add other items slowly while stirring until free from lumps; raise temperature, continuing stirring and run into cans.

4

1. Ortho Dichlorbenzol	5
Naphtha or Mineral Spirits	20
Pine Oil	4
2. Trihydroxyethylamine Linoleate	2
Tripoli or Silex	50–75
Suspendite	9
Water	260
3. Ammonium Hydroxide	12

Add "1" to "2" with stirring and then stir in "3"; allow to stand overnight and stir before packaging.

This gives a polish which does not separate if made properly. If a thicker polish or paste is desired the Tripoli is increased and the liquids decreased.

5

A. Ammonia 16°	12½	gal.
Alcohol	100	oz.
Oleic Acid	100	oz.
B. Oxalic Acid	10	lb.
Water	15	gal.
Ammonia 26°	4¼	gal.

For polish use

A	2½	gal.
B	1¼	gal.
Water	35¼	gal.
Air Floated Silex	97	lb.

Mix and run through colloid mill.

Silver Polish

1

1. Infusorial Earth	48	lb.
2. Diglycol Stearate	7	lb.
3. Soda Ash	1	lb.
4. Trisodium Phosphate	1	lb.
5. Water	70	lb.
6. Clovel	½	lb.

Heat 2 and 5 to 150° F. and stir until homogeneous. Add the other ingredients and mix to a smooth paste.

2

Castile Soap	10 parts
Water	50 parts
Tripoli Powder	10 parts
White Rouge	5 parts
French Chalk	15 parts
Petroleum	5 parts

3

Water	1 qt.
Soap Flakes	4 oz.
Whiting	8 oz.
Ammonia	½ oz.

Aluminum Polish

1. Sapinone	1
2. Water	52
3. Oleic Acid	8
4. Ammonium Hydroxide	5
5. Alcohol	4
6. Infusorial Earth	20
7. Red Iron Oxide	8

Mix (2) and (3) and stir until uniform. Mix (6) and (7) and rub into a paste with part of (1), (2) and (5). Slowly add the balance and while mixing vigorously add mixture of (2) and (3).

Brass Polish

1

Tripoli	1 lb.
Whiting	1 lb.
Prepared Chalk	1 lb.
Stearin	1 lb.
Gasoline	1 gal.
Oleic Acid	8 oz.

Dissolve the stearin in the gasoline, add the oleic acid and then stir in the powders, using care to keep them from forming in lumps. More or less stearin may be used to give any desired body, and the gasoline may be replaced in whole or in part with kerosene.

2

Petroleum Spirits	30 parts
Ammonia	4 parts
Olein	10 parts
Tripoli Powder	50 parts
Methylated Spirits	10 parts
Water	20 parts

Liquid Stove Polish

Crude Montan Wax	2
Rosin	1
Carnauba Wax	2

Heat to 90° C. with stirring and to it add slowly

Caustic Potash	2
Water (Boiling)	86
Nigrosine	3

Keep on heat and agitate vigorously until uniform. Cool and work in

Graphite Flake	5
Lampblack	3

Mix thoroughly until uniform.

Liquid Polishing Wax

Beeswax	5
Ceresin	20

Melt together and cool to 65° C. Stir in slowly

Turpentine	85
Pine Oil	2.5

Polishing Rouge

Double Pressed Saponified

Stearic Acid	50 parts
Edible Tallow	25 parts
Camphor	3 parts
Paraffin Wax	2 parts
Fine Iron Oxide	20 parts

Jewelry Polish Powder

Marble Dust	90%
Jeweler's Rouge	10%

Polishing Cloth

Crude Oleic Acid	1 lb.
Stearic Acid	½ oz.
Vaseline	1 oz.

Melt together, remove from fire and add cassia oil or methyl salicylate, or terpineol, ½ oz. Cut good weight canton flannel into desired size (½ × 1 yd.), dip in the mixture till thoroughly saturated, then run through a tight wringer. Fold and wrap in oiled paper.

Tripoli Polishing Composition

Stearic Acid	55 lb.
Edible Tallow	2 lb.
Oleo Stearine	5 lb.
Rosin	9 lb.
Petrolatum	40 lb.
Japan Wax	1 lb.
Flint	315 lb.

| Tripoli Flour, Double Ground | 93 lb. |
| Ponolith | 2 lb. |

Tripoli Buffing Stick

Double Pressed Saponified	
Stearic Acid	30 parts
Edible Tallow	25 parts
Paraffin Wax	25 parts
Tripoli Flour	20 parts

(or as much as will be absorbed)

A buffing or polishing paste may be made using the above formulae with the addition of a small amount of turpentine and of water to bring to the consistency desired.

Grease Stick for Buffing and Polishing Purposes

Single Pressed Saponified	
Stearic Acid	25 parts
Edible Tallow	70 parts
Paraffin Wax	5 parts

Buffing Nickel Polish

Double Pressed Saponified	
Stearic Acid	86 lb.
Paraffin	16 lb.
Edible Tallow	10 lb.
Japan Wax	3 lb.
Silex	376 lb.

Oil Polish

Mineral Oil	60 lb.
Naphtha	26 lb.
Turpentine	3 lb.
Stearic Acid	9 lb.
Triethanolamine	4 lb.
Methanol	4 lb.
Water	120 lb.

Preparation

Mix together the mineral oil, naphtha and turpentine and add the stearic acid. Heat the mixture to about 60° C. at which time the acid will dissolve to give a clear solution.

In a separate container mix the Triethanolamine, methanol and water and heat likewise to 60° C. Then add to this the first mixture and stir vigorously until the emulsion is smooth. Continue with gentle stirring until cool.

Properties

An oil polish of this type can be used both for furniture and automobiles. It can be rubbed dry to leave a glossy finish on the varnish or lacquer surface. Such a polish is more easily applied than a wax polish but it does not leave the same hard and permanent film.

Variations

The cleaning action of this polish can be increased with a slight alteration in formula; namely by the substitution of part of the mineral oil with kerosene or naphtha. Pine oil may also be substituted for the turpentine, or other solvent changes made. When this polish is to be used for lacquers, a fine abrasive is frequently added in small quantity.

Uses

Furniture and automobile polishes.

Floor Oil
1

Mineral Oil	92
Turpentine	5
Beeswax	1
Shellac Wax	2

Dissolve waxes in mineral oil heated to 100° C.; cool and stir in turpentine.

Floor Oil, Low Priced
2

Light Mineral Oil	5	gal.
Automobile Engine Oil	½	gal.
Paraffin Wax	2	lb.
Clovel	½	pt.

Dust-Cloth Fluid

Light Mineral Oil	3	gal.
Corn Oil	1	gal.
Clovel	3	oz.
Oil Soluble Yellow Color	to suit	

Emery Paste

Double Pressed Saponified	
Stearic Acid	17 lb.
Oleo Stearine	2 lb.
Petrolatum	38 lb.
Japan Wax	3 lb.
Paraffin	26 lb.
Emery	300 lb.
Flint	100 lb.

Emery Grease

Double Pressed Saponified	
Stearic Acid	11 lb.
Edible Tallow	1 lb.
Paraffin	3 lb.
Petrolatum	1 lb.

Razor Strops, Abrasive for

Bauxite	42
Lard	42
Powd. Emery	15
Varnish	1

Carborundum Suspension

Diglycol Stearate	4
Water	100

Heat to 60° C. and stir after turning off heat. Add with stirring

Carborundum Powder	4

Crocus Composition

Double Pressed Saponified	
Stearic Acid	11 lb.
Petrolatum	11 lb.
Edible Tallow	2 lb.
Crocus	165 lb.
Flint	23 lb.

Vienna Lime Composition

Double Pressed Saponified		
Stearic Acid	45	lb.
Edible Tallow	15	lb.
Vienna Lime	200	lb.
Ponolith	2½	lb.

White Shoe Cleaner

a.	Titanox C	30 g.
b.	Diglycol Laurate	6 cc.
	Varnolene	10 cc.
	Toluol	12 cc.

Mix a and b thoroughly.

c. Bright Drying Carnauba

Wax Emulsion	60 cc.
Water	20 cc.

Add c to ab in 4 equal portions, shaking or stirring during and after each addition.

d. Trichloroethylene	40 cc.

Add slowly with stirring.

White Shoe Dressing

Titanium White	60 g.
Diglycol Oleate	12 g.
Naphtha	20 g.

Stir the above together and while stirring vigorously add slowly

Carnauba Wax Emulsion (10% Wax)	80 g.

then stir in vigorously

Trichloroethylene	60-100 g.

Polishing Cloths

Prepare powder mixtures:
Formula No. 1

Calcium Carbonate	70 g.
Kieselguhr	25 g.
Caput Mortuum	5 g.

No. 2

Magnesia, Calcined	20 g.
English Red	40 g.
Vienna Lime	40 g.

No. 3

Calcium Carbonate	40 g.
Bolus	20 g.
Vienna Lime	20 g.
Infusorial Earth	10 g.
Magnesia Usta	5 g.

One hundred and fifty grams of these mixtures are stirred into 1000 cc. of water. Stir to keep in suspension while impregnating cloth.

Ski Finishes

For running on wet snow.
Mix:

Pine Tar	25 g.
Copal Lacquer	25 g.
Venice Turpentine	50 g.

This mixture is boiled in on the running side of the ski with a blowtorch. Before using the ski rub in a thin coating of Venice turpentine.

For running on very cold snow burn in a good coating of Pine tar and before using heat ski and rub on some spermaceti.

Shoe Dye Liquid

Carnauba Wax	2	g.
Montan Wax, Bleached	2	g.
Paraffin (50-52° C.)	4	g.
Ozokerite, Refined	1	g.
Dyestuff	1.5	g.
Thinner (Turpentine)	89.5	cc.

CHAPTER XV

RUBBER, PLASTICS, WAXES, ETC.

THE manufacture of rubber goods requires expensive, heavy machinery so that most of the formulae of this chapter are only of academic interest. These, however, give a clear idea of the composition of many types of rubber articles.

Plastics may be roughly described as that group of materials which may be molded under suitable conditions of heat and temperature to give finished products which are comparatively inert and unaffected by heat. Typical of this class is "Bakelite" and similar products. Thermoplastics differ in that they can be re-molded or deformed by heat e.g. "Celluloid."

Rubber Goods

A single rubber product may be compounded with any number of mixtures, combining various grades of rubber, reinforcing agents, pigments and vulcanizing agents. For most items, a number of different compounds will serve with equal satisfaction. All of the possible combinations cannot be included here, but the following compounds are representative and can be readily adapted to commercial factory production by slight modifications to suit specific conditions. Adjustments as to curing conditions, temperature, or time of cure may be desirable depending on prevailing factory conditions. The curing data given for the various compounds is not intended to be specific and may be modified as desired.

Rubber Goods, Non-Sticking

Sprinkling with talc prevents rubber goods and sheets from sticking.

Rubber Hospital Sheeting

Pale Crepe	100
Petrolatum	1.00
Zinc Oxide	10
Lithopone	75
Whiting	63
Color	as desired
Monex	0.50
Sulfur	2.00

Cure—In air—60 minutes, rise to 245° F. and hold 60 minutes.

Rubber Clothing

Pale Crepe	100
Plastogen	6.00
Stearic Acid	1.00
Zinc Oxide	5.00
Dixie Clay	40.00
Kalite—No. 1	40.00
Captax	1.00
Zimate	0.10
Sulfur	1.50

Cure—60 minutes rise to 260° F. and 30 to 60 minutes at 260° F.

White Rubber Tiling

Pale Crepe	15.00
Paraffin	0.3125
Whiting	50.00
Ti-Tone	25.00
Zinc Oxide	6.50
Magnesium Carbonate	1.50
10% Thionex Master Batch	0.625
Anti-Scorch-T	0.0625
Sulfur	1.00

Cure—11 to 12 minutes at 40 lb. steam.

White Rubber Tubing

Pale Crepe	100
Petrolatum	7.50
Agerite Gel	1.00
Zinc Oxide	15.00
Lithopone	130.00
Dixie Clay	40.00
Kalite No. 1	200.00

Altax	1.25
Sulfur	3.00

Cure—In talc 30 minutes at 20 lb.

Red Moulded Rubber Tube

Smoked Sheets	97.75
Medium Process Oil	1.50
Stearic Acid	1.25
Blanc Fixe	40.00
Zinc Oxide	5.00
Du Pont Rubber Orange 2R	.75
10% Thionex Master Batch	2.50
Sulfur	1.75

Cure—5 minutes at 292° F.

Passenger Car Inner Tube

Pale Crepe	50.00
Smoked Sheets	50.00
Plastogen	4.00
Stearic Acid	.50
Agerite Powder	1.00
Kalite No. 1	50.00
Zinc Oxide	5.00
Tuads	.10
Altax	.50
Captax	.50
Sulfur	1.00

Cure—3 minutes at 55 lb.

High Grade Rubber Hose Tube

Smoked Sheets	14.00
Amber Crepe	10.00
Whole Tire Reclaim	20.00
Petrolatum	2.00
Paraffin	0.50
Stearic Acid	0.25
Neozone D	0.375
Whiting	20.00
Soft Clay	20.25
Carbon Black	7.25
Zinc Oxide	3.00
Litharge	0.125
10% Thionex Master Batch	1.000
Sulfur	1.250

Cure—15 minutes at 274° F.

Rubber Fire Hose

Pale Crepe	23
Smoked Sheets	23
Zinc Oxide	32
Whiting—Precipitated	10
Litharge	10
Sulfur	2.00

Cure—45 minutes at 274° F. in steam.

Transparent Rubber

Pale Crepe	100.00
Plastogen	5.00
Rodo No. 10	0.10
Stearic Acid	1.00
Zinc Carbonate	2.00
Zimate	0.25
Captax	0.50
Sulfur	1.50

Cure—Approximately 15 minutes at 15 lb.

Tire Carcass

Pale Crepe	50.00
Smoked Sheets	50.00
Plastogen	4.00
Stearic Acid	2.00
Agerite Powder	1.00
Zinc Oxide	5.00
Tuads	.05
Captax	1.00
Sulfur	2.50

Cure—45 minutes at 274° F.

Tire Cushion Stocks

Smoked Sheets	60.00
Amber Crepe	40.00
Cumar Resin	1.00
Mineral Rubber	2.00
Stearic Acid	0.50
Neozone A	1.00
Zinc Oxide	30.00
Accelerator 808	0.6875
Sulfur	3.25

Cure—45 minutes at 281° F.

Tire Tread

Smoked Sheets	100.0
Pine Tar	4.00
Stearic Acid	2.00
Neozone A	1.25
Carbon Black	40.00
Zinc Oxide	10.00
Accelerator 808	0.875
Sulfur	3.25

Cure—60 minutes at 281° F.

Protection of Rubber Belting in Storage

Shellac	1	qt.
Alcohol	1	pt.
Ammonia	1½	qt.
Water	3	qt.

Apply with a brush.

Belt, Friction Rubber

Smoked Sheets	9.4375
Thin Brown Crepe	10.00
Whole Tire Reclaim	59.00
Paraflux	5.00
Stearic Acid	0.50
Neozone D	0.5625
Litharge	0.0625
Whiting	10.3125
Zinc Oxide	2.25
10% Thionex Master Batch	0.6250
Sulfur	2.25

Cure—15 minutes at 274° F.

High Grade Rubber Combs

Smoked Sheets	100.00
Cottonseed Oil	2.00
Beeswax	2.00
Accelerator 833	1.50
Sulfur	45.00

Cure—Approximately 6 hours in water at 274° F.

Black Rubber Footwear

Rubber	100.0
Plastogen	6.00
Agerite Powder	1.00
Zinc Oxide	5.00
Whiting	40.00
Kalite No. 1	40.00
Dixie Clay	25.00
Gas Black	2.00
Zimate	0.10
Altax	0.50
Captax	0.50
Sulfur	2.50

Cure—Dry heat. 60 minutes rise to 260° F. and one hour at 260° F. under 30 lb. air pressure.

Black Rubber Heel

Smoked Sheets	11.50
Whole Tire Reclaim	64.00
Refined Asphalt	3.00
Paraffin	0.25
Stearic Acid	0.375
Neozone A	0.50
Carbon Black	9.375
Whiting (Natural)	7.25
Zinc Oxide	1.00
Litharge	0.125
10% Thionex Master Batch	1.125
Sulfur	1.50

Cure—12 minutes at 40 lb. steam.

Hard White Rubber Sole

Pale Crepe	28.75
Stearic Acid	0.25

Magnesium Carbonate	43.00
Lithopone	21.40
Zinc Oxide	1.50
Glue	2.88
Ultramarine Blue	0.09
Diphenylguanidine	0.28
10% Thionex Master Batch	0.35
Sulfur	1.50

Cure—8 to 10 minutes at 316° F.

High Grade Black Rubber Sole

Pale Crepe	50.00
Smoked Sheets	50.00
Agerite Gel	1.25
Zinc Oxide	60.00
Gas Black	10.00
Dixie Clay	40.00
Kalite No. 1	60.00
Captax	1.25
Tuads	.0125
Sulfur	2.50

Cure—60 minutes rise and 45 to 60 minutes at 255° F. under 30 lb. air pressure.

Rubber Bathing Cap

Rubber	100.00
Stearic Acid	1.00
Cycline Oil-softener	4.00
Zinc Oxide	5.00
Whiting	15.00
Lithopone	15.00
Barytes	15.00
Ureka C	1.25
D. P. G.	.25
Sulfur	2.00

Cure—8 minutes at 40 lb. steam.

Soft Rubber Sponge

Rubber	100.00
Stearic Acid	1.00
Red Oil	1.00
Petrolatum	18.00
White Substitute	5.00
Zinc Oxide	2.50
Sodium Bicarbonate	15.00
Whiting	25.00
Ureka C	.625
Guantal	.375
Sulfur	4.00

Cure—¾ inch thick, 20 minutes at 70 lb. steam.

Rubber Packing

Smoked Sheets	35.125
Whole Tire Reclaim	10.00
Paraffin	1.00
Paraffin Oil	5.00

Stearic Acid	0.375
Clay	20.00
Whiting	20.00
Red Iron Oxide	6.00
Zinc Oxide	1.50
Beutene	0.75
Sulfur	0.75

Cure—12 minutes at 45 lb.

White Rubber Sidewall

Pale Crepe	100.00
Plastogen	4.00
Stearic Acid	1.00
Zinc Oxide	5.00
Kalite No. 1	40.00
Dixie Clay	30.00
Titanium Dioxide	25.00
Captax	1.00
Sulfur	2.25

Cure—Press Cure—Approximately 45 minutes at 30 lb. steam.

Rubber Code Wire Compound

Smoked Sheets	5.00
Blended Reclaim	48.00
Mineral Rubber	20.00
Stearic Acid	0.25
Paraffin	0.25
Neozone A	0.3125
Whiting	23.625
Zinc Oxide	1.00
Accelerator 808	0.3125
Sulfur	1.25

Cure—30 minutes rise to 275° F. plus 105 minutes at 275° F. in soapstone.

30% Rubber Wire Compound

Smoked Sheets	32.00
Paraffin	1.00
Agerite Gel	0.60
Kalite No. 1	33.00
Zinc Oxide	32.00
Carbon Black—P-33	0.20
Captax	0.20
Sulfur	0.80

Cure—Steam Cure in talc. 30 minutes at 260° F.

Electrician's Rubber Gloves

Pale Crepe	100.00
Mineral Rubber	4.50
Paraffin	0.75
Zinc Stearate	1.50
Agerite Gel	1.00
Zinc Oxide	15.50
Blanc Fixe	9.25

Tuads	3.00
Vandex	1.50

Cure—Press—15 minutes at 30 lb.

Rubber Wringer Roll Compound

Smoked Sheets	38.00
Paraffin	0.50
Mineral Oil	1.25
Du Pont Antox	0.375
Zinc Oxide	2.00
Lithopone	35.00
Whiting	21.50
Accelerator 808	0.125
Sulfur	1.25

Cure—45 minutes at 292° F.

Solution for Application on Rubber Materials to Be Embossed to Prevent Sticking on Rolls

Glycerin	5 lb.
Denatured Alcohol	95 lb.

Rubber Bands and Thread

Pale Crepe	100.00
Agerite White	1.00
Zinc Oxide (fine particle size)	2.00
Color	to suit
Zimate	0.10
Altax	0.50
Captax	0.50
Sulfur	2.00

Cure—Open steam. 10 minutes rise to 260° F. and 30 minutes at 260° F.

Raincoat Rubber Compound

Hevea Rubber	48
Litharge	10
Zinc Oxide	20.5
Mineral Rubber	5
Sulfur	1.5
Whiting	15

Anchor Rubber for Artificial Suede

Pale Crepe	40 lb.
White Reclaim	20 lb.
Tube Reclaim	15 lb.
Hard Factice (Brown)	8 lb.
Zinc Oxide	5 lb.
Lithopone	6 lb. 4 oz.
Cottonseed Oil	1 lb.
Stearic Acid	8 oz.
Sulfur	14 oz.
Captax or Ureka	14 oz.
Anti Oxidant	8 oz.

About 4 oz. per square yard of this compound is calendered onto a backing fabric. A cement of the same compound is then applied and closely followed with thorough dusting of finely divided cotton flock. The material is then festooned in an oven and cured ½ hour, rise to 250° F. and 1 hour at 250° F.

Cheap Rubber Topping Formula

Smoked Sheets	7 lb.
Boot and Shoe Reclaim	57 lb.
Cliffstone Whiting	55 lb.
Sublimed Litharge	9 lb.
Hard Mineral Rubber	3 lb.
Palm Oil	2 lb.
Tar Oil	2 lb.
Paraffin	1 lb.
Sulfur	11 oz.
Carbon Black	1 lb. 8 oz.

Cure ½ hour. Rise to 250° F., 1 hour at 250° F.

Rubber Pencil Eraser

Crepe Rubber	4
Starch	10
Petrolatum	4
Vulcanized Waste Rubber	2
Factice	1
Abrasive	2
Lithopone	3
Sulfur	0.1
Accelerator	0.05

Hot Water Bottle

Pale Crepe	34.375
Medium Process Oil	0.50
Barytes	34.00
Whiting	25.25
Zinc Oxide	3.00
Du Pont Rubber Orange AD	0.75
10% Thionex Master Batch	1.4375
Sulfur	0.6875

Cure—7 minutes at 287° F.

Coloring Rubber Articles

An oil soluble dye is dissolved in benzol or other volatile hydrocarbon. The rubber is soaked in this until slightly swollen. It is then removed and dried. The depth of color depends on the concentration of the dye and the time of immersion.

Transparent Rubber Goods

Jatex, a concentrate obtained by centrifuging latex which after evaporation to 40% gives a film as clear as glass, is used as dipping fluid. The articles are dipped at 40° C. followed by vulcanization in a bath made by dissolving 100 grams or more of the finest sulfur in 1000 cc. benzol. Part of the sulfur remains on the bottom of the vessel and maintains saturated solution when the temperature goes up, and as sulfur is taken up during the vulcanization process. To promote the reaction is used an addition of 20 grams Vulcafor ZDC (zinc di-ethylene carbamate).

Coloring Latex Black

Colloidal Micronex is a dispersed carbon black suitable for use with rubber. It does not require grinding. It is merely stirred into the latex in amounts varying with the depth of color desired.

Rubber Pyroxylin Mixture

A common solvent for rubber and pyroxylin is composed of following

Ethyl Oenanthate
Propyl Propionate
Isobutyl Butyrate

or

Ethyl Oenanthate

Thus rubber and pyroxylin may be dissolved in these to form lacquers of special properties.

Black Combining Cement for Double Texture Pyroxylin Goods

Smoked Sheets	15 lb.
Boot and Shoe Reclaim	20 lb.
Soft Factice	10 lb.
Soft Mineral Rubber	8 lb.
Carbon Black	1 lb.
Lime	1 lb. 8 oz.
By Product Whiting	65 lb.

Dissolve in petroleum naphtha to spreader consistency.

Light Color Combining Cement for Double Texture Pyroxylin Goods

Smoked Sheets	15 lb.
White Reclaim	20 lb.
Soft Factice	10 lb.
Hard Mineral Rubber	8 lb.
Cliffstone Whiting	25 lb.
By Product Whiting	50 lb.
Lime	1 lb. 8 oz.
Raw Sienna	2 lb.

Dissolve in Petroleum Naphtha.

Black Combining Cement for Double Texture Rubber Goods

Smoked Sheets	15 lb.
Boot and Shoe Reclaim	25 lb.
Soft Factice	8 lb.
Litharge	8 lb.
Cliffstone Whiting	65 lb.
Rosin Oil	2 lb.
Sulfur	8 oz.

Dissolve in Petroleum Naphtha.

Light Colored Combining Cement for Double Texture Rubber Goods

Smoked Sheets	15 lb.
White Reclaim	30 lb.
Soft Factice	8 lb.
Litharge	2 lb.
Zinc Oxide	10 lb.
Magnesium Oxide	5 lb.
Raw Sienna	4 lb.
By Product Whiting	50 lb.
Sulfur	8 oz.
Rosin Oil	8 oz.

Dissolve in Petroleum Naphtha.

Rubber Cement, Reducing Viscosity of

The addition of 2–3% alcohol reduces the viscosity of thick rubber cements.

Oil-Resisting Materials

Mention has previously been made of new products designed to resist practically all solvents, oils and fats, such as Ethanite, a reaction product of ethylene dichloride and calcium polysulfide, and Thiokol, a polymethylene polysulfide. Although different claims may be made for the individual products now on the market, in general these polysulfides may be vulcanized in a similar manner to rubber requiring no sulfur, but zinc oxide in proportions of 1 to 20% is necessary; the material in appearance is similar to rubber, being homogeneous and pliable, but the gravity is much higher, viz., 1.6. The suitable vulcanizing temperatures are similar to those with rubber mixings, such as one hour at forty pounds steam pressure. The addition of rubber is not necessary, although milling is facilitated thereby. In the case of Ethanite it is stated that the addition of 5% of rubber gives a product which is as resistant to oil as Ethanite alone, but generally speaking, the oil resistance deteriorates according to the amount of rubber present. Carbon black may be added to increase tensile strength and decrease porosity, and a mix which is stated to be resistant to practically all oils and solvents is: Ethanite 20, pale crepe 1, zinc oxide 2, carbon black 5.

When cured these products show practically no dimensional increase when immersed in such solvents as benzol, toluol, and carbon tetrachloride, and acids, with the exception of strong nitric or chromic acids, are without action. A 20% caustic soda solution or concentrated ammonia attacks the material, but the latter does not appear to suffer from aging in the usual manner of rubber goods. The particular advantages obtained are offset to some extent by the objectionable characteristic odor which, besides rendering the use of the products impracticable in many instances, for example foodstuffs, renders the general atmosphere where it is in process, particularly in the region of the mill, decidedly unpleasant. Possibly means will be found of overcoming this, at any rate to a considerable extent.

COLORS FOR PLASTICS
Mahoganies

Burnt Sienna	2.92
Black Oxide of Iron	.44
Deep Indian Red	.64
Resin	49
Wood Flour	49

Burnt Sienna, Dark	.8
Burnt Sienna, Very Dark	3.12
Black Oxide of Iron	.08
Resin	49
Wood Flour	49

Burnt Sienna	1.64
Black Oxide of Iron	.14
Deep Indian Red	.22
Resin	49
Wood Flour	49

Seal Browns

Burnt Sienna, Dark	1.85
Black Oxide of Iron	.1
Ultramarine Blue	.05
Resin	49
Wood Flour	49

Deep Indian Red	.75
Burnt Turkey Umber	1.75
Resin	49
Wood Flour	49

Red Browns

Deep Indian Red	1.75
Burnt Turkey Umber	.75
Resin	49
Wood Flour	49

Deep Indian Red	1.50
Black Oxide of Iron	.5
Resin	49
Wood Flour	49

Blacks

Nigrosine Dye	1.4
Black Oxide of Iron	.6
Resin	49
Wood Flour	49

Olive Drab

Black Oxide of Iron	1.9
Yellow Oxide of Iron	.1
Resin	49
Wood Flour	49

Bakelite Moulding Powder

Asbestos Flour	147 parts
Chalk	147 parts
Clay	147 parts
Bakelite	30 parts
"Cumarone" Resin	30 parts

"Celluloid"—Non-Inflammable

Cellulose Acetate	119–180
Acetone	33– 48
Benzol	32– 52
Alcohol	14– 20

Cork Composition Binder

Casein	45
Borax	7
Water	120
Glycerin	76

Composition Ornaments

A pattern is carved out of wood and is covered by following composition to form a "die":

Oil of Tar	3 oz.
Soapstone	4 lb.
Emery Flour	4 lb.
Orange Shellac	6 lb.
French Chalk	4 oz.

Melt the shellac and add the oil of tar. Add the soapstone, mixing thoroughly. Mix separately the (dry) emery flour and French chalk; then pour this into the melted shellac and oil of tar, stirring thoroughly and vigorously. Place the pattern or "die" in a box, flat side down, and pour this mixture over same. When cool the result will be a mould into which can be cast the materials of which the ornaments or mouldings are composed.

The following composition has been tested and found excellent for mouldings and ornaments of this kind:

White Glue	13	lb.
Rosin	13	lb.
Raw Linseed Oil	1/3	qt.
Glycerin	1	qt.
Whiting	19	lb.

This mixture is prepared by cooking the white glue until it is dissolved. Then cook separately the rosin and raw linseed oil until they are dissolved. Add the rosin, oil and glycerin to the cooked glue, stirring in the whiting until the mass makes up to the consistency of putty. Keep the mixture hot.

Place this putty mass in the die, pressing it firmly into the same and allowing it to cool slightly before removing. The finished product is ready to use within a few hours after removal. Suitable colors can be added to secure brown, red, black or any other color.

In applying ornaments made of this composition to a wood surface, they are first steamed to make them flexible; in this condition they can be glued to the wood surface easily and securely. They can be bent to any shape, and no nails are required for applying them.

Benzyl Cellulose Plastic

Asbestos (Powd.)	300
Chalk (Powd.)	300
Clay (Powd.)	300
Benzyl Cellulose	125

A moulding pressure of 30–60 lb. per sq. in. is used.

Dental Impression Wax

1.

Paraffin Wax	90
Ceresin	39
Beeswax	40
Venice Turpentine	30
Japan Wax	20

2.

Shellac	45%
Talc	30%
Glycerin	2½%

Coloring   sufficient
Tallow Fatty Acids (to
  make)   100%

3.

An impression material is prepared by mixing and heating together a mineral and drying oil mixture 2.5–4.5, a beeswax and paraffin mixture 1.5–2.5, aluminum stearate 2.5–3.5, rubber, gutta-percha or balata not more than about 0.06, starch 0.5–1.5 and glycerol not more than about 0.125 part.

Glue Composition for Toys

Indestructible mass for the manufacture of ornaments, toys, etc. A hard mass consists of 50 parts glue, 35 wax or rosin, 15 glycerin, and required quantity of a metallic oxide of mineral color. A soft mass consists of 50 parts glue, 25 glycerin, 25 parts wax or rosin. Glue is melted in glycerin with the assistance of steam and the wax or rosin added. Mass poured in liquid state into moulds. Degree of hardness of mass is increased by the addition of 30 to 35% zinc white.

Flexible Wax

Methyl Abietate	10
Gelowax	90

Heat together and stir until homogeneous. The finished product has a softening point of 58° C. and a melting point of 67° C.

Grafting Wax Solid

Lanolin	22
Rosin	44
Ceresin	13
Beeswax	8
Japan Wax	2
Rozolin	9
Pine Oil	1

Wine or Liquor Barrel Wax

Tallow	24
Paraffin	50
Japan Wax	5
Beeswax	5
Venice Turpentine	4
Rosin Oil	1
Talc	10

Thread Wax

Beeswax	40
Japan Wax	10
Paraffin Wax	150

Shoemaker's Wax, Hard

Rosin	8
Ester Gum	2
Montan Wax Crude	30
Paraffin Wax	45
Stearin Pitch	10
Beeswax	5
Oil Soluble Color	to suit

Shoemaker's Wax, Soft

Rosin	5
Paraffin Wax	65
Japan Wax	5
Stearin Pitch	20
Beeswax	5
Oil Soluble Color	to suit

Beeswax Substitute

Glyceryl Stearate	20
Beeswax	8
Japan Wax	10

Pure Stearic Acid Candles

Use Triple Pressed Saponified Stearic Acid. After melting down the Stearic Acid should be stirred or agitated until "milky" in appearance to destroy the large crystals. It should then be poured in moulds which have been heated to approximately the same temperature and cooled. A better appearance will be noted on more rapid cooling.

Standard Candle Formula

Paraffin Wax	60 lb.
Double Pressed Stearic Acid	35 lb.
Beeswax	5 lb.

The above are melted together and agitated to insure complete blending. When melted an oil soluble dye of the desired hue is added and then the combination is poured in moulds and cooled. Care in the selection of the dye should be exercised to eliminate "bleeding" or fading, but many good dyes are available. It may be desirable to make up known strength of dyes in blocks of paraffin by merely adding the dye to the melted wax and then pour in moulds, forming blocks of uniform size. This permits easy storing and somewhat facilitates the complete blending of the color when introduced to the melting kettle.

A better grade of candles is made by increasing the amount of Stearic Acid and decreasing the amount of paraffin, or vice versa.

Pure Beeswax Candles

Are made from the pure wax and range down to combinations as low as 40% Beeswax, 50% Paraffin and 10% Stearic Acid.

Tapered Candles

These are usually a hand-dipped operation entirely. The combination of waxes and color is melted in the kettle and a constant temperature maintained at slightly above the melting point. Dipping proceeds from the bottom and progresses up the wick to the desired length in order to attain the desired taper.

Candle Wicks

The matter of the selection of the wick for various compositions of candles is one of careful consideration. For instance, the wick used in a pure stearic acid candle, usually a 48 to 51 ply—meaning three strands of 16 or 17 threads each, would be entirely unsuited for a candle containing very much paraffin, which would require a smaller wick. The wick should be treated with Boracic Acid, the object of which is to prevent the wick from continued glowing and smoking when blown out. One of the strands of the wick should be woven tighter than the other two in order to force the wick into separation while burning to dissipate the ash.

Birthday Candles

Are made entirely of paraffin and the proper oil soluble dye. The procedure, though, is entirely different than in the case of other candles. The thin threads, forming the wicks are formed into endless belts and placed over two drums. These drums are spaced a few feet apart and are set up to revolve slowly, allowing the "endless belt" wicks to run through a tank of the melted wax. This operation is continued until the series of wicks have picked up the desired amount of wax and have reached the required diameter. The "belts" are then cut and laid out on tables where the candles are cut to length. The head of the candle is then inserted into a revolving cutter or a revolving hot mould to properly shape the head.

Virgil Lights

Eighty per cent Paraffin, 15% Double Pressed Stearic Acid and 5% Beeswax. This can be varied to as much as 95% Paraffin and 5% Stearic Acid.

Illumination Candles

Paraffin (50-52° C.)	79 g.
Stearin	19.5 g.
Carnauba Wax, Bleached	1.5 g.

Wax Lighting Tapers

Paraffin Wax (40-42° C. or 42-44° C.)	65-85 g.
Ceresin (58-60° C.)	30-10 g.
Beeswax	2-3 g.
Turpentine, Thickened	3-2 g.

Wick of loose cotton threads, 30 together for a size of 2-4 mm., wound on wire.

Sealing Wax for Candle Decorations

Rosin	50 g.
Ruby Shellac	3 g.
Gypsum	1 g.

Dental Wax

Stearic Acid	1 lb.
Paraffin Scale Wax	2 lb.
Glyceryl Tristereate	1 lb.
Carnauba Wax	2 lb.
Ethylene Glycol Glyceryl Stearate	2 lb.

Electrotypers' Waxes

Beeswax	5½ lb.
Paraffin Wax	3 lb.

CHAPTER XVI

SOAPS AND CLEANERS

THERE are numerous types of soaps on the market which vary in composition and appearance in accordance with their use. Thus there are toilet soaps (liquid, paste and solid); laundry soaps; industrial soaps (textile, dry-cleaner's, floor) etc.

Cleaners consist of solvents (e.g. naphtha, benzine, etc.) with or without some soap dissolved therein. For removing stains from different materials resort is had to numerous chemicals as well as solvents.

Liquid Soap

1

The soap base may be made from one-third coconut oil and two-thirds soya bean oil. The proportions used in saponification are 10.75 parts by weight of soya bean oil, crude or bleached, 5.00 parts by weight of coconut oil and about 7.87 parts by weight of 50 degrees Bé. potassium hydroxide. The soap obtained from this saponification is dissolved in 77 parts by weight of water to which a maximum of 0.5 part by weight of potash has been added.

Another soap is made from two-thirds coconut oil and one-third castor oil. The proportions used in saponification are 10.75 parts by weight of coconut oil, 5.0 parts by weight of pure castor oil and about 7.48 parts by weight of 50 degrees Bé. potassium hydroxide solution. After saponification, the soap is dissolved in 76 parts by weight of water and as above a maximum of 0.5 part by weight of potash is added.

In making the soap from coconut oil and olein, the following proportions are used: 8.5 parts by weight of coconut oil, 5.0 parts by weight of best quality oleic acid and about 7.3 parts by weight of 50 degrees Bé. potassium hydroxide solution. After saponification the soap is dissolved in 77 parts by weight of water and again up to a maximum of 0.5 part of potash is added.

It is very interesting to follow through the progress of saponification. At the beginning the temperature of the mix-

ture rises slowly, since only a small part of the mixture is saponified under the initial conditions of the process. But the rise in temperature constantly becomes greater and the principal reaction of the saponification then takes place. Hence if the mixture has been agitated at a temperature of 65 to 70 degrees C., the temperature rises slowly to approximately 75 to 78 degrees C. Thereafter the rise is more rapid until approximately 85 degrees C. is attained. At this point the greater part of the contents of the kettle is saponified and the heat of reaction liberated becomes smaller and further increase of the temperature is slower. In most cases the temperature increases to approximately 94 to 96 degrees C. and remains constant at that point for some time. Then there comes a point at which the temperature in the kettle begins to fall. Saponification reaction may then be considered as finished and it only remains to saponify residual traces of unsaponified matter. Hence the mixture in the kettle must show at this point noticeable traces of caustic alkali, so that the saponification of the residual fat and oil may be affected when the mixture is well-agitated.

As the mass in the kettle is worked up, it first becomes thick and heavy, but then soon thinner and thereafter thicker and heavier again. When this happens, agitation is best stopped and the soap mass is allowed to remain quiescent for some minutes. Then the soap is fitted and tested. If sufficient alkali were present, technically complete saponification would be obtained. Thus, the results would be as good as those obtained by hot saponification of fats.

At this point the fitting of the soap begins. The soap must have a slight but clearly perceptible acrid taste. This test may be used when the complete saponification test is not made in the works laboratory. This test is, however, very simple and should be made. A small quantity of the soap is dissolved in distilled water. The solution must not be turbid, but absolutely clear. If there is a slight turbidity, this indicates the presence of unsaponified oils or fats. However, in this case, no traces of free caustic potash could be detected in the soap, since the correctly carried out half-boil process gives absolutely good results. If too little lye has been used in the saponification process, which may also happen when the potassium hydroxide solution employed is not 50 degrees strength (this does not happen often), if the solution of potassium hydroxide is allowed to remain in storage tanks exposed to the air for too long a time so that considerable of the hydroxide is converted into the carbonate and the strength of the solution accordingly reduced, then the soap may be lacking in potash lye and in fitting the soap it then becomes necessary to add potassium hydroxide. In this case the potassium hydroxide solution is diluted with distilled or soft water to about 30 degrees Bé. concentration, so that it can be mixed with the soap more readily and more uniformly. The fitting of the soap must be repeated in this case after a short time has elapsed and the same process is carried through until a definite excess of potassium hydroxide is detectable in the soap.

Alkali in Soap Base

If the excess of alkali is found to be too large when the soap base is tested, the taste of the soap being too sharp, then there must have been an error in measuring out the alkali for saponification of the fats and oils, on the assumption that there was nothing wrong with the latter and they were completely saponifiable. However, fats and oils, which are not completely saponifiable, and hence are not of first quality (technical grade), are not suitable raw materials for making liquid soaps. However, if the soap base contains too much

alkali, then it is necessary to neutralize the same. This is accomplished by introducing a small quantity of coconut into the hot soap. Good results are also obtained with oleic acid. After the added fats or oils have been thoroughly mixed with the soap mass and saponified, the soap must be tested again after about ten to fifteen minutes and fitted.

As has been remarked above, if the soap base had a content of about 65 to 66 per cent of fatty acids, it need be dissolved only in three times its weight of distilled or soft water to give a liquid soap containing about fifteen to sixteen per cent of fatty acids. If the soap base contained only a slight quantity of alkaline excess and was used without further treatment, the liquid soap will be found to be practically neutral. On the other hand, if the proportion of excess potassium hydroxide in the soap base was quite large, then the liquid soap must be neutralized. An acid turkey red oil is used with best results for this purpose. This product dissolves rapidly and completely in the liquid soap to give a clear solution. Neutralization is therefore rapid and as complete as desired.

2

Eighty kg. palm-seed oil and 20 kg. sunflower seed oil are sapond. at 50° with 52 kg. 50 Bé. Caustic Potash. After the mixt. has stood, it is adjusted to the desired alky., and then the filling mass (consisting of 200 kg. cryst. sugar, 10 kg. Potassium Carbonate and 10 kg. Potassium Chloride dissolved in 1000 kg. water) is added.

Liquid Soaps and Shampoos

Liquid soap for use in offices, etc. (14 per cent fatty acid content) may be of similar composition to the following:

(a) Coconut oil (free fatty acid less than ½ per cent)............126 lb.
Caustic potash liquor 38° Bé... 90 lb.
Glycerin...................... 17 lb.
Water........................560 lb.

(b) Liquid soap for workshop use.
Coconut oil.................220 lb.
Caustic potash at 38° Bé....157 lb.
Glycerin................... 26 lb.
Water......................418 lb.

(c) Shampoo (good quality).
Resin (finest W.W. grade)... 1 lb. 3 oz.
Coconut oil (refined)........ 2 lb. 8 oz.
Glycerin....................16 lb. 4 oz.
Olive oil.................... 13 lb. 0 oz.
Caustic potash liquor 38° Bé. 9 lb. 10 oz.
Water...................... 57 lb. 8 oz.

This shampoo may be mentholated by adding the required amount of menthol dissolved in spirit, as before detailed.

(*d*) A pine tar type is obtained by adding 2 per cent of tar water, obtained as follows:

Take 250 gm. wood tar with 15 gm. sodium bicarbonate and 1,000 gm. of water. Place in a vessel and keep at 35° C. to 40° C. for three hours, during which time the vessel is shaken frequently. The mixture is stored for three days and then filtered.

A brown tint is imparted to the shampoo by means of a suitable water-soluble dye.

Preparation

The manufacture of shampoos and liquid soaps is best carried out in enamelled jacketted pots. 30 lb. steam pressure is adequate. Stirring devices are not really essential. A clean wooden paddle will suffice.

The pot is charged with the required amounts of the selected oils and the contents raised to 160° F. to 180° F., when the saponifying agent (which is always caustic potash standardized at 38° Bé.) is added, whilst the contents are gently stirred. From twenty to thirty minutes suffice to complete saponification, then a sample is taken and tested carefully. It is essential that there shall be at least 0.05 per cent free alkali present, expressed as Na_2O. Failure to observe this condition will result in imperfect clarification of the finished product and impair its lathering properties.

Upon the result of the test an addition of caustic potash liquor is added, if indicated. Should the free alkali lie between 0.05 and 0.15 per cent, it will be unnecessary to correct with oil.

The base will not be of a translucent pale yellow appearance of a fairly stiff consistency; this latter factor will be variable according to the amount of olive or palm oil used.

Distilled water at as high a temperature as possible is now added in the desired proportion, and the contents of the pot are gently stirred till solution is obtained. The glycerin is then added.

The product is now allowed to cool and the perfume and/or medicament incorporated with the rest.

The clarification of the shampoo or liquid soap is effected in a small diameter deep cylindrical enamelled vessel with a deep conical base. Fitted at the junction

of the straight side of the cone is a tap to facilitate the withdrawal of the clarified liquid.

Too much importance cannot be placed on the need for effective clarification at as low a temperature as possible, and the deep conical bottom influences materially the rate of settlement of the suspended matter.

An addition of potassium carbonate of the order of one-half of 1 per cent dissolved in distilled water will be found invaluable in preventing subsequent formation of cloudiness.

For *large-scale production* at least two clarification vessels should be employed, so that orders may be filled from the vessels in turn.

Coloring Soap and Other Products

Color lends a greater sales appeal to practically every product. The purpose of coloring soap and soap products, therefore, is to give them greater sales appeal, and not to cover any defects, as many people think. The soap maker, to-day, not only manufactures soaps and shampoos, but also bath salts, deodorizing blocks, sweeping compounds, floor oils, floor waxes, and emulsions. In this chapter it is our intention to consider all these products and present those dyes which we have found best suited for the particular purposes.

Before the discovery of aniline dyes and their development during the past forty years, the coloring and dyeing of all materials was done by vegetable dyes or mineral colors. With the advent of the aniline dyes, the vegetable dyes and the mineral colors were for the most part discarded. Today, there are very few vegetable colors used in any industry, let alone soap. We will, therefore, consider in this article the few vegetable dyes and mineral colors that are still being used and also the coal tar dyes, their advantages and their disadvantages.

Let us take the vegetable colors first. The one which enjoys the greatest use is Chlorophyll. Chlorophyll is a green color and is extracted from the green leaves of plants such as stinging nettles or spinach. It comes in three types—water soluble, alcohol soluble, and oil soluble. The commercial product in each case is a liquid or a paste. Of the three, the oil soluble Chlorophyll enjoys the greatest sale. It is used particularly for the coloring of olive oil soaps such as silk boil-off, and automobile soaps. It is also used in the coloring of cosmetic creams, paraffin oils and waxes. Contrary to general belief, Chlorophyll is not very fast. Liquid soap colored with Chlorophyll will not hold its color when packed in tin containers. The light fastness of Chlorophyll in certain solutions is poor. The coloring cost, when compared to a good aniline dye is excessive. There is a coal tar substitute which gives the exact shade of Chlorophyll, is faster than Chlorophyll, and has greater coloring power. It comes in a water, alcohol, or oil soluble form.

Alkinet and Bixin, yellow and red vegetable colors, are still being used somewhat for the coloring of cheese and butter, but they find no use in the coloring of soaps due to their low strength.

Carmine may be mentioned at this point although it is not a vegetable color. It is made from the dried bodies of the female insect Coccus Cacti, an insect which lives on certain cactus plants in Mexico, Central America, and South America. At one time, Carmine was used extensively. Its use today as a red color is limited.

Of the mineral colors, the two most important are Ultramarine Blue and Vermilion. Ultramarine Blue is used in the manufacture of blue mottled laundry soap. Vermilion Red is used in the manufacture of a red mottled soap. Chrome Green has been used for the coloring of soap but it has been largely replaced by aniline dyes. Spanish Oxide or Iron Oxide is used in some cases for sweeping compounds.

We now turn our attention to the coal tar derivatives or what are commonly called aniline dyes. The aniline dyes of commerce are generally powders or crystals. When considered in light of their solvents, they may be divided into these three groups: 1—water soluble, 2—spirit soluble, 3—oil and fat soluble.

Not all of the aniline dyes are fast. For most soaps and sanitary products, it is essential to have a dye or color that is fast to alkali and light. For some products, alkali fastness is essential and light fastness may be disregarded. As a general rule, the water soluble and oil soluble colors have good fastness to both alkali and light. The spirit soluble colors, with one or two exceptions, do not have good fastness to alkali and light.

We have mentioned that water, alcohol, and oil soluble colors are used. It might be advisable, therefore, to consider the preparation of each of these types of colors for use.

Water Soluble Colors: Dissolve the color in the proportion of two to three ounces of color to a gallon of hot water. Do not use a tin or an iron container as a chemical reaction will set up and will tend to decrease or destroy the coloring power. Filter to insure that all the color is dissolved. Undissolved particles of color cause spots and blotches. It is not necessary to make fresh color solutions each time. You must stir the color if you have not used it in some time as some dyes have a tendency to settle out of the color solution.

Alcohol Soluble Colors: Dissolve from two to three ounces of color to a gallon of alcohol. Filter to insure the absence of undissolved color. Alcohol soluble colors are also soluble in acetone, ethyl acetate, and some lacquers. They are also soluble in perfume oils.

Oil Soluble Colors: These are soluble in all vegetable and mineral oils and waxes, fats, oleic and stearic acid, fatty acids of all kinds, paradichlorbenzol, perfume oils, ethyl acetate, toluol and lacquers. When dissolving the color in oils, waxes, or fatty acids, heat should be employed to get the full solution of the color.

It is important to remember that soap and soap products are merely colored and not dyed, *i.e.*, the color does not combine chemically. Consequently, if too much color is used, the color will bleed out of the product.

Now, let us take the various types of materials to be colored and consider the dyes best suited for the purposes.

Milled Soaps

The primary requisite for colors here is light fastness. Water, alcohol, and even oil soluble colors are used. The water soluble colors are recommended as best for general use on milled soaps. The alcohol soluble colors have a tendency to blister. The oil soluble colors are incorporated in very few types of milled soaps. The recommended dyes are:

Pink—Rhodamine B Extra
Salmon Pink—Rhodamine 6G Extra
Green—Cyanine Green
Golden Yellow—Metanil
Blue—Alizarine Blue
Red—Cloth Red
Amber—Bismarck Brown
Lemon—Fluorescein

Canary Yellow—Fast Light Yellow
Heliotrope—Violamine
Violet—Acid Violet

These colors are all water soluble and, with the exception of the Rhodamine and the Bismarck Brown, will mix with one another to give any shade desired. They all have good fastness. For a two hundred pound batch of soap, you will require one-sixteenth of an ounce of Rhodamine B Extra; one-half ounce of Fluorescein; one-half ounce of Violamine. All the other colors require one ounce per two hundred pounds of soap. Where very delicate shades are required, these proportions can be cut down one-third. The color, in liquid form, is best added in the amalgamator. If spots or blotches form, it is a sign that some of the color was undissolved. It is, therefore, advisable to make sure that you have a clear solution.

Cold, Half-Boiled, and Boiled Soaps

The requirements for a color for these types of soaps are fastness to light, alkali, and heat. Water or oil soluble colors are used. Where a water soluble color is used, the color is dissolved in water. Where the oil soluble color is used, the color is dissolved in some of the oil or fat. In the cold soaps, the water soluble color is added in liquid form after saponification has started. In figged soaps, the color is crutched in after saponification is completed. Never add dry dye to the soap mass or to the lye unless you are looking for trouble. Dry dye causes spots and blotches. The water soluble colors recommended are:

Pink—Rhodamine B Extra
Salmon Pink—Rhodamine 6G Extra
Green—Cyanine Green
Golden Yellow—Metanil
Blue—Alizarine Blue
Red—Cloth Red
Amber—Bismarck Brown
Lemon—Fluorescein
Canary Yellow—Fast Light Yellow
Heliotrope—Violamine
Violet—Acid Violet

The oil soluble colors recommended are:

Green—Alizarine Green Oil
      soluble plus Azo Yellow
Amber—Azo Amber

Soap Bases and Liquid Soaps

The primary requisites here are fastness to light, alkali, and contact with tin. Water soluble colors are used for the liquid soaps. In the soap bases, the oil soluble colors may be used. The same colors listed for milled soaps and cold soaps can be used here. One exception, however, is Naphthol Green. Naphthol Green is used by many soap makers to get a leaf gren shade. In many cases where liquid soap or soap bases have been colored with Naphthol Green and packed in tin containers, the color has faded out due to the chemical action resulting between the dye, the soap, and the tin. We recommend a substitute which may consist of Cyanine Green and Tartrazine. The most popular shades and the colors used to obtain them are:

Pink—Rhodamine B
Yellow—Metanil Yellow
Blue—Alizarine Blue
Amber—Bismarck Brown
Opal—Fluorescein
Strawberry—a mixture of Rhodamine
      B and Bismarck Brown

Soaps

These are colored either with Chlorophyll or with a water or an oil soluble aniline dye. The replacement of the Chlorophyll with aniline dye due to the excessive coloring cost of the Chlorophyll is recommended. The dye is added after saponification is completed and before the soap is settled. A pound colors 4500 pounds of soap.

Boiled Automobile and Silk Boil-Off Laundry Soap

Where it is desired to give a laundry soap a deeper tone or a more brownish cast, an oil-soluble amber is recommended as it will not stain the clothing. For laundry powders which are already manufactured and which must be made darker in tone, or browner, a water soluble dye can be used. Do not use more than one pound of color for 16,000 pounds of soap, otherwise the dye will stain the clothing.

Medicated Soaps

These are generally colored red. Either a water soluble or an oil soluble red is used. The water soluble dye is a Rhodamine derivative. The oil dye is an Azo compound. The oil soluble dye is soluble in the cresylic acid. A pound of dye colors 2500 pounds of soap. These two colors can be used in milled, cold, or semi-cold medicated soaps.

Bath Salts

The requisites for colors for bath salts are fastness to alkali and light. There are two ways of coloring bath salts. One is to get the color and odor combined and use the proportions recommended by the manufacturer, generally a pound to one hundred to two hundred pounds of bath salts. For the small manufacturer this is the most practical and most convenient method. The other method is to use water or alcohol soluble colors and add the perfume afterwards. When you use water soluble colors, make the solution as concentrated as possible. Color some of the salt very heavily and then mix this up with the rest of the salt. This will minimize the water used. Add the perfume and then tumble or mix. The colors recommended are the water colors given at the end of this article.

Emulsions

There are two kinds of emulsions. 1—oil in water emulsions, 2—water in oil emulsions. The oil in water emulsions are best colored with water soluble dyes. The water which is used in the emulsion is first colored. If the emulsion is to be colored after completion, the color is dissolved in as little water as possible and the concentrated dye solution is added to the emulsion and stirred vigorously. The following colors are recommended. The proportions are anywhere from one pound to four hundred gallons, to one pound to twelve hundred gallons, depending upon the depth of shade desired.

Pink—Rhodamine B Extra
Green—Cyanine Green
Golden Yellow—Metanil
Canary Yellow—Tartrazine
Blue—Alizarine Blue
Red—Cloth Red
Heliotrope—Violamine
Opal—Fluorescein
Black—Nigrosine

Water in oil emulsions are best colored with oil soluble colors. The colors are dissolved in the oils before emulsification. The colors recommended are:

Yellow—Azo Oil Yellow
Red—Azo Oil Red
Black—Oil Black
Orange—Azo Oil Orange

Blue—Alizarine Oil Blue
Violet—Alizarine Oil Violet
Green—Oil Green

With the exception of the black, one pound colors 200 gallons. One pound of the black colors 50 gallons.

Wax Emulsions

Wax emulsions, wax polishes and wax pastes are colored with oil soluble colors. The same colors as for the water in oil emulsions are recommended. In the case of shoe polish pastes, both the water and oil soluble colors are used. The water soluble color is dissolved in the water and the oil soluble color is dissolved in the wax before emulsification takes place. In the case of the wax floor pastes, the color is added to the solvent for the wax.

Cosmetic Creams and Lotions (See Emulsions)

Nail Polishes

Either basic dyes soluble in acetone and ethyl acetate are used, or oil soluble dyes. The basic dyes generally used are Rhodamine B (Pink) Safranine Y (Red). They are used alone or mixed with Auramine (Yellow) or Chrysoidine (Orange) to give all desired shades. The oil reds given below are used, or shaded with oil yellow.

Deodorizing Blocks

These are colored in the same manner as bath salts. Buy combinations of colors and perfumes and use the proportions recommended, or use the oil soluble colors mentioned at the end of the article. If the blocks are molded, dissolve the color in the molten paradichlorbenzene. If the blocks are pressed, dissolve the color in the perfume oils and spray.

Mineral Oils

Brilliantines and mineral oils are colored with oil soluble colors given at the end of this article. A pound of color generally colors 1600 gallons.

Washing Powders

Dish washing and cleaning compounds made from Trisodium Phosphate, modified soda, soda ash, or combinations of same are generally colored with water soluble Fluorescein. The proportion used is one pound of color to 1250 pounds of compound. The color of the dyed compound is peach. When dissolved in water, a greenish fluorescein is given. The use of this color in the cleaning compounds is covered by patents. For coloring washing compounds made from the above chemicals where a fluorescein is not desired, use any of the water colors given at the end of this article.

Sweeping Compounds

Water or oil soluble red or green is used. The water soluble green is Malachite Green. The water soluble red is Croceine Scarlet. The best oil soluble green is an Alizarine Oil Green, the red is an Azo Oil Red. Where the water soluble colors are used, the color is dissolved in water and the sawdust is colored first. Then the oil and sand are added afterwards. If the oil soluble colors are used, the oil is colored first and then mixed thoroughly with the sawdust and the sand is added afterwards. The whole mass is then thoroughly mixed.

Washing and Bluing Powders or Tablets

These tablets may be soap, cleansing agent like Trisodium Phosphate, and Bluing. Three kinds of bluing may be used. 1—Ultramarine Blue, 2—Soluble Prussian Blue, 3—Aniline Blue. The Aniline Blue is the best to use as it gives a more attractive finish. The Prussian Blue may cause rust spots when the laundered material is ironed. The proportion of the Aniline Blue is one pound to 2000 pounds of compound.

Glycerin Anti-Freeze

These mixtures are colored with basic colors. The shades used are a scarlet which is a mixture of Safranine and Phosphine, a green which is a mixture of Malachite Green and Auramine, a blue which is Methylene Blue zinc free, or Victoria Blue. The proportions are one pound to 1500 to 3000 gallons depending upon the depth of shade desired.

Silicate of Soda Compounds

Three coloring media are used. Two are dyes and one is an indicator. The indicator is Phenolphthalein. It gives a color from a pale pink to a deep wine

color depending upon the amount of indicator added. The others are Fluorescein which gives a yellow with a greenish fluorescein, and Eosine which gives a red with a yellowish fluorescein. In both cases, the water soluble Fluorescein and Eosine should be used. The silicate will act as a solvent. Stir thoroughly to insure perfect solution.

In closing, we list the colors recommended for quick references.

Water Soluble Dyes

These can be used for coloring milled, cold, semi-boiled soaps, liquid soaps and bases, shampoos, toilet waters, bath salts and emulsions.

Pink—Rhodamine B Extra
Salmon Pink—Rhodamine 6G Extra
Green—Cyanine Green
Golden Yellow—Metanil
Blue—Alizarine Blue
Red—Cloth Red
Amber—Bismarck Brown
Lemon—Fluorescein
Canary Yellow—Tartrazine
Heliotrope—Violamine
Violet—Alizarine Violet

Alcohol Soluble Dyes

These can be used for coloring nail polishes, anti-freeze glycerin, denatured alcohol, and shellac.

Pink—Rhodamine B
Red—Safranine Y
Blue—Methylene Blue ZF
Violet—Methyl Violet
Green—Malachite Green
Yellow—Auramine
Black—Nigrosine
Orange—Chrysoidine
Brown—Bismarck Brown

Oil Soluble Dyes

They can be used for coloring emulsions, nail polishes, waxes, wax pastes, oleic and stearic acids, lacquers, acetone, toluol, cello-solve, creams, mineral oils, petrolatum paradichlorbenzene and orthodichlorbenzene.

Red—Azo Oil Red
Yellow—Azo Oil Yellow
Orange—Azo Oil Orange
Black—Oil Black
Blue—Alizarine Oil Blue
Violet—Alizarine Oil Violet

Green—Oil Green
Brown, Amber—mixtures of the above

---◆---

Pine Oil Liquid Hand Soaps

Liquid Soaps usually are made with cocoanut oil-potash soaps, or a combination of palm-kernel oil and vegetable oil-potash soaps.

These soaps are diluted with water, depending upon the price the consumer wishes to pay for such a product. When high percentages of water are present large percentages of ethyl (or grain) alcohol, glycerol or sugar are added to lower the freezing point. Consequently, there is less chance for the soaps to solidify out of solution and cause a subsequent clouding of the finished product. A cloudy product causes sales resistance while a clear, transparent product does not.

Manufacturers of liquid soaps have found that the addition of pine oil increases the cleaning action of the soap. In addition, pine oil imparts a piney fragrance to the soap. The following formula was developed for use in a washroom dispenser:

Parts by Weight

Coconut Oil (Saponification No. 257)	160.0
Potassium Hydroxide (89% Pure)	46.0
Pine Oil	40.0
Water	754.0
	1000.0

It is prepared in the following manner:

Coconut oil of Ceylon Grade is added to a vat and heated to a temperature of 80°–85° C. The potassium hydroxide is then dissolved in a sufficient amount of the water to make a 15% to 20% solution. One-half the solution is then added to the coconut oil and stirred in slowly. The balance of water is then added followed by the balance of potassium hydroxide solution which is stirred in slowly. The temperature of the mix is then kept at 80°–85° C. for a period of from two to three hours with good agitation. After complete saponification, the solution is then cooled, chilled and filtered in this chilled state. The Pine Oil is then added by stirring in very slowly. A sufficient amount of water is then added to balance water loss during sus-

tained heating to bring product to original weight.

---◆---

Pine Oil Liquid Scrubbing Soaps
1.

The scrubbing soaps on the market are either liquid or powder. The former are principally composed of soaps and solvents with lesser percentages of alkali, whereas, the latter are mostly alkali with slight traces of soap and solvent.

Pine Oil Liquid Scrubbing Soap is recommended for general use and is widely used in many institutions to preserve costly surfaces and for its deodorizing properties.

The following is a good formula for a liquid scrubbing soap:

Parts by Weight

Oleic Acid (Acid Number—194)	61.6
"I" Wood Rosin (Acid Number—165)	61.6
Sodium Hydroxide (100%)	16.3
Pine Oil	133.0
Tri-sodium Phosphate	26.7
Water	700.8
	1000.0

It is prepared in the following manner:

The Oleic Acid and "I" Wood Rosin are added to a vat and heated to a temperature of 80° C. The sodium hydroxide is then dissolved in a sufficient amount of the water to make a 15% to 20% solution. One-half of the alkali solution is then added to the mass and stirred in slowly. The remainder of the water together with the tri-sodium phosphate is then added by stirring in slowly. After temperature has been dropped to 60° C. the balance of the sodium hydroxide solution is added with vigorous agitation and continued for 15 minutes. After complete saponification the Pine Oil is added by stirring vigorously for several minutes.

The finished or completed product is light red to dark brown in color, dependent upon the type of rosin or oleic acid used.

Such a pine liquid scrub soap is especially adapted for fine tile, cork, rubber, linoleum, mastic, terrazzo and painted floors.

1. It is a powerful solvent.
2. It does not contain any injurious ingredients.

3. It is an efficient cleanser.
4. It removes grease and stains.
5. It deodorizes.
6. It repeats.
7. It is economical to manufacture.
8. It is a concentrated product and effects a great economy.
9. Use 4 oz. in a 10 quart pail of (preferably hot) water and then apply in usual manner.

Pine Oil Powder Scrubbing Soaps

The pine powder scrubbing soaps are specialty products since they are manufactured for specific use rather than for general use.

Manufacturers have found that cleaners may be recommended for many purposes; in addition, however, pine powder scrubbing soaps are invaluable to the public garage owner and filling station manager for dissolving grease and dirt from concrete flooring. Its light sudsing property is a great advantage in that it does not leave a slippery film. In addition its searching piney fragrance excellently dispenses many obnoxious odors.

The following is representative of the best grades:

Parts by Weight	
Oleic Acid (Acid Number —195)	50
"I" Wood Rosin (Saponification Number—165)	50
Sodium Hydroxide (100%)	13.3
Pine Oil	100
Soda Ash (58%)	737
Water	4.7

It is prepared in the following manner:

The oleic acid and "I" Wood Rosin are added to a vat and brought to a temperature of 80° C. The sodium hydroxide is dissolved in the specified amount of water. Temperature of the mass is then dropped to 60° C. and the sodium hydroxide solution is added by stirring in slowly. After complete saponification the Pine Oil is added by stirring in slowly. Add the soda ash to the previous mass and mix it in a mechanical stirring device similarly constructed to a cement mixer. The resultant product is free flowing.

The pine powder is sprinkled over the greasy floors and wet down with a hose. The usual scrubbing practice is followed. Or it may be dissolved in a bucket of hot water and applied in usual manner.

Laundry Soap

"Manila" Type Coconut oil, or coconut oil fatty acids, 50 pounds. Stearic Acid, single pressed, 450 pounds. Caustic Soda, solid or flake, 75 pounds.

The caustic soda is to be dissolved in sufficient water to make a solution of 15° Baumé (1.116 specific gravity). The fatty material is placed in the soap kettle, the caustic soda solution is run in slowly, steam being turned into the coils or jacket of the kettle and the entire mass boiled vigorously. As the caustic soda solution is taken up by the fat, more and more is added until saponification is complete.

The soap is then grained out with dry salt, re-boiled and finished as usual. If the resultant product is too hard it can be made softer by increasing the proportionate amount of coconut oil.

This formula is based upon production of about 800 pounds of finished soap containing 30% moisture. The amounts can be varied proportionately for any capacity charge.

Concentrated Liquid Soap for Silk Goods, Silk Stockings, Etc.

Water	55 parts
Solid Caustic Potash	5 parts
Diethylene Glycol	20 parts
Red Oil or Oleic Acid	20 parts
Yield	100 parts

Dissolve the caustic in the water, add the diethylene glycol, bring to a boil and add the red oil. Adjust either with red oil or alkali until the sample dissolved in alcohol is neutral to phenolphthalein.

Saddle Soap

Beeswax	500
Caustic Potash	80
Water	800

Boil for 5 minutes while stirring. In another vessel heat

Castile Soap	160
Water	800

Mix the two with good stirring; remove from heat and add

Turpentine	1200

while stirring well.

Soap, Castor Oil

To obtain a transparent, amber-colored castor-oil soap (A), mix 30 cc. caustic potash of 80% with 15 cc. industrial alc. and 99.4 g. castor oil. The resulting opaque jelly when put into a warm place will be clear after 10 minutes. To prep. from this a *compound soln. of cresol*, add further 142 g. cresol, shake, then add water to make 300 cc. To prep. a more dil. soln. of *A*, add to the above quantity of *A* sufficient water to make 225 cc. This soln., *liquid castor-oil soap* (*B*), is miscible with water in all proportions, is permanent and may be used as a stock soln. for other prepns.

------◆------

"Waterless" Soap

A soap which may be used to clean hands without water consists of

Agar-Agar	2
Psyllium	3
Glycerol	50
Soda Ash	50
Soft Soap	50
Ammonium Hydroxide	25
Javelle Water	5
Water	815

------◆------

Hand Wash, Mechanics Antiseptic

Chloride of Lime Powd.	175 gm.
Sodium Bicarbonate	359 gm.
Boric Acid	35 gm.
Water	30 oz.

For use on grimy hands to prevent dermatitis dilute with 10 times water and follow by thorough rinsing with mild soap and water.

------◆------

Cleaning Paste for Mechanics

Stearic Acid	100 lb.
Caustic Soda Soln. 32° Bé.	54 lb.
Soda Ash	10 lb.
Water	836
	———
	1000 lb.

Heat at 85° C. for about 10 minutes, stirring until uniform. Fine pumice stone may be incorporated as an abrasive if desired.

------◆------

Mechanics' Hand Soap Paste

Water	1.8 qt.
White Soap Chips	1.5 lb.
Glycerin	2.4 oz.
Borax	6 oz.
Dry Sodium Carbonate	3 oz.
Coarse Pumice Powder	2.2 lb.
Safrol	enough to scent

Dissolve the soap in ⅔ of the water by heat. Dissolve the last three in the rest of the water. Pour the 2 solutions together and stir well. When it begins to thicken, sift in the pumice, stirring constantly till thick, then pour into cans. Vary amount of water, for heavier or softer paste. (Water cannot be added to the finished soap.)

------◆------

Powdered Scouring Compound

Rosin Soap	5%
Oleate Soap	5%
Steam-distilled Pine Oil	10%
Soda Ash	75%
Water	5%

This product makes a very efficient scouring compound for cleaning concrete floors, tile, marble, granite, etc. The pine oil content insures good penetration and is essential for the efficient removal of greasy and oily dirt. Yamor Pine Oil is an excellent solvent for grease, oil, etc.

------◆------

Sweeping Compounds

Although there are many sweeping compounds on the market made of sawdust, sand, ground feldspar, oil, wax emulsions, coloring matter, disinfectant, etc., it is believed that in many cases fine sawdust moistened with water at the time of use will prove satisfactory. Some prefer a compound containing sand, oil, etc.; for example, the Treasury Department at one time used a compound made up according to the following formula:

Sand	10 parts by weight
Fine Sawdust	3½ parts by weight
Salt	1½ parts by weight
Paraffin Oil	1 part by weight

Mix thoroughly.

Certain Government offices have advised us that a compound conforming to the following formula has been satisfactory in service:

Fine Sand	35%
Pine Sawdust	40%
Paraffin Oil	15%
Water (dye if coloring is desired)	10%

The Navy Department has used a compound consisting of a uniform mixture of clean, fine sand and finely ground sawdust properly impregnated with a refined heavy mineral oil and water. Such a compound must show on analysis: not more than 20% of water, not more than

50% of clean sand, not less than 5% of refined heavy mineral oil, and the remainder finely ground sawdust. Some of the commercial compounds are colored with iron oxide or other pigment and contain naphthalene flakes.

Essential oils, such as oil of eucalyptus, oil of sassafras, etc., are frequently added to impart a pleasant odor to the compound or to mask any unpleasant odor that may be due to the ingredients used.

Dry Cleaning Soap

	Parts
1. Oleic Acid-white	10
2. An alcohol solution of potassium Hydroxide (2 oz. by wt. of potassium Hydroxide in 10 oz. of denatured alcohol)	10
3. Carbon Tetrachloride	50

Mix 1 and 2 then add 3.

Use plain then rinse article with gasoline or better still with carbon tetrachloride allow to dry.

Dry Cleaning Liquid Soap (Non-Alkaline)

Diglycol Oleate	130
Tetralin	28
Naphtha	30

Rug Cleaning Soap

Oleic Acid	28 lb.
Butyl Cellosolve	5 lb.
Ethylene Dichloride	13 lb.
Triethanolamine	15 lb.
Water	125 lb.
Isopropanol	14 lb.

The oleic acid, ethylene dichloride and Butyl Cellosolve are mixed and then added to a solution made of the Triethanolamine and water. The mixture is well stirred and sufficient isopropanol is added to form a clear solution. The product emulsifies in water, and the emulsion made with an equal volume of water is recommended for cleaning rugs.

Rug Cleaner

Di-Glycol Oleate	44
Butyl Cellosolve	5
Ethylene Dichloride	12
Alcohol	15
Oleic Acid	11

Ammonium Hydroxide	11
Water	45

This may be made thinner by increasing the amount of water.

Dry Cleaner

Glycol Oleate	2 parts
Carbon Tetrachloride	60 parts
Varnoline	20 parts
Benzine	18 parts

An excellent cleaner that will not injure the finest fabrics.

Gasoline Cleaning Cream

1. Cocoa Soap	5 gm.
Ammonia Water	8 cc.
Solution Potassa	4 cc.
Water, enough to make	30 cc.

Dissolve the soap, by the aid of heat, in 10 cc. of water, add the ammonia and solution of potassa, and sufficient water to make 30 cc. To this saponaceous cream carefully add, in small portions at a time, 5000 cc. of gasoline. This is stated to be an excellent cream for removing grease spots from clothing.

2. Spirit of Ammonia	20 gm.
Ether	50 gm.
Gasoline	150 gm.
Oil Lavender	5 gm.
Tincture Soapbark	225 gm.
Alcohol	500 gm.

3. Oleate Ammonia	2 oz.
Solution Ammonia	2 oz.
Ether	1 oz.
Benzine	5 oz.
Chloroform	1 oz.

Mix the solution and oleate; shake well and add the ether; shake, and add 5 ounces of benzine; agitate thoroughly; then add 1 ounce of chloroform and shake again. Allow to stand a few minutes and shake at intervals, when a mixture having the consistency of cream and showing but little tendency to separate will result.

Kerosene Jelly Cleaner

1. Trihydroxyethylamine Stearate	5
2. Kerosene	16
3. Cresylic Acid	1
4. Water (Boiling)	45

Heat (1) and (2) until dissolved; add (4) slowly while stirring with high speed mixer then add (3).

The above makes an excellent antiseptic cleaner for woodwork, tile, porcelain, etc.

"Nitro" Solvent for Cleaning Guns

Mix

Amyl Acetate	4 oz.
Benzol	4 oz.
Motor Oil (S.A.E. 50)	8 oz.

Apply to cloth and pull through gun barrel, repeating with fresh cloth until the cloth comes through unstained. This leaves a thin coating of oil as a protective film in the gun barrel.

Wall Paper Cleaner

Whiting	10 lb.
Magnesia Calcined	2 lb.
Fuller's Earth	2 lb.
Pumice Powd.	12 oz.
Lemenone	4 oz.

Dry Cleaning Fluid

(Non-inflammable and quick acting)

Butyl Cellosolve	1
Diglycol Oleate	1
Water	1
Isopropyl Alcohol	10
Carbon Tetrachloride	14

Non-Inflammable Dry Cleaner

Naphtha 200–300° F.	40
Carbon Tetrachloride	60

Household Cleanser

Soap Powder	2%
Soda Ash	3%
Trisodium Phosphate	40%
Finely Ground Silica	55%

Mix well and put up in the usual containers.

Solvent for Grease in Drain Pipes

Potassium Hydroxide, dry flake	99 parts
Aluminum, fine powder	1 part

Mix well and keep dry.

Javelle Water

Bleaching Powder	20 lb.
Soda Ash	20 lb.
Water	60 gal.

Mix well until reaction is completed. Allow to settle over night and siphon off the clear liquid.

Laundry "Sour"

Oxalic Acid	3 lb.
Water	3 gal.

Heat with stirring until dissolved. Cool and add

Acetic Acid (56%)	8½ lb.

One pint of this sour is used per 200 lb. of goods.

Laundry Blue

Ultramarine Blue	35
Aniline Blue Soluble	1
Soda Ash	30
Corn Syrup	7

Make into a paste with water and press in forms.

Liquid Laundry Blue

Prussian Blue	1
Distilled Water	32
Oxalic Acid	¼

Window Cleanser

Castile Soap	2 parts
Water	5 parts
Chalk	4 parts
French Chalk	3 parts
Tripoli Powder	2 parts
Petroleum Spirits	5 parts

Cleaning Straw Hats

1

1. Hats made of natural (uncolored) straw, which have become soiled by wear, may be cleaned by thoroughly sponging with a weak solution of tartaric acid in water, followed by water alone. The hat after being so treated should be fastened by the rim to a board by means of pins, so that it will keep its shape on drying. Packets containing some of the acid in powdered form and wrapped in wax paper may be put up and sold for this purpose. Of course, printed directions for the use of the acid should accompany the packet.

2. Sponge the hat with a solution of:

Sodium Hyposulfite	10 parts
Glycerin	5 parts
Alcohol	10 parts
Water	75 parts

Lay aside in a damp place for 24 hours and then apply:

Citric Acid	2 parts
Alcohol	10 parts
Water	90 parts

Press with a moderately hot iron after stiffening with gum water if necessary.

3. If the hat has become much darkened in tint by wear the fumes of burning sulfur may be employed. The material should be first thoroughly cleaned by sponging with an aqueous solution of potassium carbonate, followed by a similar application of water, and it is then suspended over the sulfur fumes. These are generated by placing in a metal or earthen dish, so mounted as to keep the heat from setting fire to anything beneath, some brimstone, and sprinkling over it some live coals to start combustion. The operation is conducted in a deep box or barrel, the dish of burning sulfur being placed at the bottom, and the article to be bleached being suspended from a string stretched across the top. A cover not fitting so tightly as to exclude all air is placed over it, and the apparatus allowed to stand for a few hours. Hats so treated will require to be stiffened by the application of a little gum water, and pressed on a block with a hot iron to bring them back into shape.

------♦------

Straw Hat Cleaner

2

(Poison!)

1. Powdered oxalic acid,   2 g.
2. Powdered sodium bisulfite   2 g.

Put up separately in waxed paper then tinfoil. *To use:* Dissolve each separately in ½ teacup of water. (Remove band of hat and clean with ammonia and water before replacing.) Wash the hat with soap and water to remove dirt. With a sponge apply solution 1 till well saturated. Then apply solution 2 in the same way. Let remain on from 2 to 20 minutes; then rinse well in clear water and dry.

------♦------

Stains, Removing

Argyrol stains can be removed by applying potassium iodide solution followed by hypo crystals.

Blood stains can be removed in water with ammonia.

Candle drippings are removed with lard and benzol.

Cod liver oil stains are removed with soap dissolved in amyl acetate.

Enamel stains are removed with amyl acetate and acetone.

Fruit stains are removed by pouring boiling water through the garment from a height of several feet. Use peroxide of hydrogen.

Grass stains are removed with ether or soap and alcohol.

Gum stains are removed with carbon tetrachloride, benzol.

To remove ink stains apply hydrogen peroxide and hold in steam issuing from a kettle until yellowish. Repeat. Then apply oxalic acid solution and wash with water. Repeat if needed.

To remove iodine stains use sodium thiosulfate.

Lacquer stains can be removed easily with amyl acetate (banana oil), lacquer thinner.

To remove mercurochrome stains, 1st, boil ¾ hour in soapy water, and, 2nd, apply benzaldehyde, then a 25% hydrochloric acid solution. Rinse thoroughly afterward.

Mildew is removed in one minute with Javelle water, but *not* from silk or wool.

Paint or varnish is removed with carbon tetrachloride, benzol, Stoddard's Solvent, amyl acetate; *not* for Rayon, which should be scrubbed with two parts carbon tetrachloride, two of alcohol, one part of oleic acid.

Perfume can be removed with alcohol.

Perspiration stains are removed with soapy water and hydrogen peroxide.

Scorched stains are removed with potassium permanganate followed by hydrogen peroxide.

Shoe polish stains are removed the same as candle drippings, or use benzol.

------♦------

Removing Stains

Stain   Treatment

Albumen.—Soak for a few hours in Pepsin 25, Hydrochloric Acid (25%) 50, Water 100 at 45° C.

Antimony Compounds.—Ammonium Sulfide solution.

Arsenic Compounds.—Ammonium Sulfide solution followed by ammonium hydroxide if necessary.

Asphalt ⎱ Soften by rubbing with
Gilsonite⎰ warm petrolatum or mineral oil or tetralin and dissolve with following: Benzol, 1, Carbon tetrachloride 1, Trichlorethylene 1, Ethylene Dichloride 1.

Balsams.—Ether, Toluol or Chloroform.

Beer } Ammonium Chloride 2,
Champagne} Glycerin 2, Alcohol 2, Water 7 followed by water.

Blood.—Sodium Hydrosulfite or Trisodium Phosphate and Hydrogen Peroxide.

Burnt Sugar.—Glycerin 10, Water 10, Isopropyl Alcohol 20.

Cadium Compounds.—Potassium Cyanide (poisonous) and thorough removal with water.

Chromic Compounds } Sodium Bisulfite
Chromates } or Sodium Hyposulfite and dilute sulfuric acid.

Cobalt.—Potassium Cyanide (poisonous) Solution followed by water.

Copper.—Warm 25–30% Potassium Iodide Solution.

Egg Yolk.—Soften with glycerin and treat with Alcoholic soap solution.

Grass.—Alcohol or Chloroform or Zinc Chloride 2% solution.

Henna.—Hydrogen Peroxide 10% 20, Ammonium Chloride 4, Water 20.

Iodine.—10% Potassium Iodide followed by 10% Sodium Thio Sulfate followed by water.

Iron Salts.—Sodium Hydrosulfite 8% solution.

Lacquer.—Trichlorethylene 5, Paraffin Wax 1, Acetone 1, Benzol 1, Tetralin 1, Methanol 1.

Lead Compounds.—Stain with Tinc. Iodine; dry and dissolve with concentrated potassium iodide solution.

Manganese.—10% Ammonium Sulfate Solution followed by dilute Hydrochloric Acid then water.

Mercury.—5–10% Solution Potassium Cyanide (poisonous) followed by water.

Milk.—Ether or Ethylenedichloride followed by warm borax solution.

Mold.—3% Hydrogen Peroxide, Ammonium Chloride 4, Alcohol 10, Water 70.

Nickel.—10% Solution Potassium Cyanide (poisonous) then water.

"Nicotine."—On skin—Sodium Sulfite 25, Water 100, Hydrochliroc Acid 2 or 10% Hydrogen Peroxide 10, Ammonium Chloride 1, Alcohol 5.

Oil or Fat.—Glycol Oleate 1, Hexalin 2,

Carbon Tetrachloride 1 followed by any dry cleaning solvent.

Perspiration.—10% Borax Solution or 10% Ammonium Carbonate Solution.

Picric Acid.—20% Solution Sodium Sulfate followed by soap and water.

Rust.—Potassium Binoxalate 1, Water 44, Glycerin 1, allow to remain for a few hours and wash.

Silver.—10% Solution Sodium Hydrosulfite (warm) for 15 minutes followed by soap and water.

Urine.—Citric Acid 10% followed by hot water.

Varnish.—Rosin Oil 1, Ethyl Acetate 1, Tetralin 1, Amyl Alcohol 1, Ammonium Hydroxide 1, Alcohol 1.

Vomit.—Ammonium Chloride 10% solution, followed by alcoholic soap and then water.

Water.—Rub with flannel wet with 5% White Mineral Oil and 95 Toluol.

Wine } Acetic or Tartaric Acid (10%)
Fruit } or Hydrogen Peroxide (10%) 5, Ammonium Chloride 20, Water 75.

Mercurochrome Stains, Removing

It is stated that two treatments with benzaldehyde, followed with a 25 per cent hydrochloric acid applications and an alcohol rinse, with a final bath in water will remove fresh mercurochrome stains for silk. Glacial acetic acid followed by ether is also recommended as a remover of mercurochrome stains, as is phosphoric acid in rubbing alcohol.

Marble and Concrete Stain Removal

While practically every type of stain can be removed from concrete without appreciable injury to either the texture or color, the eradication of old stains which have been long neglected may require considerable patience. It is often a matter of repeating the treatment day after day until the desired results are attained. It is not always possible to determine what the staining matter is, and hence the treatment sometimes has to be a matter of experimentation. Usually the staining matter will be found to exist in a stable form, and its removal may require several applications of a solvent which does not appreciably affect the surface. A considerable variety of chemicals may be applied to concrete without appreciable injury, but acids or those chemicals which develop an acid condition should be carefully avoided. Even weak acids, such as oxalic and acetic, may show their effects on the surface if left on concrete for a considerable length of time.

Usually stains penetrate to such an extent that they cannot be readily removed by merely applying the proper chemical to the surface or by scrubbing the stained part and it is necessary to resort to a poultice or bandage. A poultice is made by mixing one or more chemicals with a fine inert powder to a pasty consistency. This is applied to the stain in a thick layer. The bandage treatment consists of a layer of cotton batting or a few layers of cloth soaked in a chemical solution and pasted over the stain. A stain may be eradicated, first by dissolving the staining matter and drawing it out by capillary suction or driving it back from the surface; and, second, by converting the coloring matter into a form which does not show as a stain. In removing an oil stain it is usually necessary to apply a solvent and draw the dissolved oil out. An iron stain is more satisfactorily treated by applying a reducing agent, although means must be taken to prevent the re-oxidation of the iron and the reappearance of the stain. This is accomplished by an application of sodium citrate solution. Some chemicals used for removing stains are very unstable and decompose under certain conditions, producing stains of their own which may be more troublesome than the original. This is particularly true of the hydrosulfite ($Na_2S_2O_4$) used in removing iron stains, but unless the method of application described is rather closely followed a yellow stain will result. If the poultice is left on several hours, a black stain may develop, which is probably due to the formation of a sulfide of iron. Some staining matter is easily dissolved by a surface scrubbing and apparently removed, but as the area dries the stain may reappear. Tobacco stains scrubbed with a solution of washing soda may disappear in this way, but reappear stronger than before due to the solvent driving the staining matter into the surface in stronger concentrations. The chief function of a poultice is to draw dissolved staining matter out of the surface. In some cases a porous paper or blotter pasted to the stained surface after the proper solvent has been applied may be made to answer the purpose. When a stain has to be treated with a very volatile solvent, such as benzol, ether, acetone, etc., it is best to use a slab of stone or brick over the solvent. This prevents a rapid evaporation of such solvents, prolonging their action and affording a capillary action similar to a poultice. When so used, the stone or brick should be thoroughly dry.

In some cases it may not be possible to determine the type of stain. Many stains are yellow or brown, resembling iron rust. Oil stains when new resemble the oil itself, but after a considerable period of time they are apt to become yellow or dark brown. Copper and bronze stains are usually green, although, due to the iron or manganese content, or due to the alteration of fine particles of pyrites in the concrete, bronze sometimes causes a brown stain. In experiments on copper stains, made with a solution of copper sulfate, a brown stain was found on the surface after the copper stain had been removed. This yielded readily to the treatment for iron stains, indicating that it was caused by the alteration of some element in the surface, since the copper salt applied was "chemically pure."

Concrete in certain parts of buildings is apt to become stained from the perspiration or oil from the hands. Such discolorations sometimes become very prominent and resemble iron stains. This stain is not as difficult to remove as those caused by lubricating or linseed oils.

Under damp conditions, wood will rot and finally produce a chocolate-colored stain. When pine wood burns, pitch from the wood may penetrate the surface and produce a stain which is almost black. The eradication of such stains is a slow process, but in many cases it may be entirely practical.

1. Treatment of Iron Stains

Iron stains can usually be recognized by their resemblance to iron rust or by their position with respect to steel members of the structure.

Method No. 1.—Dissolve 1 part sodium citrate in 6 parts of water and mix this thoroughly with an equal volume of glycerin. Mix a part of this liquid with whiting to form a paste just stiff enough to adhere in a thick coating to the surface. Apply this to the stained area with a putty knife or trowel. This will become dry in a few days and it should then be replaced with a new layer or softened by the addition of more of the liquid. While this treatment has no injurious effects, its action may be too slow to be practical in cases of intense stains. Ammonium citrate may be used instead of sodium citrate to obtain somewhat quicker results, but, due to the development of an acid condition, it may injure a polished surface slightly.

Method No. 2.—For deep and intense iron stains it is more satisfactory to employ sodium hydrosulfite $(Na_2S_2O_4)$. Before applying the hydrosulfite to the stain the surface should be soaked for a few minutes with a solution of sodium citrate made by dissolving 1 part of the citrate crystals in 6 parts of water. To apply the citrate solution, dip a white cloth or piece of cotton batting into the solution and paste it over the stain for 10 or 15 minutes. If the stain is on a horizontal face, sprinkle a thin layer of the hydrosulfite crystals over it, moisten with water, and cover with a stiff paste of whiting and water. If the stain is on a vertical face, place a layer of the whiting paste on a plasterer's trowel, sprinkle on a layer of the hydrosulfite, moisten slightly, and apply it to the stain. Remove after one hour. If the stain is not all removed, repeat the operation. Unless the stain is deep, one treatment will be sufficient. When the stain disappears, rinse the surface thoroughly with clear water and make another application of the citrate solution as at first. Although the polish is apt to be dimmed somewhat by this treatment, it is not a difficult matter to repolish the treated portion.

2. Copper or Bronze Stains

Such stains are found where the wash from bronze, copper or brass runs over concrete. The stain is nearly always green, being due to the formation of the carbonate of copper, but bronze apparently causes a brown stain in some cases. The green stains may be eradicated in the following way:

Method No. 1.—Mix dry 1 part of ammonium chloride (sal ammoniac) and 4 parts of powdered talc. Add ammonia water and stir into a paste. Place this over the stain and leave until dry. A stain of this kind that has been collecting for several years may require several repetitions of this procedure to completely remove it. Sometimes aluminum chloride is employed instead of sal ammoniac.

Method No. 2.—Dissolve 8 ounces of potassium cyanide in 1 gallon of water. Saturate a thick white cloth in the solution and place it over the stain. When the cloth has become dry, soak it again in the cyanide solution and repeat the operation until the stain disappears. Sometimes it may be advantageous to combine this and the method above; that is, remove the greater part of the stain with the poultice and finish with the cyanide solution. This solution is very poisonous if taken into the system.

3. Ink Stains

Inks are of various compositions, and require different treatments.

Ordinary writing inks usually consist of gallotannate of iron, a blue dye, a mineral acid, phenol and a gum or glycerin. Such an ink may etch the surface of concrete due to the acid content. To remove a stain of this type, make a strong solution of sodium perborate in hot water. Mix this with whiting to a thick paste, apply in a layer ¼-inch thick, and leave until dry. If some of the blue color is visible after this poultice is removed, repeat the process. If only a brown stain remains, treat it by Method No. 1 for iron rust. Sodium perborate can be obtained from any druggist. Repolish the surface if necessary.

Synthetic Dye Inks.—Many of the red, green, violet, and other bright colored inks are water solutions of synthetic dyes. These contain no acid and do not etch concrete. Stains made by this type of ink can usually be removed by the sodium perborate poultice described above. Often the stain from such inks can be removed by applying ammonia water on a piece of cotton batting. Javelle water may also be effectively used in the same way as ammonia water or mixed to a paste with whiting and applied as a poultice. A mixture of equal parts of chlorinated lime and whiting reduced to a paste with water may also be used as a poulticing material.

Prussian Blue Inks.—Some blue inks

contain Prussian blue, which is a ferro-cyanide of iron. Stains from this type of ink cannot be removed by the perborate poultice, Javelle water, or chlorinated lime poultice. Such stains yield to a treatment of ammonia water applied on a layer of cotton batting. A strong soap solution applied in the same way may also be effective.

Indelible Ink.—This type of ink often consists entirely of synthetic dyes. Stains from dye inks may be treated as outlined above for that type. However, some indelible inks contain silver salts which cause a black stain. This may be removed with ammonia water applied on a layer of cotton batting. Usually several applications will be necessary.

4. Tobacco Stains

Method No. 1.—The grit scrubbing powders, commonly used on marble, terrazzo, and tile floors are usually satisfactory for application as a poulticing material on this type of stain. Stir the powder into a pail of hot water until a mortar consistency is obtained. Mix thoroughly for several minutes, then apply to the stained surface in a layer about one-half inch thick. Leave this on until dry. In most cases two or more applications of the poultice will be necessary.

Method No. 2.—If the scrubbing powders called for in Method No. 1 are not at hand, the following procedure may be used. Make up a soap solution by dissolving about 1 cubic inch of soap in a quart of hot water. In another vessel dissolve 1 large tablespoonful of soda ash or 2 tablespoonfuls of washing soda in one pint of water. Combine equal parts of these two solutions and apply a portion of it to the stained surface with a mop, or saturate a piece of cotton batting in the liquid and place it over the stain for a few minutes. Make up a poultice by mixing a portion of the soap and soda solution with powdered talc or whiting. Apply this to the stain and leave until dry. Scrape it off and repeat if necessary. Powdered talc is preferable to whiting, since it holds the moisture longer and thus prolongs the action of the active chemicals. It also has the advantage of being easier to remove from the surface after it has dried. Whiting is apt to cling so firmly that it has to be moistened before it can be scraped off. This is an undesirable feature, since the

dried poultice contains the staining matter, and if it has to be soaked loose from the surface some of the staining matter is apt to be driven back into the concrete. If the paste is made of the proper consistency, it can be applied with a paint brush. A whiting paste has the desired brushing properties, but in order to make the talc poultice work well as a brushing coat it is necessary to add a teaspoonful of sugar to each pound of talc. Powdered talc in the raw state is of low cost, but is not always easily obtained. When only a small amount is required, one may employ the cheaper grades of talcum powders or purchase the unscented grades from automobile tire distributors.

Method No. 3.—The following formula will be found to be somewhat more efficacious than either of the foregoing: Dissolve 2 pounds of trisodium phosphate crystals in 1 gallon hot water. Mix the contents of a 12-ounce can of chlorinated lime to a paste in a shallow enameled pan by adding water slowly and mashing the lumps. Pour this and the trisodium phosphate solution into a stoneware jar and add water until approximately 2 gallons are obtained. Stir well, cover the jar, and allow the lime to settle. For use add some of the liquid to powdered talc until a thick paste is obtained, and apply as a poultice ¼-inch thick with a trowel. If it is desired to apply this with a brush, add about one teaspoonful of sugar to each pound of powdered talc. When dry scrape off with a wooden paddle or trowel. This mixture is a strong bleaching agent and is corrosive to metals, hence in using it care should be taken not to drop it on colored fabrics or metal fixtures.

This formula is also valuable for treating other stains and will be frequently referred to in the following methods. Trisodium phosphate may be purchased at most drug stores, at chemical supply houses, or laundry supply houses.

5. Urine Stains

Use Method No. 3 as outlined above for tobacco stains. Should some part of the stain prove stubborn, saturate a layer of cotton batting in the liquids and paste over that part of the surface. Resaturate the cotton if necessary.

If the polish has been injured, moisten a piece of felt cloth or chamois skin with water, dip it into some FF carborundum or emery flour and rub the surface until it appears smooth and glossy. Then polish with putty powder in the same manner until the desired finish is obtained. When applying the putty powder, use a new piece of felt or chamois skin.

◆

Marble, Cleaning

A solution of potassium permanganate about ½ per cent strength is made, the permanganate being dissolved in a little hot water. This is a product which can be obtained from almost any chemist; this is then brushed into the marble until uniform penetration is obtained. Before it is allowed to dry, it is treated with a solution of ammonia and a little sodium hydrosulfite in warm water. When making up this solution it is essential to add the ammonia first as otherwise the hydrosulfite will be decomposed; this is then sponged on to the marble when the violet coloration of the permanganate will entirely disappear leaving a clean white product. This method can be applied efficiently on floors which become discolored through age, etc. If one application is not enough it can easily be repeated without harming the marble in any way whatsoever. If the floor is very greasy an initial washing with soda ash may be resorted to being well rinsed with clean water before applying the permanganate solution.

◆

Cleaning Copper Stained Limestone

Copper stains are occasionally observed on limestone surfaces below copper roofs or gutters, adjoining copper down-spouts, or around copper, bronze, or brass name plates, lamp standards, and the like. The following methods of removing copper stains have been developed in our laboratory. A potassium cyanide solution will wash off this stain very satisfactorily but must be used with caution because of its poisonous nature.

◆

Cleaning Stained Limestone

1. Scrub surface with

Washing Soda 5–10% Solution

using a bristle brush according to the intensity of the stain. After half an hour use a steam jet, applying the treatment uniformly to remove the stain.

After this treatment the stone usually appears clean and fresh, but if left to itself the stain tends to come back. To prevent this the surface should be scrubbed uniformly with the 10% formic acid solution.

2. A poultice method has been worked out which can be used advantageously under certain conditions for indurated stains, especially for localized or interior stains. The material for poultices can be conveniently prepared by shredding old newspapers or similar paper stock under a steam jet, sufficient fireclay being added to make the mass plastic. Washing soda is then added, according to the intensity of the stain, in amounts of from 5 to 10 per cent, and the whole is plastered over the stained surface with a trowel. The alkaline poultice is easily stripped off after 24 hours and a similar poultice containing 10% formic acid is applied in the same way and removed after another 24 hours. If the wall is dry at the start this treatment is usually successful if carried out by a workman experienced in its use.

◆

Cleaning Colored Concrete

Colored concrete surfaces may be cleaned and made more impervious by washing with liquid soap. When this treatment is used the soap should be applied and allowed to stand overnight, being washed off thoroughly the next morning.

The application of ordinary floor wax once a month after the concrete is dry and clean will produce deep colors, improve the wearing surface and make it easy to keep clean. After the first two or three waxings, unless the surface is to be subjected to unusually severe wear, waxing twice a year will be sufficient.

◆

Copper Cleaner

Oxalic Acid	1	oz.
Rotten Stone	6	oz.
Gum Arabic	½	oz.
Cottonseed Oil	1	oz.

Water sufficient to make paste.

Apply to small portion and rub dry with flannel.

◆

To Clean Bronze

Saturate a 5% acetic acid solution (or household vinegar) with ordinary table salt. This solution will clean bronze or brass; and if the metal is immediately polished and lacquered with clear lacquer, a reasonably permanent finish will result.

General Spot Remover (Egg, Blood, Candy, General Dirt)

2% Liquid Soap Solution

Wet the spot and place folded cloth underneath. Dip clean cloth in soap solution and gently rub spot until lather forms. Remove suds by rubbing with wet cloth. Repeat if necessary.

Grass, and Fruit Stain Remover

Immerse spot in 95% denatured alcohol and then follow with 2% soap solution.

Removing Oil and Grease Spots

Immerse the goods for one hour in a warm saturated solution of sodium aluminate, diluted to about ½ strength. Then rinse in warm water, extract and dry. Much better results are obtained when the solution is lukewarm, although it can be used cold.

Solutions made by this same formula may also be bottled and used for removing small spots, as it leaves no fringe or ring. Put a piece of blotting paper under the spot and apply solution with a cloth.

Grease, Oil, Paint and Lacquer Spot Remover

Alcohol	10 lb.
Ethyl Acetate	20 lb.
Butyl Acetate	20 lb.
Toluol	20 lb.
Carbon Tetrachloride	30 lb.

Paint and Tar Solvent

Xylene	140 lb.
Trichlorethylene	47 lb.
Ethylene Dichloride	61 lb.
Oleic Acid	40 lb.
Sulfonated Castor Oil	24 lb.
Isopropanol	33 lb.
Triethanolamine	16 lb.

Cleaning Compound, Bottle

Sodium Metasilicate	10
Soda Ash	20
Trisodium Phosphate	25

Hectograph Stains from Skin, Removing

Sodium Hydrosulfite	5–10
Water	95–90

Press-Marks on Celanese-Garments

In order to remove such lustrous spots from dull finish Acetate rayon often a good result is obtained (in case of plain colored garments) by soaking the whole garment for 1 hour in pure Methanol with addition of a little Castor Oil. The amount of liquid should be just enough to perfectly penetrate the garment without any excess liquid. Thus bleeding of colors is avoided. The spots will disappear due to swelling action. Sometimes it is advisable to rub and slightly pull the parts having marks, to loosen the fibers, melted by the heat, from each other. Then the garment is dried on a hanger with a fan.

Scorch Remover

Slight scorch spots can be removed by immersing for about an hour or more in a 3% Hydrogen peroxide solution.

Paint Brush Cleaner

Mix (1)

Kerosene	2 pt.
Oleic acid	1 pt.

Mix (2)

Strong liquid ammonia, 28%	¼ pt.
Denatured alcohol	¼ pt.

Slowly stir 2 into 1 till a smooth mixture results. Directions: To clean brushes, pour into a can and stand the brushes in it overnight. In the morning, wash out with warm water.

Printers' Form Cleaner

Sodium Metasilicate	20 lb.
Water	50 gal.

Rust and Ink Remover

Immerse portion of fabric with rust or ink spot alternately in Solution A and B rinsing with water after each immersion.

Solution A

Ammonium Sulfide Solution	5%
Water	95%

Solution B

Oxalic Acid	5%
Water	95%

Stains, Blacking Removing

The following will probably be effective:

Nitrobenzene (Oil or Mirbane)	1 part
Phenol (Carbolic Acid, U. S. P. 90% Solution)	7 parts

After application, rinse well with alcohol.

Cleanser for Hunting Calf Leather

Trioxymethylene	70 g.
Cleaning Naphtha	30 cc.
Oxalic Acid	5 g.
Liquid Soap	20 cc.

Mix thoroughly.

Cleanser for Sporting Leathers

Water	75 cc.
Acetic Acid (80%)	5 cc.
Alcohol (95%), Denatured	30 cc.

Radiator Cleaner

Compound for use in hot force pump automobile radiator flushing tanks.

76% Flake Caustic Soda	60 lb.
Sal Soda	30 lb.
Rosin	10 lb.

Use about 40 lb. to 75 gal. water.

Laundry Bleach

Chlorinated Lime	1	lb.
Washing Soda	1½	lb.
Water	1	gal.

Allow to stand for a few days and filter.

Laundry Blue

Good Quality

Formula No. 1

Ultramarine	60 lb.

Cigarette Stain Removal

The following method removes cigarette stains from fingers.

A. Potassium Permanganate (2% Soln.)	

B. Sodium Bisulfite	10
Orris Root, Powd.	10
Perfume	to suit

Apply solution A with a swab and after a few minutes rub with B moistening with water if necessary. Wash well with soap and water.

Bicarbonate of Soda	40 lb.
Glucose	12 lb.

No. 2

Cheap Quality

Ultramarine	18 lb.
Kiln-Dried Blue Earth	20 lb.
Terra Alba	15 lb.
Bicarbonate of Soda	45 lb.
Glucose	10 lb.

No. 3

Lime	5 oz.
Water	10 oz.

Stir until smooth and mix with a hot solution of

Dextrin, Yellow	5 oz.
Water	3 oz.
Glycerin	5 oz.
Phenol	0.2 oz.
Ultramarine Blue Powder	75 oz.

Ultramarine Blue Paste, Laundry

a.	Glue	5 oz.
	Water	10 oz.

Soak cold, then warm to dissolve.

b.	Yellow Dextrin	5 oz.
	Water	3 oz.
	Glycerin (sp. g. 1.23)	5 oz.

Mix both parts warm, preserve with 0.2% Moldex or phenol, etc., and grind now with

Ultramarine Blue	75 oz.

CHAPTER XVII

TEXTILES AND FIBRES

TEXTILES and their treatments are many and diverse. The principal natural textile materials are cotton, wool and silk. The chief synthetic textile materials are "rayon and viscose and other cellulose products. Each type of textile requires a different treatment for production, sizing, dyeing and finishing." Specialty effects are produced by intricate processes.

Textile Materials, Identifying

	Vegetable Fibres					Artificial Fibres			Animal Fibres	
	Cotton	Linen	Jute	Hemp	Ramie	Viscose	Chardonnet	Acetate Silk	Wool	Silk
Burning..............	Burn rapidly with pungent smell					Burn rapidly with pungent smell		Forms beads	Burn slowly with characteristic smell	
Caustic soda, 76° Tw...	Insoluble	Insoluble	Brown. Insoluble	Yellow. Insoluble	Insoluble	Unchanged	Disintegrated and partly dissolved	Fibre swells	Soluble cold	Soluble hot
Alkaline lead.........									Black	
Sulphuric acid, 168° Tw.	Dissolves rapidly	Dissolves slowly	Dissolves slowly	Dissolves slowly	Dissolves slowly	Rapidly dissolve			Insoluble	Dissolves
Nitric acid.............	Insoluble	Insoluble	Brown. Insoluble	Yellow. Insoluble	Insoluble	Dissolve rapidly with yellow coloration			Yellow. Insoluble	Yellow. Dissolves
Ammoniacal copper solution..............	Soluble	Soluble	Insoluble	Insoluble	Insoluble	Swells, disintegrates and is partly dissolved		Unchanged	Insoluble cold	Soluble cold
Aniline sulphate.......			Yellow	Yellow						
Acetone..............						Unchanged	Unchanged	Dissolves rapidly		
Iodine and sulphuric acid	Blue	Blue	Yellow	Yellow	Blue					
Diaphenylamine and sulphuric acid.........							Blue			

Sizing of Textiles

For this service hide glue finds extensive use because of absence of the most objectionable impurity sulfur dioxide or sulfites. As the colors employed for dyeing fabrics are much more delicate than those used in paper and are usually soluble, the absence of traces of mineral acids or alkalies is also indicated.

Hide or extracted bone glue is used on cotton goods to stiffen and give body to the material. If solution of this glue is too thin it will penetrate the pores of cotton fibre to such a degree that the latter will be altogether too stiff to use, while if it is too viscous it will not be absorbed at all and will fail to dry out during passage through drying chamber. The desired results are obtained when a very dilute solution of this glue is treated with a solution of alum. The alum thickens the solution and is satisfactory because no precipitation will result.

Carpets, tapestries, burlap wall covering are all heavily sized with this grade of glue.

In the case of shade cloth where firmness with flexibility is desired—strong high grade glue is used.

All straws used in the manufacture of hats are sized. In this case a product that is more or less resistant to the action of water and also light in weight is desired. A final bleaching is given the material, by the use of oxalic acid, or lead acetate. Many manufacturers bleach their glue before sizing.

Metallic Printing on Textiles

A certain number of fabrics are adorned with metallic powders printed with the aid of hot solutions of glue or gelatin, containing powders of aluminum, copper, bronze or brass in suspension, which remain fixed on the material after cooling. Cylinders of copper, aluminum or brass are used for applying the paste and are hollow so that steam or hot air may be introduced. The color-feed rollers are also heated. The trough for the metallic paste has a double bottom and it, too, is heated. All the heating elements are maintained at about the same temperature.

The printing completed, the cotton fabrics are passed through a drying machine.

It has been found by experience that the use of a glue or gelatin paste at a high temperature has the great advantage of causing the metallic powder to adhere more easily to the surface of the fabric. But, to increase the fixation still more, the cloth is submitted, immediately after drying, to a certain pressure by passing it through a pair of calender rolls, which at the same time give it a slightly glazed finish.

If the metallic powder used is sufficiently fixed, the designs are very smooth and glossy, and if they are geometrical shapes they form a collection of fine lines almost imperceptible to the eye, but giving more attraction to the cloth. It is the impression of the rollers which produces this effect.

(1) *Dress goods with metallic effects.* —Certain garments for daily use gain much from the discreet use of metallic fabrics, and as these give a rather exclusive air their use has developed of late. The printing of these fabrics must be done with greater care than of those destined for carnival wear. The fixation of the powders must be absolutely complete, to the point of being able to resist a soaping without risk of the powder bleeding, even partially.

The designs used are most frequently flowers or leaves on a background of accentuated lines, to which a very special finish is obtained by pressure. The cheapness of the powders permits their use for muslins, tulles and voiles. When these more common fabrics are manufactured with care there is not much to choose between them and the older and more expensive goods. Their appearance in light, after they have passed through the calender, is remarkable.

(2) *The Printing Pastes.*—The printing pastes employed for the manufacture of these goods are very varied, but the majority of them permit the ordinary use of the metallic powders just enumerated. These are finally fixed with albumen, casein, rubber, or even with resin, bakelite or cellulose acetate.

One can, in this case, obtain very good results by printing in the cold, followed by drying and steaming. The goods produced in this way have sufficient resistance to washing and rubbing.

Sometimes, in the preparation of the pastes blood albumen (*e.g.,* 10 parts of the commercial quality, inodorous as far as possible) is used. It is wetted with 15 parts of water and mixed with a wooden rod 12 hours later, until a uniform mass is formed. This is then filtered through a sieve and ¾ part of essence of terebenthine and 1 to 3 parts of bronze, brass, aluminum or other powder are added. This mixture is used for direct printing from engraved rollers.

The smell left by blood albumen in the fabric sometimes gives rise to complaints. It is avoided by mixing an egg albumen with the blood albumen, or by using the former exclusively. This leads to a marked economy, but the results are less certain and sales more difficult. One or other of these albumens is sometimes replaced by casein dissolved in a weak ammonia solution. In these various cases, the fixation of the powders is not so good. When it is wished to use rubber for the fixation, 150 to 200 parts of the powder are mixed with 1,000 parts of a solution of this substance in benzine; fixation takes place after the solvent has evaporated.

(3) *Production of Metallic Designs with the Aid of Acetyl Cellulose.*—Solutions of cellulose or of its esters give excellent results, when it is a question of producing fine designs. The cellulose is dissolved in ammoniacal copper oxide, the metallic powder is added in the desired proportion, and the paste used on a color printing machine. The copper oxide in the fabric is eliminated with the aid of acid. The objection to this procedure is its high cost.

Instead, one may use acetyl cellulose dissolved in an appropriate solvent. The paste is prepared by mixing 12 parts of the acetyl cellulose solution, 24 parts of resorcine, 16 parts of water (added later), and 48 parts of denatured alcohol. The mixture is agitated, allowed to stand until the constituents are entirely dissolved and 15 parts of fine metallic powder are then added. One hundred parts of this paste are used on the roller printing machine, together, if wished, with pastes containing basic colors or others.

If colors are being used, one proceeds as in the following example: 1 part of Rhoduline Blue 3GO, 2 parts of Rhoduline Yellow 6G, 3 parts of a good commercial acetic acid; to this mixture add 20 parts of iron-free water and, later, 10 parts of hydrolite dissolved in 10 parts of water. After mixing these substances well, 10 parts of aniline oil, 10 parts of alcohol and 12 parts of tannin powder are added. The paste is then ready for use.

When the designs have been printed on the cotton fabric, this is dried, steamed for 4 minutes, and passed through a tartar emetic bath, if the color must possess good fastness; finally, the fabric is rinsed in running water, dried and calendered.

It is simple to vary the effect by mixing color of various kinds with the powders, so as to shade or modify these. Interesting effects are also obtained by confining the powders to certain parts of the print, obtained with basic colors or others on cotton, and by limiting the print to points, circles and so on, with lines of gold or silver, applied on the bench and giving the appearance of original oriental goods.

Concentrated Warp Sizing
(For Cotton Warps)

Sulfonated Tallow (75%, if 50% used increase proportion)	36–42 lb.
Raw Beef Tallow — good quality preferred, otherwise size may be discolored slightly	18–24 lb.
Dry Gum Tragacanth	14–20 oz.
Water	38–45 lb.

The gum tragacanth should be placed in separate vessel and heated up to boil and allowed to stand until complete jell has been reached, then it is ready to add to mix.

Mix the sulfonated tallow and raw tallow in kettle and heat while mixing until thoroughly blended and syrupy.

Add the gum tragacanth jell and mix until blended.

Add the necessary amount of preservative and place in closed barrels until ready for use.

Finish for Fancy Woven Goods

1. Composition of the finish:

Dextrin	150	parts
Epsom Salt	80–90	parts
Monopole Soap	6– 7	parts

per 1000 parts paste or brought up to the required degree of Tw.

2. Thicker finish:

Dextrin	200	parts
Epsom Salt	110–130	parts
Glucose	50	parts
Monopole Soap	6– 7	parts

per 1000 parts paste or brought up to the required degree of Tw.

3. Cheap finish:

Potato Flour	50	parts
Epsom Salt	50	parts
Monopole Soap	5–66	parts

per 1000 parts paste.

Dissolve the different constituents separately in water and mix them together by good stirring. In cases where the products cannot be dissolved separately owing to want of accommodation, dissolve the dextrin or potato flour together with the Epsom Salt and boil, then add the glucose and finally the Monopole Soap. The latter is dissolved with direct steam in a small quantity of water, but before adding it to the finish, dilute the dissolved soap with as much water as possible in order that the fatty matter may be finely and uniformly divided and thus render same particularly stable. The dissolving of a little dextrin (4–5 oz. dextrin per 1 lb. of soap) together with the Monopole Soap will be found advantageous.

It is not necessary to boil the finish again after the addition of the soap, although a boiling is not detrimental. The temperature of the size ready for use should be 95–115° F.

Concentrated Finishing Compound (For Cotton Piece Goods)

Sulfonated Tallow (75%)	22–26 lb.
Japan Wax	12–15 lb.
25% Tri-Sodium-Phosphate Solution	20–24 lb.
Water	50–60 lb.

The Japan wax should be emulsified in a separate vessel.

Mix the tallow, ⅓ of the Japan wax (emulsified) and required amount of T. S. P. solution until thoroughly blended.

Add the remainder of the Japan wax emulsion, agitate and heat; it is best not to boil.

Stir until a creamy mix is secured.

The Dyeing of Cotton

The preparation of the fibre for dyeing depends upon the form in which it comes into the dyehouse and differs in the handling as well as processing.

Skeins are boiled out under pressure of 2–3 pounds with 0.25–0.5% calc. soda and 1% of a sulfonated oil or suitable wetting-out agent for 3 hours. The boiling liquor should be at least 15–20 inches above the check-chain before the kier is closed.

Piece-goods must be thoroughly desized before dyeing to prevent ''Landscapes'' or cloud effects. An addition of 0.1–0.2% of Activin based upon the weight of the goods will aid in a rapid and more complete desizing of the material. Piece-goods which must be bleached are best boiled out with 3% caustic soda, 2% calc. soda, 1% of a wetting agent and 0.1–0.2% Activin for 4 hours under 3 pounds of pressure. It may be said here that the degree of desizing can be successfully tested with a solution of potassium iodide.

When piece-goods are to be dyed with vat colors it is well to note that the ends of the pieces when sewn together should lie over one another, somewhat in the manner of roof-shingles. Pieces sewn together side by side, i.e., against each other, will show ''airstripes'' after dyeing, evidenced by a deeper shade.

Tubular knit goods (jersey) and delicate materials are, of course, not boiled out under pressure, but are boiled out on the reel with 1% of calc. soda and 1% of sulf. oil for 1 hour.

Raw cotton, slubbing, cops, bobbins, and warp on the beam are usually handled in mechanical apparatus and are boiled out with 1% calc. soda and 1% sulf. oil for 1 hour.

Preferred and often used is the cold-wetting-out method for raw cotton and slubbing, which has the advantage of preserving the spinning qualities of the fiber. During the packing of the material attention should be paid that no channels develop, as this will interfere not only with the proper boiling-out process but also will give unsatisfactory results in dyeing.

Bobbins and warps on beams can, of course, be dyed with vat colors in mechanical apparatus, however, certain irregularities must be overlooked, and the same is true when dyeing skeins in apparatus which employ so-called ''Hang-systems.''

Dyeing skeins with vat colors in the dye kettle offers, of course, also certain difficulties such as unevenness, and an aid to good results are levelling and protecting agents such as Tetracarnit, Glue, Sulfite-cellulose-waste liquors, Soap, Sulfonated oils, etc. It must, however, be remembered that Soap or Sulfonated oils can be used only to limited amounts in the dyebath, as they will induce the material to swim and thereby only hinder the dyeing process. An addition of Glucose to the dyebath will often aid in overcoming unevenness, however, the amount of caustic soda must be increased about 30%, as the Glucose will use up this amount. A further aid to level uneven dyeings is to remove the lot from the dye liquor, squeeze, and return to the dyebath under addition of more sodium hydrosulfite, and raising the dyeing temperature from 60–100° F. It must be mentioned, however, that most of the vat color types will lose their brilliancy and also give up part of their fastness qualities should the temperature be raised above their regular dyeing temperature.

It is perhaps more advisable, providing the dyeing qualities of the dyestuffs are accurately known to the dyer, to begin dyeing at a lower temperature and gradually raise to the dyeing temperature, as in this manner no complications will have to be feared, provided the condition of the vat is constantly observed.

After dyeing the material is squeezed and hung on sticks to oxidize. Should oxidation be too sluggish the process can be hastened by passing the lot through a bath made up with 0.3–0.5 cc. per liter of 30% Hydrogen Peroxide, at a temperature of 80–100° F. Sodium Perborate (1–3% from the weight of the goods) can be used instead of Hydrogen Peroxide. After the material has been handled in such a bath for 10–15 minutes, the temperature can be raised to the boil and the subsequent soaping be carried out without fear of complications, as the perborate will give up its oxygen quickly at a temperature of 150° F. It may be pointed out that such a method is also more economical as it eliminates one extra handling of the material.

Dyeing Knit Fabrics

Using direct colors. For light shades dissolve dyes separately and strain into bath. Dye goods for 10 minutes at 80° F. Add glauber salts (5% of weight of goods) and raise temperature to 120° F. Shade should be reached in 15 minutes. For dark shades increase glauber salts to 150% and increase temperature to 160° F.

½ of 1% neutral olive oil soap may be used for improving feel of finished goods. Dry at 100° F.

Dyeing Silk Black (Lyons)

About 10 to 20% yellow prussiate of potash is used in proportion to the weighting with oxide of iron which the silk has received previously. In addition, a quantity of hydrochloric acid, equal to the prussiate, is required. Prepare the bath with the prussiate and half the hydrochloric acid. Enter at 30° C., turn the silk about 10 times, heat to 45° C., turn a few times, add the other half of the acid and heat to 50 to 55° C., turn again a few times, wring out and wash well in water.

A weighting of 16 to 24% is obtained; or by a threefold treatment with nitrate, etc., the loss sustained by the discharging is recovered, and the silk brought to "pari." A further weighting of 4% may be added by one more treatment with "nitrate of iron" after the blue dyeing, and subsequent rinsing with water to precipitate the ferric hydroxide (hot soaping would affect the Prussian blue). Work the silk after these treatments 1 hour in an old bath of catechu (gambier) standing at 4 to 7½° Tw., the temperature of which should not exceed 50° C., so that the Prussian blue may not be decomposed and the shade become too dark; rinse and hydro-extract. The silk acquires in the catechu bath an over-charge (over pari) of 15% and becomes more greenish.

Dyeing Straw Green

The light green which is so popular on straw hats is produced with basic colors in a bath made up of 5% acetic acid and 5% Malachite green crystals. The dyeing is continued at about 160° F. for an hour or until the shade is acquired, after which the straw is removed, rinsed, hydro-extracted and dried at a low temperature.

Removing Dye from Woolen Goods

To strip all dyes from old woolen garments, mainly sweaters, first remove the loose color by heating to 180° F. with ammonia 28 degree using from 3% to 5% on weight of goods. Drain thoroughly. Wash with warm water until clear, repeat if necessary using 2% ammonia. In a wooden barrel mix 3% to 5% basic zinc sulfoxalate formaldehyde with an equal amount of 28% acetic acid and about 5 gal. hot water, stir and mix thoroughly; add this solution to the material to be stripped. Heat to boiling and boil 20 minutes. Add 3% more acetic acid and boil 15 minutes; drain and wash thoroughly to completely remove the stripping compound.

Scouring Cotton-Rayon Fabrics

Turkey Red Oil	5
Olive Oil Soap	5
Soda Ash	1
Water	100 gal.

Use at 200° F. for 1–2 hours. If fabric contains celanese keep temperature below 175° F. and leave out soda ash.

Scouring Knit Goods

Scour at 160° F. for 20 minutes in	
Trisodium Phosphate	1

Olive Oil Soap	2
Water	97

Rinse well in soft water.

Cotton, Coloring

Cotton and cotton materials are generally dyed with *direct* dyes, sometimes called substantive dyes. They do not need any chemical to develop or lock the dye into the fibre. Common salt, however, is used as an auxiliary to aid dyeing.

Dyeing instructions: Prepare dye bath using about four gallons of water to each pound of material.

Add five pounds of salt for each pound of dye used.

Bring temperature up to 140° F. Introduce the material. Bring temperature up to a boil and keep at boiling point three-quarters of an hour. Rinse and dry.

Average Yellow requires
    1 lb. of dye to 100 lb. material
Average Red requires
    2 lb. of dye to 100 lb. material
Average Blue requires
    2 lb. of dye to 100 lb. material
Average Green requires
    2 lb. of dye to 100 lb. material
Average Black requires
    5 lb. of dye to 100 lb. material

Representative dyes are:
    Direct Fast Yellow NN
    Chrysophinine (Yellow)
    Direct Blue 2B
    Direct Sky Blue 5
    Direct Orange 2R
    Direct Green
    Congo Red
    Direct Black E
    Direct Pink E
    Direct Violet N
    Direct Brown

Wool, Coloring

Wool and woolen materials, for the most part, are dyed with acid dyes; the acid used is Sulfuric. In some cases acetic acid is used. Glauber salts are added as an auxiliary in dyeing.

Dyeing instructions:
    For each 100 lb. of material
    use 4 gallons of water.
    add 3 lb. of Sulfuric Acid.
    add 10 lb. of Glauber Salts.
    add 1 to 5 lb. of color depending
    on shade and color strength.

Yellow generally requires 1 lb.
Red, blue, green generally require 2 lb.
Black generally requires 5 lb.
Bring temperature of dye bath to 140° F. Immerse material, bring to boil and boil three-quarters of an hour and rinse.

Representative dyes are:
    Yellow—Tartrazine
    Lemon Yellow—Erio Flavine
    Orange—Orange II
    Red—Ponceau 2R
    Red—Crocein Scarlet
    Magenta—Acid Magenta B
    Violet—Acid Violet 6 BN
    Green—Patent Blue A
    Black—Acid Black J
    Black—Acid Black 10 BX

Silk, Coloring

Silk may be colored with Direct, Acid, or Basic colors. The Direct colors are dyed in a neutral bath. Some direct colors require the addition of Acetic Acid to the dye bath toward the end of the operation. Temperature 180 to 200° F. Time about 30 minutes.

Acid Colors.—Dyed in bath acidulated with Sulfuric Acid. Temperature 180 to 200° F. Time about 30 minutes.

Basic Colors.—Dyed in bath acidulated with Acetic Acid. Temperature start at 100° F., go to 140 to 175° F. slowly. For Auramine, temperature must not exceed 1404° F.

Direct dyes (see dyes for cotton).

Acid dyes (see dyes for wool).

Basic dyes:
    Yellow—Auramine
      2 lb. per 100 lb. material
    Orange—Chrysoidine Y
      2 lb. per 100 lb. material
    Brown—Bismarck Brown
      2 lb. per 100 lb. material
    Pink—Rhodamine B
      2 lb. per 100 lb. material
    Blue—Methylene Blue 2B
      2 lb. per 100 lb. material
    Violet—Methyl Violet
      2 lb. per 100 lb. material
    Green—Malachite Green X
      2 lb. per 100 lb. material
    Black—Basic Black
      2 lb. per 100 lb. material

Bleaching Cotton

The goods to be bleached are impregnated with a solution of Turkey-Red Oil of from 5 to 10% strength, according to the natural color of the cotton, wrung and centrifuged to get rid of the excess, and then dried. The goods are next boiled for 6 hours under pressure with from 1½ to 2% of caustic soda, rinsed, slightly soured, rinsed again, passed through a very weak soap bath, again rinsed, and then dried. If the cotton is very pure and easily bleached the process may be simplified by putting the Turkey-Red Oil into the boiler with the lye. The process has special importance for bleaching makkoyarn, as that yarn, so largely used for finer counts, has been hitherto very difficult to bleach, requiring strong baths of chloride of lime.

Turkey-Red Oil may also be used to advantage in bleaching cotton by the usual chloride of lime method, as follows:

Goods may be treated with the oil before bleaching. Pad goods in a 5% solution of the oil, and steam without pressure. The oil may also be added to the contents of the kier, whether this consists of lime, soda, or caustic soda. Two litres of Turkey-Red Oil per cubic meter of caustic soda at 3° Tw. are sufficient. The oil is added to the saturated liquor, which is afterwards introduced into the kier. There is no change required in the bleaching operation.

When lime is used, the oil is added to the lime after slaking, and then the necessary quantity of water is added. A milky liquid is thus obtained, which only settles very slowly, and which penetrates the goods perfectly, especially when tepid. The use of the oil in the lime boil gives better results than in the caustic soda boil.

Before the anti-chlorine bath it is advisable to wash well in soft water, in order to remove any undecomposed oil. Goods bleached with the aid of Turkey-Red Oil are much softer than those bleached without. The chemicking is easier and quicker, while at the same time less bleach may be used.

Sodium Hypochlorite Bleach

To prepare Sodium Hypochlorite. Dissolve 100 lb. of 33% Bleaching Powder in 40 gallons of water.

Dissolve 60 lb. of Soda Ash in 20 gallons of boiling water, afterwards diluting with 10 gallons of cold water.

The Soda solution is then to be mixed with the bleaching powder paste and well stirred for one-half hour and allowed to settle over night.

In the morning the clear solution is to be drawn off.

The residue should be washed with clear water, allowed to settle, and the top liquor added to the main solution.

The washing may be done for economy, several times, each time letting the solution settle and adding the top to the main solution.

Use only sufficient wash waters to bring the main solution to stand at 6° to 7° Tw.

Now add 1½ to 2 lb. Soda Ash. Dissolve and let stand over night, when all the lime will have been thrown out of solution.

It is then ready for use by simple dilution in water to the desired strength for bleaching.

Sodium Hypochlorite has advantages over the old-time Chloride of Lime solution. The goods come out softer. They rinse cleaner, and this insures better strength of the fibre and a more permanent white.

Viscose Manufacture
For Rayon and Cellophane

Steep 2 lb. cotton or pure wood pulp fibre in 18% caustic soda solution at 20° C. for 1 hour

Press excess caustic out till pulp weighs 6.5 lb.

Keep in a closed container for 70 hours at 20° C.

Place in large mason jars, first breaking pulp up. Add ¾ lb. Carbon Bisulfide; close jar and shake for 2 hours till orange color appears.

Dissolve this xanthate in a 3½% caustic soda so as to finally have 7% cellulose in solution, approximately use 16 lb. to 18 lb. of 3½% caustic soda solution.

Keep this viscose for 3 days at 18° C.

For coagulation use a spin bath of following specifications:

Sulfuric Acid	9%
Glauber's Salts	18%
Zinc	1%
Glucose	5%
Temperature	45° C.

Then rinse acid out of thread.

For transparent films spread very thin on a plate of glass. Place glass in a solution of 30% ammonium sulfate. Then place in saturated salt solution. Then place in 3% sulfuric acid solution till film is clean. Wash acid free and dry.

Carbonizing Wool in Cotton Mixture

Some kinds of burnt out embroideries which consist partly of pure cotton and partly also of artificial silk and cotton, are prepared on a ground of wool or cotton. The ground is then usually carbonized before the dyeing, that is to say, removed so that the actual embroidery alone remains standing out.

For cotton embroidery, a wool ground is usually used, and is carbonized by a hot treatment or by boiling for 20 to 30 minutes with caustic soda lye or 3°–5° Tw. The embroidery is then rinsed thoroughly, scoured off and dried, the destroyed wool then being removed by heating.

Boiling Off Silk

Raw silk consists chiefly of two substances, the true silk fibre, called "fibroin," and an outer layer of material known as "sericin." It also contains a very small amount of wax, fat, coloring matter and ash. Most of the coloring matter is in the outer sericin layer.

Sericin is a substance resembling gelatin in its properties, and is soluble in water only by prolonged boiling.

Fibroin is a proteid and is not noticeably affected by prolonged boiling in water, but is somewhat readily attacked by caustic alkalies even in weak solutions, their action rendering it more brittle and rough and diminishing its gloss. Fibroin is also attacked by soap solutions if boiled for a long time, but it is not acted upon by weak acid solutions.

In preparation of silk for the dye bath it has been customary to "boil off." This process consists in boiling in a bath of soap and water, sometimes with the addition of Carbonate of Soda, the purpose of such treatment being to remove the outer layer of sericin, whereby the silk becomes lighter in color and the luster is developed, and it becomes softer and more suitable for dyeing.

During the process of boiling off, the sericin first swells up, making the silk sticky. It then dissolves, leaving the lustrous and internal thread exposed.

In treating piece goods which are composed partly of cotton or wool, the boiling off process serves the further purpose of cleansing from the material whatever dust may be adhering to the silk.

It tends also to improve the quality of the cotton or wool mixture. It is customary to put the goods through a washing process after boiling off. The boiling off and washing processes consume much time and labor, and employ materials which, while not expensive in themselves or in small quantities, become expensive when used in large quantities, as they must be used in the customary practice of the art.

It is claimed by users of Sulfonated Castor Oil that silk left to soak in a bath made up to consist of:

One part of the Oil to 1000 parts of water, with the addition of sufficient soda ash, or about two parts, to make the bath slightly alkaline at a temperature of about 98° C. for one-half hour, the degumming process will become complete during the dyeing.

The solution is very mild in its action upon the fibroin, leaving it coated with a very thin layer of nitrogenous material which is repellent to water, though soluble on prolonged boiling. The protective layer is of extreme thinness, and is removed in whole or in part in the ordinary operations to which silk goods are subjected subsequent to boiling off. This layer also probably protects the fibroin from weakening not only during the time that it is in the bath, but during the subsequent operation of dyeing.

Olive Oil Finishing Emulsion

May be used for finishing blankets, hosiery, mercerized cottons, etc.

25% Tri-sodium-phosphate Solution	50	parts
Olive Oil	30	parts
50% Sulfonated Tallow	10–15	parts

Add half of olive oil and mix thoroughly in trisodium phosphate solution then boil and agitate until saponification takes place and add in the remaining half; then add in sulfonated tallow and mix until a smooth blended emulsion is formed. Test—10 cc. in 100 cc. lukewarm water; should emulsify and not separate out in oily spots, etc. Should have consistency of soft lard or butter.

Anti-Seize Compound

Used in threads to prevent seizing.

Petrolatum	50%
Zinc Dust	50%

Solubilizing Starch

The starch is mixed with required amount of water and 1% Aktivin S on amount of starch used.

A wooden vat with mechanical agitator preferred, copper can be used but wood keeps solution hot the longest. Direct steam may be used in boiling up starch. A thick paste is made first, this becomes thinner and after boiling 20 minutes or longer the starch becomes thin flowing. Do not fail to actually boil starch and covering to prevent splashing.

Starch	100 lb.
Water	150 gal.
Aktivin S	1 lb.

Stirring and boiling is discontinued when desired thinness is reached.

------◆------

Cotton Good Softeners

The saponified coconut oil softeners are easily made by heating the melted oil with the required amount of a concentrated caustic soda solution until saponification is complete, following which the mixture is diluted to approximately 20% fat content.

------◆------

Coconut Oil Softener

Coconut Oil	2060 lb.
Soda Ash	135 lb.
Caustic 39° Bé.	1090 lb.
Dilute to produce	9000 lb.

These products are finished off alkaline or neutral as desired and are exceptionally well suited for use in hard water or in mixes containing excessive amounts of salts, such as Epsom and others. Their excellent solubility, moreover, permits of easy removal on washing when this is necessary. Coconut oil soaps almost invariably become rancid with age, although this can be retarded by complete saponification. Softeners made from the completely neutralized fatty acids are less liable to this fault than those made from the oil itself. The great fluidity of the soap with its capacity for holding water enhances the value of this material as a softener, as well as for the lustrous sheen imparted on calendering. A shirting formula containing this oil is given here:

Shirtings

Wheat Starch	1 lb.	10 oz.
Potato Starch	15 lb.	

Talc		60 lb.
Stearic Acid Softener	2 lb.	8 oz.
Coconut Oil Softener		13 oz.
40 gal. Mix		

------◆------

Waterproofing Composition

To thirty parts of commercial petrolatum fifteen parts, by weight, of aluminum palmitate are added and the mixture kneaded into a smooth paste free from lumps. Or the petrolatum may be heated to about 130° F., whereupon the consistency of the petrolatum is such that a smooth mixture is produced by introducing the palmitate and stirring. To this mixture is added fifty parts of commercial yellow beeswax and one hundred five parts of soft paraffin wax, such as white scale wax, and the resulting mixture agitated in a steam heated container. The temperature is brought up to 250° to 270° F. and the agitation continued until a smooth, homogeneous mass is obtained. The mixture is then allowed to cool to about 220° F. and about eight hundred parts by weight of a petroleum thinner having a boiling range in this instance of 275° to 450° F. added. It will be found that the resulting product is stable and homogeneous, of proper viscosity for application by hand or machine, and extremely suitable as a saturant for waterproofing fabrics. It acts as a preservative to fabrics to which it is applied and forms a water-repellent and impervious coating on each of the fibers making up the material.

------◆------

Waterproofing Liquid (Cloth or Wood)

Paraffin	⅔ oz.
Gum Damar	1⅕ oz.
Pure Rubber	⅛ oz.
Benzol	13 oz.

Carbon tetrachloride q. s. 1 gallon. Dissolve rubber in benzol; add other ingredients and allow to dissolve. (Inflammable).

------◆------

Waterproofing Liquid

This may be used on fabrics, paper and other fibrous bases. It penetrates quickly and leaves a flexible, odorless product which is highly water repellent.

Example 1.—Use of high melting paraffin wax and plasticizer for the cellulose nitrate.

	Per cent
Nitrocotton (15–20 seconds)	1.0
High Melting Paraffin Wax	4.0
Naphthene Base Mineral Oil	6.0
Butyl Stearate	2.0
Butyl Acetate	4.0
Ethyl Acetate	25.0
Gasoline	13.0
Toluol	40.0
Alcohol (Denatured)	5.0
	100.0

Example 2.—Use of Japan wax and no plasticizer for the cellulose nitrate.

	Per cent
Nitrocotton (15–20 seconds)	1.0
Japan Wax	3.0
Naphthene Base Mineral Oil	3.0
Toluol	30.0
Ethyl Acetate	33.0
Butyl Acetate	30.0
	100.0

The compositions of the above examples are prepared by a simple mixing operation. Preferably the wax is added to the toluol in a mixer and agitated until dissolved, and the cellulose nitrate is separately dissolved in the ester solvents and alcohol, the other materials then being added to the nitrocellulose solution, which is then combined with the wax solution.

The compositions may be applied to fabrics by a number of known methods but it is preferred to apply these compositions simply by immersing the fabric, or paper, or material to be treated until it is thoroughly saturated and then wringing out the excess coating material by squeeze rolls or centrifuging. This process is conducted at room temperature generally, although in using the composition in Example 1, it is preferred to carry out the process at a temperature not lower than 73° F., since there is some tendency for the high melting paraffin wax to precipitate out if the operating temperature is below 73° F. In the case of the composition in Example 2, it is not necessary to observe this temperature requirement since the Japan wax does not show any tendency to precipitate out. After the excess coating material has been removed the volatile solvents of the composition are then removed by drying the fabric, or paper, at ordinary or slightly elevated temperatures.

Canvas Waterproofing

1

Raw Linseed Oil	1 gal.
Beeswax Crude	13 oz.
White Lead	1 lb.
Rosin	12 oz.

Boil the above and apply warm to upper side of canvas, wetting the canvas with a sponge on the underside before applying.

2

Gilsonite	80 lb.
Stearine Pitch	62 lb.
Scale Wax	34 lb.
Mineral Oil	10 lb.
Creosote Oil	10 lb.
Copper Linoleate	9 lb.

Melt together.

Apply at a temperature of 300° F. Scrape off excess while hot.

3

Beeswax	25 lb.
Glyceryl Stearate	5 lb.
Stearine Pitch	102 lb.
Copper Oleate	15 lb.
Castor Oil	48 lb.
Naphtha	50 lb.

4

For canvas paulins or large portable covers:

Formula 1

Petrolatum Dark or Amber	8½ lb.
Beeswax, Yellow Refined	1½ lb.
Earth Pigment, Dry (Ochre, Sienna, or Umber)	5 lb.
Volatile Mineral Spirits (Painters' Naphtha)	5 gal.

Formula 2

Petroleum Asphalt, Medium Hard	7½ lb.
Petrolatum, Dark or Amber	2½ lb.
Lampblack, Dry	1 lb.
Volatile Mineral Spirits (Painters' Naphtha)	5 gal.

The quantities specified are sufficient to treat about 40 square yards of canvas on one side.

A mixture of 3 gallons of gasoline and 2 gallons of kerosene can be substituted for the volatile mineral spirits, but will evaporate more slowly. Canvas treated according to the first formula will be colored buff by ochre, khaki by raw sienna, drab by raw umber, and brown by burnt umber. If a white treatment is preferred, use dry zinc oxide in place of earth pigment. For some purposes, Formula 1 with a light-colored pigment will be preferable to Formula 2, because canvas treated with the latter will absorb more heat from sunlight, owing to its black color. For permanently fixed canvas covers:

Formula 3

Boiled Linseed Oil	1 gal.
Lampblack, Ground in Linseed Oil	2 lb.
Japan Drier	1 pt.

Formula 4

Boiled Linseed Oil	1 gal.
Aluminum Bronzing Powder	1 lb.
Japan Drier	½ pt.

For lightweight fabrics not continuously or frequently exposed to sunlight:

Formula 5

Beeswax, Yellow Refined	½ lb.
Spirits of Turpentine	1 gal.

Mixing the Materials

In the preparation of waterproofing solutions according to Formulas 1, 2, and 5, place the specified weights of waterproofing materials in a suitable metal container and melt slowly and carefully at as low a temperature as possible, with constant stirring. Then remove to a place where there is good ventilation and no fire or open flame and pour the melted material into the solvent while stirring. When a pigment is used, thin the pigment in a separate container by mixing with it small additions of the liquid, and when the pigment mixture is sufficiently thinned strain it through fine-mesh wire screen or several thicknesses of cheesecloth into the waterproofing liquid. In Formulas 3 and 4 the pigments should be thinned in a similar manner with linseed oil before they are added to the bulk of the oil.

When the waterproofing material settles to the bottom of the container or thickens, it will be necessary to warm the mixture just before applying it to the canvas. This must be done in the open air by placing the container in a tub or can of hot water. Be sure that the container is open, and *never place it over or near a flame.*

Application

The mixture must be thoroughly stirred before and during application, in order to keep the undissolved material in suspension. These preparations may be applied to the canvas by means of a paint brush or by spraying. Wagon covers, shock covers, etc., may be treated best by stretching the canvas against the side of a barn or attaching it to a frame and applying the material with a brush. Once the canvas is fixed in position, no more time is required to treat it than is necessary to apply a first coat of paint to a rough board siding having the same area. Much time may be saved in treating large paulins and standing tents by applying the material with a spray pump, with which a pressure of at least 50 pounds is developed. Some loss of material, however, results from this method.

The experience has been that one coat applied to one side of the canvas usually is sufficient. With one coat applied to one side, using the strength of solution as given in the formulas, there will be an increase in weight of approximately 40 to 50 per cent when Formula 1 or 2 is used. When Formula 3 or 4 is used the fabric will gain about 75 per cent in weight. When Formula 5 is used the gain in weight will be around 10 per cent.

When canvas is treated with linseed-oil preparations it should be allowed to dry thoroughly (for two or three weeks) while freely exposed to the air. If folded and stored in a warm place before drying is complete the accumulated heat from continued oxidation may result in spontaneous combustion.

Waterproofing Cloth
1

The process is carried out in two padding machines.

The first padder contains a soap emulsion made up as follows:

Twenty-five pounds Soap (stearic acid type) is dissolved in 100 gallons boiling water. Twelve pounds Japan wax is added a little at a time with stirring so that an emulsion is obtained.

The second padder contains the following solution:

Fifty pounds Lead Acetate and 40 lb. Aluminum Acetate are dissolved in 100 gallons water. The clear solution is siphoned off the lead sulfate which is formed in the reaction and is run into the second padder.

The cloth is entered into the first bath at the rate of about 15 yards per minute so that it is in contact with the emulsion for about 12 seconds. This rate has to be varied with the type of cloth treated. The cloth is squeezed between rollers and without rinsing is passed into the second bath. It is squeezed between rollers again and dried.

2

Aluminum acetate is used for waterproofing cloth, the usual procedure being to immerse the well cleaned material in a solution of aluminum acetate of 4 to 5 degrees Baumé strength. The material is soaked for a period of about twelve hours and then dried in a warm room. The cloth is then introduced into a soap solution made up of about five pounds of soap in 13 gallons of water, the excess liquid wrung out and the cloth then given a bath in a 2% alum solution, followed by drying. This latter process precipitates aluminum stearate into the fibers of the cloth.

Another process, somewhat similar to the one above, consists in first immersing the cloth in a solution of:

White Soap Chips	10 lb.
Dextrine	20 lb.
Water	16 gal.

To cause thorough solution, the above is heated. After passing the cloth into this first solution, it is hung to drain and while still wet immersed in:

Zinc Sulfate (White Vitriol)	6 lb.
Dissolved in Water	9 gal.

The material is then removed after thorough penetration by the second solution, and dried, any coarse precipitated particles being brushed out.

Another method uses the following formula:

Lead Acetate (Sugar of Lead)	1 lb.
Tannic Acid	2 oz.
Sodium Sulfate (Glauber's Salts)	1 oz.

Alum	10 oz.
Water	1 gal.

3

Naphtha	100
Rubber Cement	45
Ester Gum	20
Cumar	4
Paraffin Wax (128°)	32

Moisture-Proofing Clothing

Clothing may be made moisture repellent in the course of dry cleaning by the addition to the dry cleaning solvent of 1 per cent aluminum palmitate (or certain other metallic soaps such as magnesium stearate or oleate which have been used successfully for this purpose). The addition of rosin or paraffin is preferred by various cleaners since it seems to prevent any further deposit of white film on the fabric. The proportion of these latter materials is quite small and varies, depending upon shop conditions and temperature of the solvents used. It is very difficult to put metallic soaps into solution, and is therefore far preferable to make about a 10 per cent solution in cleaners naphtha by heating the soap and naphtha together in a kettle for several hours and pouring the resulting solution into the system. After it is allowed to cool, it will set to gel, but may be redissolved very easily on reheating.

Straw Hats, Waterproofing for

Bleached Shellac	75 parts
White Rosin	15 parts
Venice Turpentine	15 parts
Castor Oil	2 parts
Alcohol (Denatured)	250 parts

Gum Sandarac	135 gm.
Gum Elemi	45 gm.
Castor Oil	11 gm.
Rosin, Bleached	45 gm.
Alcohol (Denatured)	1,000 cc.

White Shellac	4 oz.
Gum Sandarac	1 oz.
Gum Thus	1 oz.
Alcohol (Denatured)	1 pt.

Waterproofing Textiles

1

Fabrics may be rendered waterproof with glue and tannin. Both should penetrate the fabric. If fabric is dipped in strong solution of glue and than in tannin, the glue only will become insoluble on the outside, and that which has penetrated deeper in fibre will be unchanged. Treatment is thus commenced with a very weak solution composed of 5 parts of glue in 100 parts of water and fabric immersed 10 to 15 minutes.

Fabric wrung out and when nearly dry passed into tannin solution. This solution can be strong as only so much of it is taken up as corresponds to glue present. Tannin reacts quickly with glue so that only a short period of immersion is necessary. The fabric again hung to dry and then washed in water to remove excess tannin. Process is twice repeated. Fabric is now passed through a stronger glue solution, 5%, and then again tannin. By repeating the process as many times as desired the coating can be made as thick as desired.

Another Method: Potash alum 100 lb. dissolved in 10 gallons of boiling water in one pot; in another pot 100 lb. glue, 200 lb. water. Solution is affected when glue is hot, add 5 lb. tannin and 2 lb. sodium silicate. Two solutions are boiled together with constant stirring. When mixture is complete, allow to jell. To waterproof: 1 lb. jelly to 1 lb. water is boiled, bath cooled to 176° F. and fabric soaked ½ hour and then stretched out horizontally for 6 hours to drain. If drying room is used keep temperature below 122° F.

Another Method: Dissolve 10 lb. gelatin, 10 lb. tallow soap in 30 gal. boiling water and mix solution in 4 gal. water in which 15 lb. alum has been dissolved. The whole is boiled for ½ hour and cooled to 104° F. At that temperature fabric is soaked in it, dried, rinsed, dried, and finally calendered. In this process the alum partially decomposed the soap, forming either free fatty acid or an acid alumina soap. The gelatin form an insoluble compound with the alum. The free fatty acid or acid soap is mostly carried down on the fibre by the precipitate formed by the alum and gelatin.

2

Rubber Cement	46
Ester Gum	22
Cumar	2
Paraffin	31
Naphtha	100

Protecting Silk Stockings against "Runs"

Silk stockings may be protected against runs by washing them as usual in soap, squeezing them as dry as possible, and afterwards rinsing them in ½ to 1 per cent alum solution. It is quite immaterial whether the aluminum salt used is potash or ammonia alum or aluminum sulfate.

◆

Fireproof Coating for Curtains, etc.

For coating curtains, paper, etc., the L.C.C. recommends 1 lb. of ammonium phosphate and 2 lb. of ammonium chloride to 1½ gallons of water, or alternatively 10 oz. borax and 8 oz. boracic acid per gallon of water. The second formula is stated to be better for delicate articles. The fabrics should be dried without rinsing, and in all cases a small piece of the cloth should be treated first, in order to find the effect on color and texture.

◆

Fireproofing Canvas

Ammonium Phosphate	1 lb.
Ammonium Chloride	2 lb.
Water	1½ gal.

Impregnate with above; squeeze out excess and dry.

◆

Fireproofing Light Fabrics

Borax	10 oz.
Boric Acid	8 oz.
Water	1 gal.

Impregnate; squeeze and dry.

◆

Fireproofing Textiles

1

The cloth is impregnated with

Borax	70
Boric Acid	30
Water	600

and dried.

2

Ammonium Chloride	20 kg.
Zinc Chloride 30 per cent	300 l.
Ammonia 28 per cent	350 l.
Water	100 l.

3

The Paris Municipal Laboratory recommended the following process: Prepare a 2 per cent solution of aluminum sulfate and a 5 per cent solution of silicate of soda. Mix and enter the cloth. After squeezing and drying the aluminum silicate formed is insoluble.

4

Another method consists in padding the fabric in a solution of ammonium phosphates, then steeping in an ammoniacal solution of magnesium chloride. The compound formed on the fiber is insoluble in water. The fabric is rinsed to remove the excess of magnesium chloride and dried.

5

Tungstate of zinc resists washing, and this makes it preferred at times to tungstate of alumina. The most usual method consists in padding in a solution of stannate of soda at 14 deg. Bé., and then drying. The gods are then entered into a bath of the following composition:

Tungstate of Soda, 35° Bé.	4 parts
Acetic Acid, 9° Bé.	1 part
Ammonium Hydrochloride, 4° Bé.	3 parts
Acetate of Zinc, 17° Bé.	2 parts

After centrifuging and drying the fabric is hot-calendered to evaporate the acetic acid.

6

Perkin recommends the following method: Pad with a solution of stannate of soda at 26 deg. Bé., and dry, then treat with a solution of ammonium sulfate at 10 deg. Bé., squeeze dry and wash in water to remove the excess of ammonium sulfate. This last step is not indispensable, as the sulfate has flame-proofing properties. The stannate of soda combines intimately with the fiber and the ammonium sulfate precipitates the oxide which combines also with the fiber.

Asbestos Dope

Asbestos.—The cloth is painted with a dope containing asbestos which hardly interferes with suppleness. An interesting composition is:

Asbestos	350 gr.
Silicate of Soda, 36° Bé.	350 gr.
Water	1000 gr.

The particles crumble and shrink. Continue heating for about 12 hours. Crush and screen to uniform sizes; replace in pans and reheat at 185–195° F. until proper state of dryness is reached (about 8 hours). The dried material is of a granular glassy light yellow color. This material is air-cooled and sifted through No. 6 and No. 8 screens.

Embroidery Treatment

Cotton cloth is saturated with

Alum	1
Aluminum Chlorate (20%)	3
Water	17

Dry in air and embroidery is then worked on cloth. Then dry in oven at 80° F. Chlorine is liberated and attacks cotton so that latter may be brushed off from embroidery.

Felt Hat Stiffener

Carnauba Wax Emulsion (Bright Drying)	90 lb.
Shellac (Ammonia Water Solution)	10 lb.

Fireproofing Solutions

The following is the formula of a solution used in theatrical work for rendering materials non-inflammable:

Tungstate of Sodium	17½ oz.
Water	1½ pt.

Dissolve in the cold and add:

Sodium Phosphate	2½ oz.
Water	1 pt.

or a sufficiency of water to make the solution sp. g. 1.140.

Dip the material in the solution, wring out with the hands, dry, and iron if necessary.

CHAPTER XVIII

MISCELLANEOUS

Freshening Cut Flowers

Dissolve half an aspirin tablet in two quarts of water and pour this into vase holding flowers.

◆

Preserving Cut Flowers

For those whose flowers are supplied from private gardens, advise cutting the flowers in the early morning or late evening when the stems are turgid. A sharp knife is recommended. The sharper the cut the less is the bruising of the conducting vessels and the greater the absorption of water. The elimination of ragged edges will lessen the chances of bacterial action.

The proper stage of the flower's development should be observed when cutting. Gladioli are best for cutting when the first floret is open; peonies, when the petals are unfolded; roses, before the buds open; dahlias, when fully open; poppies, the night before they open.

Flowers after they are cut should be plunged stem-deep in water. All arranging should be postponed until after the stems have been thoroughly soaked.

Flowers should be kept in a humid room and never in sunshine. This reduces the evaporation to a minimum. It is well to keep them at 45° F. If they are kept cooler than that during the night, the lasting quality is improved. Containers which permit a free entrance of air through the top are recommended. Narrow-necked vases should be avoided. Stems should be cut each day with change of water. The aspirin treatment may be used with each change to prolong the freshness of the flowers. In cutting stems, a slanting cut will prevent the ends from resting squarely on the bottom of the vase. All leaves which are submerged should be removed to prevent decomposition and fouling the water.

Wilted flowers may be revived by cutting their stems short, plunging them deep in water and storing in a cool dark place for ten hours or more. The so-called "hot water" treatment is also useful in restoring wilted flowers. Immerse the stems in hot water (not boiling) for half an hour, keeping them in the dark, and then change to cooler water. Usually several hours are re quired for restoration.

◆

Coloring Artificial Flowers
(Made from Cotton, Muslin, Silk, Velvet)

Material is colored in two ways.

1. Before cutting to shape.
2. After cutting to shape.

Method (1). Material is put in frames and backed with a starch sizing to give body. Dye is then brushed on. Dye may also be added to the sizing. Dried and die cut to shape.

Method (2). After backing coat is put on, the material is die cut and then dipped into the dye solution.

Dye solutions prepared as follows:

Yellow

Auramine O	1 oz.
Denatured Alcohol	4 oz.
Water	4 oz.

Rose

Rhodamine B	1 oz.
Water	4 oz.
Denatured Alcohol	4 oz.

Purple

Pylam Purple	1 oz.
Water	4 oz.
Denatured Alcohol	4 oz.

Peacock Blue

Patent Blue	1 oz.
Water	2 oz.
Denatured Alcohol	2 oz.

Green

Pylam Brilliant Green	1 oz.
Water	4 oz.
Denatured Alcohol	4 oz.

Pink

Eosine	1 oz.
Water	2 oz.
Denatured Alcohol	2 oz.

Cerise

Rose Bengale	1 oz.
Water	2 oz.
Denatured Alcohol	2 oz.

Flower Gardens (Chemical)

Salt	6 tblsp.
Bluing	6 tblsp.
Water	6 tblsp.
Ammonia Water	1 tblsp.

and pouring, after thorough mixing, over a clinker, a piece of coke or of brick in a broad bowl or dish. After the clinker (or coke or brick) has been wet with the liquid, drop on it a few drops of mercurochrome solution or of red ink or green ink. But do *not* use iodine, because this reacts with ammonia water to form the dangerously explosive nitrogen iodide, a black powder which is safe as long as it is wet but explodes with a loud report from very slight shock when it is dry. After the materials have been brought together, a coral-like colored growth soon begins to appear on the clinker. This increases rapidly.

The growth also tends to form on the edges of the dish and will climb up and over them unless they have been rubbed with vaseline. The growth will not extend beyond the vaseline.

The "depression flower garden" is a capillary phenomenon involving the tendency of ammonium salts to "creep." The saturated solution deposits crystals around its edges and upon the clinker where the evaporation is greatest. The crystals are porous and act like a wick, sucking up more of the solution by capillary action. The solution thus sucked up evaporates to produce more crystals, more wick, and more growth. The addition of a little more ammonia water to the dish will produce more growth after the first growth has stopped. Or the whole may be allowed to dry and may then be kept without further change.

The "mineral flower garden" which florists sometimes sell or display in their windows, depends upon an entirely different principle, that of osmosis or of osmotic pressure. A solution of sodium silicate or "water glass" is poured into a jar or globe, and the crystals of readily soluble salts of certain metals which form colored and insoluble silicates are

thrown in and allowed to sink to the bottom. Growths resembling marine plants spring up from these crystals and in the course of a few minutes climb rapidly upward through the liquid, often branching and curving, producing an effect which might lead one to believe that he sees exotic algae growing in an aquarium. The experiment works best if the solution of water glass is diluted to a specific gravity of about 1.10.

Ferric chloride produces a brown growth; nickel nitrate, grass green; cupric chloride, emerald green; uranium nitrate, yellow; cobaltous chloride or nitrate, dark blue; and manganous nitrate and zinc sulfate, white.

Colored Waters (Non-Fading)

These are for filling bottles which are exposed to sunlight.

Amethyst

Sodium Salicylate	10	gm.
Tinc. Ferric Chloride	½	dr.
Distilled Water	2½	gal.

Blue

Copper Sulfate	4 oz.
Ammonia	sufficient to dissolve precipitate
Distilled Water	2½ gal.

Green

Nickel Sulfate	3	oz.
Sulfuric Acid	6	oz.
Distilled Water	2½	gal.

Garnet Red

Potassium Bichromate	16	oz.
Sulfuric Acid	16	oz.
Water	2½	gal.

Rose Red

| Cudbear | 2 oz. |
| Water | 10 oz. |

Macerate for two days and filter; dilute with water to the proper shade and add ½ oz. Ammonium Hydroxide to each gallon.

Orange

Potassium Bichromate	16	oz.
Nitric Acid	8	oz.
Distilled Water	2½	gal.

Solidified Gasoline

Gasoline	0.5 gal.
White Soap (Fine Shaved)	12 oz.
Water	1.0 pt.
Household Ammonia	5 oz.

Heat the water, add soap, mix and when cool add the ammonia. Then work in slowly the gasoline to form semi-solid mass.

Gasoline Carbon Looseners

There are in the market a number of gasoline addition agents for the removal of carbon. These are used in the following manner:

Add 4 oz. to five gallons of gasoline in tank or supply through manifold by attached cup.

The formulas for a few of these are:

1. Medium Oil	50%
Varnoline	50%
2. Medium Heavy Oil	50%
Light Paraffin	50%
Wintergreen Odor	0.2%
3. Kerosene or Varnoline	80%
Vaseline	20%

Radiator Solder

Flaxseed Meal	100
Aluminum Powder	1–2

Mix together until all the flaxseed is covered with Aluminum. When this is added to the water in a leaky automobile radiator, it swells and plugs up all leaks as the water circulates.

Copper Tubing, Bending

Fill tubing completely with molten lead and bend around wood form. When bent heat and drain out lead.

Ultra Violet Filter

A filter useful for absorbing ultra violet light in connection with fluorescence photography consists of a 2% solution of Sodium Nitrite in a glass cell 1 cm. in thickness.

Anti-Rot Compound for Wood

Sodium Fluoride	2 lb.
Water	98 lb.

X-Ray Screen, Fluorescent

Sodium Tungstate	29 gm.
Calcium Chloride	11 gm.
Sodium Chloride	58 gm.

Boiler Scale, Removal of

8–10% Hydrochloric Acid is most suitable for Copper or brass app.; 5–10% Oxalic Acid for Aluminum or tinned metals; 15% Acetic Acid for Zinc or galvanized iron.

Boiler Compound

Soda Ash	87
Trisodium Phosfate	10
Starch	1
Tannic Acid	2

Demulsifier

Concentrated turkey red oil is a very efficient demulsifier and is used quite extensively in the oil fields for breaking petroleum emulsions. This material is made by slowly adding 10% of sulfuric acid (66° Baumé) to pale blown castor oil. The above is allowed to stand for two hours. It is then added to four times its volume of a half of one per cent water solution of sodium chloride and mixed thoroughly. After about 24 hours the water will be precipitated, whereupon the same is decanted and the remaining sulfonated castor oil is neutralized with ammonium hydroxide.

Embalming Fluid

Glycerin	250
Formaldehyde	1565
Potassium Nitrate	150
Borax	40
Boric Acid	120
Dark Red Dye	0.4
Water	2800

Embalming Fluids

Solution of Formaldehyde	11	lb.
Glycerin	4	lb.
Sodium Borate	2½	lb.
Boric Acid	1	lb.
Potassium Nitrate	2½	lb.
Solution of Eosin, 1%	1	oz.
Water	enough to make 10	gal.

The sodium borate, boric acid and potassium nitrate are dissolved in 6 gallons of water; the glycerin is added, then the solution of formaldehyde, and lastly the solution of eosin, and the necessary amount of water.

Another formula in vogue is as follows:

Thymol	15	gr.
Alcohol	½	oz.
Glycerin	10	oz.
Water	5	oz.

Gems, Synthetic

Titanium Tetrafluoride	0.2
Beryllium Oxide	0.5
Iron Oxide	10
Aluminum Oxide	500
Magnesium Powder	100

Fuse together in a crucible and allow to cool slowly.

Biological Fixing Fluid

These new fluids have been developed as the result of intensive research and are more or less free from difficulties. Materials fixed in them remain soft and will not harden when placed in a 70 per cent alcohol solution. In addition, all common stains may be used.

Two of the solutions are given as follows:

Cupric-paranitrophenol Fixing Solution

60 per cent Alcohol	100 cc.
Nitric Acid, sp. gr. 1.41–1.42	3 cc.
Ether	5 cc.
Cupric Nitrate, Crystals	2 gm.
Paranitrophenol, Crystals	5 gm.

This fluid is perfectly stable and is not limited as to duration of fixation, but has a slow penetration rate.

Cupric-phenol Fixing Solution

Stock Solution A

Distilled Water	100 cc.
Nitric Acid (as above)	12 cc.
Cupric Nitrate (as above)	8 gm.

Stock Solution B

80 per cent Alcohol	100 cc.
Phenol, Crystals	4 gm.
Ether	6 cc.

These solutions are perfectly stable and may be kept in glass stoppered bottles, but the mixture does not keep and for this reason the duration must not exceed forty-eight hours. For use, take: Solution A—one part; Solution B—three parts. In using either fixing solution wash the material in several changes of 70 per cent alcohol.

Anti-Freeze

Pints of anti-freeze per gal. of water for protection at:

	+10° F.	0° F.	−10° F.	−20° F.
Denatured alcohol 180° proof	3.4	4.9	6.5	8.3
Denatured alcohol 188° proof	3.3	4.7	6.0	7.7
Glycerin (USP) 95%.	3.8	5.3	7.1	9.0
Radiator glycerine 60%	10.0	18.7	39.0	106.5
Ethylene glycol 95%.	2.7	4.0	5.1	6.5

Specific gravity for protection at:

	+10°F.	0°F.	−10°F.	−20°F.	−30°F.
Denatured alcohol.	0.968	0.959	0.950	0.942	0.921
Glycerin	1.090	1.112	1.131	1.147	1.158
Ethylene glycol	1.038	1.048	1.056	1.064	1.069

Anti-Freeze Solution

Denatured Alcohol	50
Methanol	10
Glycerin	30
Water	10

Anti-Mist Liquid
(For Use on Glass)

Potash Coconut Oil Soap	120
Glycerin	60
Turpentine	8
Naphtha	3
Clovel	1

Anti-Fog Windshield Liquid

Glycerol	10	oz.
Alcohol	⅛	oz.
Rose Water	6	oz.
Salt	0.06	oz.
Sulfur Powd.	0.06	oz.

Algae Removal

In a swimming pool one pound of copper sulfate, or blue stone, to two million pounds of water destroys algae. This material is likely to be fatal to fish. The solid is placed in a sack and dragged back and forth across the pool to secure proper mixing. In computing amount needed, one gallon of water weighs eight and one-third pounds, or one cubic foot of water weights 62.5 pounds.

Brake Lining, Composition for

Crepe Rubber	14
Litharge	10
Barytes	34
Zinc Oxide	5
Carbon Black	3
Graphite	4
Sulfur	4
Asbestos Yarn	12
Brass Wire	14

Gasket Compound

Asbestine Powd.	56
Copal Varnish 9 lb. Cut	44

Grind in ball mill for 3 hours

Puncture Preventive, Tire

Bentonite	100
Magnesium Oxide	2
Asbestos Fiber	50
Water	suitable quantity

Battery Terminals, Coating for

Diglycol Stearate	10
Water	300

Heat until melted and stir until dispersed. Run in slowly with stirring

Graphite Powd.	30–100

Stiffeners for Toes of Shoes

Cumarone	12 lb.
Petroleum	1 gal.
Pine Oil	2 fl. oz.

To Drill Holes in Glass

By taking a good steel drill and wetting with a saturated solution of camphor in oil of turpentine, holes may be rapidly and easily drilled through the thickest plate glass.

Radiator Scale-Remover

Use 6 ounces of tri-sodium phosphate to 5 gallons of water, and place in radiator. Run motor slowly for ten minutes. Draw off the water, and flush out with hose.

Pyrotechnics
Red Fire

Strontium Nitrate	66 parts
Potassium Chlorate	25 parts
Powdered Orange Shellac	9 parts

Strontium Carbonate	16 parts
Potassium Chlorate	72 parts
Orange Shellac Powdered	12 parts

Potassium Chlorate	37 parts
Strontium Nitrate	50 parts
Shellac Powd.	13 parts

Strontium Nitrate	8 oz.
Sugar	4 oz.
Potassium Chlorate	1 oz.
Potassium Perchlorate	15 oz.
Strontium Nitrate	80 oz.
Flowers of Sulfur	20 oz.
Wood Charcoal (powdered)	1 oz.
Gum Kauri (red gum)	2 oz.
Vaseline-sawdust Mixture	10 oz.

The sawdust and vaseline mixture is made by rubbing 8 oz. of sawdust with 6 oz. of melted vaseline.

Potassium Perchlorate	4½	oz.
Strontium Nitrate	20	oz.
Sulfur	5½	oz.
Rosin	½	oz.
Sugar	½	oz.
Antimony, Powdered	¼	oz.
Vaseline-sawdust Mixture	10	oz.

Perchlorate Potash	12½	parts
Nitrate Strontia Powdered	50	parts
Powdered Charcoal	1	part
Powdered Sugar	4	parts
Red Gum	15	parts

Potassium Chlorate	6	parts
Strontium Nitrate	2	parts
Strontium Carbonate	1½	parts
Gum Kauri (red gum)	2½	parts

Green Fire Composition

Barium Chlorate	90	gm.
Powdered Orange Shellac	10	gm.

This mixture is made by mixing the above two ingredients together.

Barium Chlorate	23	parts
Barium Nitrate	59	parts
Potassium Chlorate	6	parts
Orange Shellac	11	parts
Stearic Acid Powd.	1	part

Barium Chlorate	55	parts
Barium Nitrate	33	parts
Shellac	12	parts

Barium Nitrate	6	parts
Potassium Nitrate	3	parts
Sulfur	2	parts

Barium Nitrate	18	parts
Shellac	4	parts
Mercurous Chloride	4	parts
Potassium Chlorate	2	parts

Barium Nitrate	3	parts
Potassium Chlorate	4	parts
Gum Kauri (red gum)	1¼	parts

Blue Fire Composition

Potassium Chlorate	6	parts
Ammonio-sulfate of Copper	8	parts

| Shellac | 1 part |
| Willow Charcoal | 2 parts |

Potassium Chlorate	40 parts
Copper Sulfate	8 parts
Rosin	6 parts

White Fire Compositions

Potassium Nitrate	24 parts
Sulfur	7 parts
Charcoal (wood)	1 part

Potassium Nitrate	7 parts
Sulfur	2 parts
Powdered Antimony	1 part

Potassium Perchlorate	3½ oz.
Barium Nitrate	17 oz.
Powdered Sulfur	3½ oz.
Finely Powdered Aluminum	5 oz.

Potassium Perchlorate	7 oz.
Barium Nitrate	34 oz.
Flowers of Sulfur	7 oz.
Aluminum Bronze (dust)	2 oz.
Aluminum Flakes	7 oz.

Pyrotechnic

A nonhygroscopic successively exploding composition consists of

Pot. Chlorate	35 lb.
Magnesium Oxide	35 lb.
Phosphorus Trisulfide	12 lb.
Gum Arabic	1 lb.
Pot. Dichromate	5 lb.
Clay and Sand	8 lb.

Eggs of Pharaoh's Serpents

Take sixty-four parts of Mercuric Nitrate in solution and add thirty-six parts of Potassium Thiocyanate. The resulting precipitate of Mercuric Thiocyanate should be filtered off, and washed three times with distilled water. The residue then should be placed in a warm place to dry.

To make the powder into pellets, gum tragacanth serves as the binder. For every pound of the powder an ounce of gum is required. This is softened by soaking in hot water to form a paste. The dried precipitate is gradually mixed into the paste by constant and thorough stirring. Add a little water if necessary, so as to present a somewhat dry pill mass, from which pellets of the desired size can be made by hand. Place on a piece of plate glass and allow to dry. They are then ready for use. On ignition they burn forming a voluminous ash, in the form of snake like tubes—the so-called "Pharaoh's Serpents."

Showers of Fire

Potassium Nitrate	18 parts
Sulfur	8 parts
Lampblack	5 parts

This composition burns with a yellowish color, throwing out streamers of golden sparks, due to the lampblack which is used. The mixture burns slowly and is suitable for filling paper tubes.

Potassium Nitrate	10 parts
Sulfur	2 parts
Charcoal	2 parts
Iron Filings (fine)	7 parts

For loading into ordinary paper cases.

Potassium Nitrate	36 parts
Sulfur	2 parts
Charcoal (wood)	10 parts

For loading into paper cases.

Light Sticks

Fill thin paper tubes of about ⅜" outside diameter and 1' long with the colored fire compositions, alternating. One end of the tube should be closed tightly to a depth of 3" with clay or sand. Fill with powder of the desired color and close end by pasting a piece of tissue paper around it, after inserting a fuse.

Boil a handful of sawdust or wood shavings in a cup of water containing a teaspoonful of potassium nitrate. When dry, it will burn with a whitish yellow flame, sizzling as it burns. Add ½ teaspoon of strontium nitrate to the water before boiling the sawdust in it. When the sawdust is then immersed and dried it will burn with a red flame. Barium nitrate will make the flame green; copper sulfate, blue.

Homemade Sparklers

Potassium Chlorate	10 oz.
Granulated Aluminum	2 oz.
Charcoal	1/16 oz.

Mix to consistency of thick cream with a solution of 2 oz. of dextrin in a pint of water and coat upon wires or slender wooden sticks.

For red sparkler add 1½ oz. powdered strontium nitrate.

For green sparkler add 2 oz. powdered barium nitrate.

Smoke Composition

White: Powdered Potassium Nitrate	4 oz.
Powdered Soft Coal	5 oz.
Sulfur	10 oz.
Fine Sawdust	3 oz.
Red: Potassium Chlorate	15 parts
Paranitraniline Red	65 parts
Lactose (powdered)	20 parts
Green: Synthetic Indigo	26 parts
Auramine Yellow O	15 parts
Potassium Chlorate	33 parts
Lactose (powdered)	26 parts
Yellow: Precipitated Red Arsenic Sulfide	55 parts
Powdered Sulfur	15 parts
Potassium Nitrate	30 parts

Smoke, Composition for Producing

Tetrachlorethane or Chloronapthalenes	40–50
Zinc Filings	55–25
Potassium Nitrate	
Sodium Nitrate	
Calcium Silicide	5–15%
Pitch	

Smokeless Flashlight Powder

Zirconium	28
Zirconium Hydride	7
Magnesium	7
Barium Nitrate	30
Barium Oxide	25
Rice Starch	5

Fuel Oil, Smokeless

Fuel Oil	460 cc.
Degras	5 gm.

Dissolve by vigorous stirring; run in slowly following solution

Pot. Nitrate	6½ gm.
Borax	2½ gm.
Water	38 cc.

Finally pass through colloid mill.

The above mixture ensures perfect, rapid and complete combustion.

Special High Temperature Fuel

Aluminum Powder	95
Sulfur Powder	5

Match, Repeatably Igniting

From the following is molded a match which ignites on rubbing and may be blown out and used repeatedly.

Pyroxylin	50
Potassium Chlorate	20
Powd. Glass	10
Camphor	8
Pyridine	4
Am. Oxalate	2

Fire Extinguisher

A fire extinguisher is absolutely necessary in the laboratory if the workers are to be protected. Manufactured extinguishers are rather expensive, but the following substitute is very efficient. The metal part of a burned out electric light bulb is removed. The tube used to seal the bulb is dipped in carbon tetrachloride, and the tube broken. The vacuum draws the liquid into the bulb. The break is sealed with wax. Fire extinguishing "bombs" of this type may be put in convenient places about the laboratory.

Dry Fire Extinguisher

Ammonium Sulfate	30 lb.
Sodium Bicarbonate	18 lb.
Ammonium Phosphate	2 lb.
Red Ochre	4 lb.
Silex	46 lb.

Fire Extinguishing Fluid

Carbon Tetrachloride	94–95
Solvent Naphtha	5
Ammonia Gas	0.5–1

The above minimizes production of toxic fumes when extinguishing fires.

Soldering Solutions

Zinc Chloride made by completely neutralizing hydrochloric acid with zinc is most universally used. In addition to this rosin, ammonium chloride and a mixture of 15% zinc chloride, 25% glycerin and 60% water are satisfactory for copper, brass, steel, terne plate, tinned steel, monel metal, etc. Hydrochloric acid is necessary on galvanized steel.

A well-made soft-soldered joint will develop 5000 to 6000 lb. per sq. inch in shear.

Silver solders consist of silver 20% to 70%, copper 50% to 18%, zinc balance, Borax or Boric acid mixture used for fluxes. Melting points of silver solder vary according to composition usually 200 to 300 degrees F. below those of the usual brazing—brasses and about 1100 to 1200 degrees F. above ordinary soft solder.

Aluminum solder is a 12% silicon and 88% aluminum melting at about 580 degrees C. (1076 degrees F.).

___

Soldering Solution for Stainless Steels

Zinc Chloride, Commercial	37 gm.
Glacial Acetic Acid, 99.9%	23 gm.
Hydrochloric Acid, Com. 34.5% Hcl.	40 gm.

___

Soldering Solution for Rustless Irons

Hydrochloric Acid, specific gravity 1.18	60 gm.
Ferric Chloride, Lump Form, Pulverized	33 gm.
Nitric Acid, Specific Gravity 1.42	2 gm.

Add in order named.

Tinning Flux—Zinc chloride stick from saturated solution in water.

___

Non-Corrosive Soldering Flux

Rosin	1 oz.
Denatured Alcohol	4 oz.

___

Solder, Aluminum

Aluminum	30
Zinc	20
Tin	15
Copper	5
Bismuth	10
Silver	10

___

Solder, Brazing

Copper	40–55
Zinc	60–45

___

Flux, Soldering

Zinc Chloride	71
Ammonium Chloride	29

___

Zinc Solder Flux

Cadmium Chloride	40
Lead Chloride	40
Ammonium Bromide	16
Sodium Fluoride	4

Pewter, Soldering

The surfaces are cleaned thoroughly. As a flux there is used a mixture of rosin and olive oil. A good solder consists of

Bismuth	50
Tin	25
Lead	25

___

Solder, Silver

Silver	20
Copper	45
Zinc	30
Cadmium	5

___

Solder, Stainless Steel

1

Tin	66
Lead	34

___

"Stainless Steel"

2

Manganese	20
Copper	25
Nickel	5
Silver	49
Gold	1

___

Soldering Paste

Water	10 parts
Zinc Chloride	25 parts
Ammonium Chloride	2 parts
Dark Petrolatum	65 parts

Dissolve the salts in the water and stir into the petrolatum.

___

Solder (Powder Form)

Iron Filings	100 parts
Ammonium Chloride	50 parts
Sulfur in Powder Form	25 parts

Mix well.

___

Welding Rod Composition

1

Tungsten	1 –12%
Chromium	1 –10%
Nickel	0.1 – 5%
Aluminum	0.1 – 8%
Vanadium	0.1 – 2%
Carbon	1.75– 4%
Manganese	0.5 – 5%
Silicon	0.2 – 3%
Molybdenum	0.1 – 6%
Iron	Balance

2

Carbon	0.60–0.85%
Manganese	11 –13.5 %
Nickel	2.5 – 3.5 %
Silicon	< 0.60%
Iron	Balance

+

Solidified Alcohol

1

Alcohol	1000.0 cc.
Stearic Acid	60.0 gm.
Caustic Soda	13.5 gm.

Dissolve the stearic acid in 500 cc. of the alcohol, and the caustic soda in the remaining alcohol. Warm to 60° C., mix, and allow to solidify.

2

Denatured Alcohol	1000 cc.
Soap Chips (Well Dried)	28–30 gm.
Gum Lac	2 gm.

Heat alcohol to 140° F., add soap and lac, mix till completely dissolved, allow to cool.

+

Fire Starters

Rosin or Pitch	10
Sawdust	10 or more

Melt and mix and cast in forms.

+

Fire Kindler

1

1. Cork Dust	50
2. Sawdust	50
3. Paraffin	80
4. Potassium Chlorate	10
5. Sugar	10

Dissolve (4) and (5) in a minimum amount of water and mix thoroughly with (1) and (2). Place in heated dough mixer and pour in melted (3); mix until uniform and cast in blocks.

2

Paraffin Crude	30
Rosin Pitch	10
Wood Flour	60

Compress strongly into bricks.

3

Rosin, Dark	30
Petroleum Oil, Thin	5
Sawdust	65

Mix and compress strongly into bricks.

4

Distillery Waste	20
Paraffin Crude	10

Mix in a heated dough mixer. Mix in

Sawdust	60
Charcoal or Coal Dust	10

Compress strongly into bricks.

+

Alcohol Motor Fuel

Alcohol	15
Benzol	20
Gasoline	65

This mixture will not separate and has high anti-knock properties.

+

Freezing Mixture

A mixture of 230 g. of ammonium sulfocyanate, 30 g. of sal ammoniac, and 300 cc. of water produces a fall of temp. from 15° to −19°. Increase of sal ammoniac content reduces the cooling effect, which is thus well under control.

Wood Floor Bleaching

Water used in the bleaching solutions raises the grain of the wood, consequently, after the surface has been allowed to thoroughly dry, it must be sandpapered to remove the raised wood fibres. A thin coat of shellac—about 2-lbs. of shellac gum to 1-gal. of denatured alcohol—brushed on and allowed to dry, makes these fibres easy to remove.

After bleaching, the surfaces may retain some of the chemicals; if so, immediate sponging with clean water will be necessary. A coat of ordinary vinegar will assist in neutralizing any traces of alkali left by the solutions, and will put the surfaces in suitable condition for the finishing process. At least 12 hours should be allowed for a thorough drying of the surface before finishing begins.

Many chemical solutions are used for bleaching; some are more effective on certain woods; others succeed better on other woods. Oxalic acid solutions are used by the majority of house painters. Before using any solution, clean and scrub the surface, using hot water to which soap and a small quantity of sal soda has been added. Use No. 2 or No. 3 steel wool for scrubbing, washing well with clean water and a sponge.

Make up a saturated solution, dissolving as much of the oxalic acid in a gallon of hot water as the water will take up. Apply the solution hot, using a flat wall brush, and let it dry on the surface. For bleaching dark sap stains and weather stains, about 8-oz. of oxalic acid to 2-qts. of water is sufficient. When the first application does not give the desired results in color, apply the solution two or more times, as may be found necessary. For greasy surfaces, rub with denatured alcohol and let dry before the bleaching solution is applied.

Dissolved in water, chlorinated soda makes an effective bleach, especially if it is followed by a solution of peroxide of hydrogen. This solution is used as follows:

Solution No. 1

Sal soda	5¼ oz.
Water	10 oz.
Dissolve.	

Solution No. 2

Chloride of lime	2½ oz.
Water	6 oz.

Mix well and allow to settle.

Pour the clear liquid of solution No. 2 in another container and a sediment will be found in the bottom of the first container. Add to this 6-oz. of water and stir well. Let settle and pour off the clear fluid as before. Add 1 or 2-oz. of water to the remaining sediment; stir well and strain into the second container through filter paper.

Mix solutions No. 1 and No. 2, and a clear liquid bleach of green color, and having a faint odor of chlorine and a strong alkaline taste, results. Use this solution hot and brush on with a flat wall brush; let dry and wash the surface with clean water.

For bleaching walnut, ordinary chloride of lime dissolved in water and brushed on the wood has been found satisfactory. It will work on many other woods too.

Permanganate of potash dissolved in water and used in varying strengths makes an excellent bleach. The wood will have a purple tint when dry. The solution is to be applied with a brush and, when dry, a second coating of a saturated solution of hyposulfite of soda in water will "fix" the tint. A 5% solution of oxalic acid in water has been found an effective second coat treatment over the permanganate of potash.

Hydro-sulfite of soda, when used as a 10% solution in water, has been found to be a satisfactory bleach. One or two coats brushed on and allowed to dry thoroughly following each application, then washing with clear water, has proved to be the required treatment.

To Ebonize Veneer

For ebonizing veneer for inlays before they are laid: Dissolve 6 to 8-oz. of nigrosine to the gallon of alcohol, and place this solution in a tank large enough for dipping the veneers. Use steam for heat, and heat the solution to the boiling point. Leave the veneer in for about one hour; then remove the veneer and dry it, using a plate dryer if possible. Repeat this procedure until the veneer is properly colored clear through. This cycle of steaming and drying may have to be repeated four or five times, depending upon the thickness of the veneer. When the color is clear through and the veneers are dry, they are ready for use. Black inlay will harmonize effectively with practically any kind or color of veneers.

Some finishers claim they get an effective ebony black with a single dipping of the inlays by using a strong water solution of jet black nigrosine heated to a temperature under the boiling point. Inlays are seldom more than 1/16-in. thick, and they are generally very narrow. A warm water solution of jet black nigrosine will penetrate deep and quickly—except possibly on some of the hardest of close-grained woods like maple, for example, which might require a second dipping or a longer period of immersion.

Removing Plaster Casts

Plaster casts may be cut readily if hydrogen peroxide is applied along the line one wishes to open and then continually applied as the cutting progresses. In place of hydrogen peroxide vinegar, or ordinary acetic acid may be used in same way.

To Remove Plaster from Hands

Rub either a little sugar syrup or a little moistened sugar into the hands. The plaster will disintegrate and wash off easily.

To Remove Teeth from Vulcanite Dentures

Boil the dentures in glycerin until the latter smokes. The teeth which will come away clean and not discolored are placed back in the glycerin for tempering.

When they are cool, wash them with soap and water.

------♦------

Removing Odors from Cutlery

Odors of garlic, onions, etc., may be removed from cutlery by washing with dilute hydrochloric acid.

------♦------

Canary Bird Food

Yolk of eggs, dried	2 parts
Poppy heads, coarse powder	1 part
Cuttlefish bone, coarse powder	1 part
Granulated sugar	2 parts
Soda crackers, powdered	8 parts

------♦------

"Art" Gum Type Eraser

Corn Oil	4 lb. 6	oz.
Light Magnesium Oxide	2½	oz.
Sulfur Chloride	16	oz.
Ground Pumice (milled to 200 mesh)	1 lb. 8	oz.

Add the magnesium oxide and the pumice to the corn oil, and stir until the solids are uniformly distributed throughout the mass. Now add slowly, and with vigorous agitation, 8 ounces of sulfur chloride at such a rate that the temperature does not rise above 90° F. The mixture is now allowed to stand for 12 hours, whereupon the balance of the sulfur chloride may be added somewhat more rapidly than the initial quantity, but with equally vigorous agitation.

Upon the addition of the second quantity of sulfur chloride the mass thickens somewhat and is poured into wooden channels, open at the top, and having a cross section approximately one inch square. The mixture is allowed to stand for several hours and is removed from the molds when solidified, cut into three-inch lengths, imprinted and packed.

------♦------

Paper Barometer

Paper barometers are made by impregnating white blotting paper with the following liquid, and then hanging up to dry:

Cobalt chloride	1	oz.
Sodium chloride	½	oz.
Acacia, gum	¼	oz.
Calcium chloride	75	grn.
Water	3	fl. oz.

The amount of moisture in the atmosphere is indicated by the following colors:

Rose red	rain
Pale red	very moist
Bluish red	moist
Lavender blue	nearly dry
Blue	very dry

------♦------

Easter Egg Dyes

Blue

Marine Blue	1	dr.
Citric Acid	10	dr.
Dextrin	2	oz.

Chocolate Brown

Vesuvin	1	oz.
Citric Acid	10	dr.
Dextrin	1	oz.

Green

Brilliant Green	0.5	oz.
Citric Acid	5	dr.
Dextrin	2	oz.

Orange

Azo Orange	2.5	dr.
Citric Acid	5	dr.
Dextrin	2.5	oz.

Rose

Eosin	75	gr.
Dextrin	3	oz.

Violet

Methyl Violet	1	dr.
Citric Acid	5	dr.
Dextrin	2.5	oz.

Yellow

Naphthol Yellow	0.5	oz.
Citric Acid	10	dr.
Dextrin	2.5	oz.

Red

Diamond Fuchsin	1	dr.
Citric Acid	5	dr.
Dextrin	2.5	oz.

One-twentieth of each of the above formulae, dissolved in a half pint of boiling water, is sufficient to color a dozen eggs. After the eggs have been immersed in the solution and allowed to dry, they should be polished with a little olive oil.

------♦------

First Aid for Chemical Injuries

Acetic Acid.—Use chalk, emetics, magnesia, oil or soap.

Acetylene.—Remove to fresh air immediately and call for pulmotor. Then apply artificial respiration for at least one hour or until pulmotor arrives. The administration of oxygen containing 5%

of carbon dioxide is beneficial; inhalation of ammonia or amyl nitrate is often of value.

Arsenic, Rat Poison, Paris Green.— Use flour and water, lime water, milk, raw egg or sweet oil.

Carbolic Acid.—Use any soluble nontoxic sulfate, after provoking vomiting with zinc sulfate; ice, milk of lime, olive or castor oil with magnesia in suspension, saccharate of calcium, uncooked white of egg in abundance, washing the stomach with equal parts of water and vinegar; or give alcohol or whiskey or about four fluid ounces of camphorated oil at one dose.

Carbon monoxide.—Same as for Acetylene.

Chloroform, Chloral, Ether.—Dash cold water on head and chest, and apply artificial respiration.

Ethylene.—Same as for Acetylene.

Gas (illuminating).—Same as for Acetylene.

Hydrochloric Acid.—Use albumen, alkali carbonates, ice or milk of magnesia.

Hydrocyanic or Prussic Acid.—Use hydrogen peroxide internally and artificial respiration; breathing ammonia or chlorine from chlorinated lime, ferrous sulfate followed by potassium carbonate, emetics, warmth.

Iodine.—Use emetics, sodium thiosulfate, starchy foods in abundance, or stomach siphon.

Lead Acetate.—Use albumen, emetics, milk, potassium or magnesium sulfate, sodium, or stomach siphon.

Mercuric Chloride.—Use castor oil, chalk, emetics, milk, reduced iron, stomach siphon, table salt, white of egg, or zinc sulfate.

Nitrate of Silver.—Use salt and water.

Nitric Acid.—Same as for hydrochloric acid.

Opium, Morphine, Laudanum, Paregoric, etc.—Drink strong coffee and take a hot bath. Keep awake and moving at any cost.

Phosphoric Acid.—Same as for hydrochloric acid.

Sodium Hydroxide or Potassium Hydroxide.—Drink lemon juice, oil, milk, orange juice, or vinegar.

Sulfuric Acid.—Use albumen, alkali carbonates, ice, magnesia, oil, or soap.

Sulfurous Acid or Sulfur Dioxide.—

Use expectorants or narcotics; use mustard plaster on chest.

Wood Alcohol (Methyl Alcohol or Methanol).—Use emetic or wash out stomach (stomach tube) with a solution of 10 grains sodium citrate per ounce of water. Give flour in water, milk or white of egg; purgative of magnesium sulfate (15 grams); stimulate and combat collapse. In case of cardiac or pulmonary failure use artificial respiration. Physicians may administer atropine, digitalin or strychnine as stimulants; to cause perspiration and elimination of the poison use 0.1 grain of pilocarpine hydrochloride.

------------◆------------

Burns and Scalds

Exclude air by baking soda, flour or thin paste of starch. Ordinary oils such as lard, olive or castor oil, or vaseline may also be used. Lime water mixed with an equal part of raw linseed oil makes an excellent dressing. An especially valuable material for all burns is picric acid gauze which may be applied in the form of a compress.

After treatment with any of the above materials, cover with a cloth or with cotton and hold in place with a light bandage.

------------◆------------

Acid and Alkali Burns

Wash off as quickly as possible with a large quantity of water. Water from tap may be allowed to flow over burns.

------------◆------------

Acids

While the injury is being washed, have procured, lime water or lime water and raw linseed oil mixed together in equal proportions or a mixture of baking soda and water or soap suds and apply freely. For acid in the eye wash as quickly as possible with water and then with lime water. Or use a 5% sodium bicarbonate solution.

------------◆------------

Alkalis

Wash with a large quantity of water as for acid burns. Neutralize with hard cider, lemon juice or weak vinegar. For lime or other strong alkali burns in the eye wash with weak solution of vinegar or with olive oil or a saturated solution of boric acid.

CHAPTER XIX

TABLES

WEIGHTS AND MEASURES
ENGLISH SYSTEM

Avoirdupois and Commercial Weights

16 drams, or 437.5 grains	=1 ounce, oz.
16 ounces, or 7000 grains	=1 pound, lb.
28 pounds	=1 quarter, qr.
4 quarters (English)	=1 hundredweight, cwt.—112 lbs.
20 hundredweight	=1 ton of 2240 lbs., gross or long ton
2000 pounds	=1 net, or short, ton
2204.6 pounds	=1 metric ton =1000 kilos

1 stone = 14 pounds; 1 quintal = 100 pounds

Troy Weights

24 grains	=1 pennyweight, dwt.
20 pennyweights	=1 ounce, oz. =480 grains
12 ounces	=1 pound, lb. =5760 grains
1 carat	=3.168 grains =0.205 gram

Troy weight is used for weighing gold and silver. The grain is the same in Avoirdupois, Troy and Apothecaries' weights.

Apothecaries' Weights

20 grains	=1 scruple
2 scruples	=1 drachm, ℨ =60 grains
8 drachms	=1 ounce, ℥ =480 grains
12 ounces	=1 pound, lb. =5760 grains

Apothecaries' Measures

60 minims (min.)	=1 fluid drachm (fl. dr.)
8 fluid drachms	=1 fluid ounce (fl. oz.)
20 fluid ounces	=1 pint (O) +
8 pints	=1 gallon (C) +

Relations of Apothecaries' Measures to Weights

(All liquids to be measured at 62° Fahr.)

1 minim is the measure of	0.0115 grains of distilled water		
1 fluid drachm	" "	54.687	" " " "
1 fluid ounce	" "	437.5	" " " "
1 point	" "	8750	" ?: " "
1 gallon	" "	70000	" " " "

Linear Measure

12	inches	=1 foot
3	feet	=1 yard
6	feet	=1 fathom

5½	yards	=1 rod pole, or perch
4	poles	=1 chain
40	poles	=1 furlong
8	furlongs	=1 mile=1760 yards

Square Measure

144 square inches = 1 square foot
9 square feet = 1 square yard
30.25 square yards or 272.5 sq. feet = 1 square rod
160 square rods or 4840 sq. yards or 43560 sq. feet = 1 acre
640 acres = 1 square mile
An acre equals a square whose side is 208.7 feet

Cubic Measure

1728 cubic inches = 1 cubic foot
27 cubic feet = 1 cubic yard
1 cord of wood = a pile 4×4×8 feet = 128 cubic feet
1 perch of masonry = 16.5×1.5×1 foot = 24.75 cubic feet
1 cubic inch of water at 62° Fahr. weighs 252.286 grains
" " " " " " " " 0.57665 oz. (av.)
" " " " " " " " 0.036041 lb.
1 cubic foot " " " " " " " 996.458 oz. (av.)
" " " " " " " " 62.2786 lb.
1 cubic yard " " " " " " " 0.75068 tons

Capacity Measure (Liquid)

4 gills = 1 pint
2 pints = 1 quart
4 quarts = 1 gallon

Conversion Factors

1. Grams per litre (g./l.) multiplied by 0.134 = avoirdupois ounces per gallon (oz./gal.).

2. Avoirdupois ounces per gallon (oz./gal.) multiplied by 7.5 = grams per litre (g.l.).

3. Grams per litre (g./l.) multiplied by 0.122 = troy ounces per gallon (troy oz./gal.).

4. Troy ounces per gallon (troy oz./gal.) multiplied by 8.2 = grams per litre (g./l.).

5. Grams per litre (g./l.) multiplied by 2.44 = pennyweights per gallon (dwt./gal.).

6. Pennyweights per gallon (dwt./gal.) multiplied by 0.41 = grams per litre (g./l.).

7. Amperes per square decimeter (amp./dm.^{2}) multiplied by 9.29 = amperes per square foot (amp./sq. ft.).

8. Amperes per square foot (amp./sq. ft.) multiplied by 0.108 = amperes per square decimeter (amp./dm.^{2}).

Thermometer Readings:

Degrees Centigrade × 1.8 + 32 = deg. Fahr.

Degrees $\dfrac{\text{Fahrenheit} - 32}{1.8}$ = deg. Cent.

Degrees $\dfrac{\text{Reamur} \times 9}{4}$ + 32 = deg. Fahr.

Degrees $\dfrac{(\text{Fahrenheit} - 32)4}{9}$ = deg. Reaumur.

Degrees $\dfrac{\text{Reamur} \times 5}{4}$ = deg. Cent.

Degrees $\dfrac{\text{Centigrade} \times 4}{5}$ = deg. Reaumur.

SPECIFIC GRAVITY
WEIGHT REQUIRED TO MAKE A GALLON

	Specific Gravity	Pounds to Gallon
Litharge	9.3	77.5
Red-Lead	8.7 to 8.8	72.5
Orange Mineral (orange lead)	8.6 to 8.7	73.0
White-Lead	6.7	55.8
Basic Lead Sulphate	6.4	53.3
Chrome Yellow (medium)	6.0	50.0
Zinc Oxide (white zinc)	5.6	46.6
Basic Lead Chromate	6.8	56.6
English (mercury) Vermilion	8.2	68.3
Bright Red Oxide of Iron	4.9 to 5.26	42.0
Indian Red Oxide of Iron	5.26	43.8
Brown Oxide of Iron (Prince's)	3.2	26.6
Ultramarine	2.4	20.0
Prussian Blue	1.85	15.4
Chrome Green (blue tone)	4.44	37.0
Chrome Green (yellow tone)	4.0	33.0
Lithopone	4.25	35.4
Ochre	2.94	24.5
Barytes	4.35 to 4.46	35. to 37.0
Blanc Fixe	4.25	35.4
Gypsum (terra alba)	2.3	19.0
Asbestine (magnesium silicate)	2.75	23.0
China Clay (aluminum silicate)	2.6 to 2.7	22.5
Whiting	2.65	22.0
Silica	2.65	22.0
Natural Graphite	2.1 to 2.4	18.0
Acheson's Graphite	2.2	18.3
Lampback	1.85	15.4
Carbon Black	1.85	15.4
Keystone Filler (ground slate)	2.66	22.0
Titanox	4.3	35.8
Titanium Oxide	3.9 to 4.0	33.3
Drop Black	2.5	20.8

To this table the following data may be added: The weight of one gallon of paste made with

	Pounds
Red-Lead	44.8
White-Lead (heavy paste)	34.0
White-Lead (soft paste)	30.8
White Zinc	25.0
Chrome Yellow (medium)	24.0
Chrome Green	24.0
Venetian Red	19.0
French Ochre	15.0
Prussian Blue	10.0
Lampblack	9.1
Drop Black	11.7

Dry

2 pints =1 quart
8 quarts=1 peck
4 pecks =1 bushel
1 U. S. standard bushel (struck) = 2150.42 cubic inches.
0.80356 U. S. bushels (struck) = 1 cubic foot.

METRIC EQUIVALENTS
Linear Measure

1 centimeter=0.3937 in.
1 decimeter=3.937 in.=0.328 ft.
1 meter=39.37 in.=1.0936 yds.
1 decameter=1.9884 rods
1 kilometer=0.62137 mile
1 inch=2.54 centimeters
1 foot=3.048 decimeters
1 yard=0.9144 meter
1 rod=0.5029 decameter
1 mile=1.6093 kilometers

(The meter, as used in Europe, is 39.370432 inches.)

Square Measure

1 sq. centimeter=0.1550 sq. inch
1 sq. decimeter=0.1076 sq. foot
1 sq. meter=1.196 sq. yards
1 are=3.954 sq. rods
1 hectare=2.47 acres
1 sq. kilometer=0.386 sq. mile
1 sq. inch=6.452 sq. centimeters
1 sq. foot=9.2903 sq. decimeters
1 sq. yard=0.8361 sq. meter
1 sq. rod=0.2529 are

1 acre=0.4047 hectare
1 sq. mile=.259 sq. kilometer

Weights

1 decigram=0.003527 oz.=1.5432 grains
1 gram=0.03527 oz. avoir., or about 15½ troy grains
1 kilogram=2.2046 lbs. avoir.
1 metric ton=1.1023 English short tons
1 ounce avoir.=28.35 grams
1 pound avoir.=0.4536 kilogram
1 English short ton=0.9072 metric ton

Approximate Metric Equivalents

1 decimeter=4 inches
1 meter=1.1 yards
1 kilometer=⅝ of a mile
1 hectare=2½ acres
1 stere, or cu. meter=¼ of a cord
1 liter=1.06 qt. liquid, 0.9 qt. dry
1 hectoliter=2⅝ bushels
1 kilogram=2⅕ lbs.
1 metric ton=2200 lbs.

Comparison of Avoirdupois and Metric Weights

Grains	Drams	Oz. Av.	Lbs. Av.	Deniers	Grams
1.000	. . . . .	. . . .	. . . . .	1.296	0.065
27.340	1.000	. . . .	. . . . .	35.437	1.772
437.500	16.000	1.000	. . . . .	566.990	28.350
7000.000	256.000	16.000	1.000	9071.840	453.592
0.772	. . . .	. . . .	. . . . .	1.000	0.050
15.432	. . . .	0.03527	. . . . .	20.000	1.000

pH Values of Chemicals

Solution Strength	Reagent	pH
1%	Commercial Olive Oil Soap (Neutral)	10.1 –10.3
1%	Commercial Olive Oil Soap (Neutral)	10.1 –10.3
1%	Commercial Olive Oil or Tallow Soap Containing 20% Soda Ash	10.75–10.88
1%	Commercial Olive Oil or Tallow Soap Containing 5% Caustic	12.0 –12.2
½%	Commercial Olive Oil or Tallow Soap	10.0 –10.2
¼%	Commercial Olive Oil or Tallow Soap	9.9 –10.1
1%	Sulfonated Oils (Neutral)	6.0 – 7.0
1%	Sulfonated Oils Containing Free Acid	Below 6.0
1%	Sulfonated Oils Containing Soap or Alkalies	Above 7.0
¼%	Trisodium Phosphate	12.3
¼%	Sodium Silicate	12.2
¼%	Sodium Carbonate	11.3
¼%	Sodium Sulfite	9.7
¼%	Disodium Phosphate	8.9
¼%	Borax	8.8
¼%	Monosodium Phosphate	5.0

pH Ranges of Common Indicators

	Useful pH Range
Thymol Blue	1.2–2.8
Bromphenol Green	2.8–4.6
Methyl Orange	3.1–4.4
Bromcresol Green	4.0–5.6
Methyl Red	4.4–6.0
Propyl Red	4.8–6.4
Bromcresol Purple	5.2–6.8
Brom Thymol Blue	6.0–7.6
Phenol Red	6.8–8.4
Litmus	7.2–8.8
Cresol Red	7.2–8.8
Cresolphthalein	8.2–9.8
Phenolphthalein	8.6–10.2
Nitro Yellow	10.0–11.6
Alizarin Yellow R	10.1–12.1
Sulfo Orange	11.2–12.6

◆

Melting Points of Resins, Etc.

Material	Melting Point ° C.
Amber	250–325
Benzoin	75–100
Copal (Zanzibar)	280
Copal (Congo)	220
Copal (Kauri)	165
Copal (Manila)	120
Cumarone	127–142
Dammar (Batavia)	100
Dammar (Singapore)	95
Dragon's Blood	120

Elemi	75–120
Ester Gum	120–140
Gilsonite	123
Guiac	85–90
Indene	127–142
Mastic	105–120
Pontianak	135
Rosin (Colophony)	100–140
Sandarac	135–150
Shellac	120

◆

* Melting Points of Common Waxes

Wax	Melting Point ° C.
Bayberry Wax	40–44
Beeswax White	63–70
Beeswax Yellow	61
Candelilla Wax	67–70
Carnauba Wax	83–86
Ceresine	74–80
Chinese Insect Wax	80–83
Cocoa Butter	21.5–27.3
Japan Wax	54.5–59.6
Montan Wax	76–130
Myrtle Wax	47–48
Ozokerite	65–110
Paraffin	55–65
Spermaceti	44–47.5
Tallow (Beef)	42.5–44

* Very often there is considerable difference between the melting and solidifying point Natural and commercially adulterated articles will also show variations.

CONVERSION OF THERMOMETER READINGS

F°	C°	F°	C°	F°	C°	F°	C°	F°	C°	F°	C°
−40	−40.00	30	−1.11	80	26.67	250	121.11	500	260.00	900	482.22
−38	−38.89	31	−0.56	81	27.22	255	123.89	505	262.78	910	487.78
−36	−37.78	32	0.00	82	27.78	260	126.67	510	265.56	920	493.33
−34	−36.67	33	0.56	83	28.33	265	129.44	515	268.33	930	498.89
−32	−35.56	34	1.11	84	28.89	270	132.22	520	271.11	940	504.44
−30	−34.44	35	1.67	85	29.44	275	135.00	525	273.89	950	510.00
−28	−33.33	36	2.22	86	30.00	280	137.78	530	276.67	960	515.56
−26	−32.22	37	2.78	87	30.56	285	140.55	535	279.44	970	521.11
−24	−31.11	38	3.33	88	31.11	290	143.33	540	282.22	980	526.67
−22	−30.00	39	3.89	89	31.67	295	146.11	545	285.00	990	532.22
−20	−28.89	40	4.44	90	32.22	300	148.89	550	287.78	1000	537.78
−18	−27.78	41	5.00	91	32.78	305	151.67	555	290.55	1050	565.56
−16	−26.67	42	5.56	92	33.33	310	154.44	560	293.33	1100	593.33
−14	−25.56	43	6.11	93	33.89	315	157.22	565	296.11	1150	621.11
−12	−24.44	44	6.67	94	39.44	320	160.00	570	298.89	1200	648.89
−10	−23.33	45	7.22	95	35.00	325	162.78	575	301.67	1250	676.6
− 8	−22.22	46	7.78	96	35.56	330	165.56	580	304.44	1300	704.44
− 6	−21.11	47	8.33	97	36.11	335	168.33	585	307.22	1350	732.22
− 4	−20.00	48	8.89	98	36.67	340	171.11	590	310.00	1400	760.00
− 2	−18.89	49	9.44	99	37.22	345	173.89	595	312.78	1450	787.78
0	−17.78	50	10.00	100	37.78	350	176.67	600	315.56	1500	815.56
1	−17.22	51	10.56	105	40.55	355	179.44	610	321.11	1550	843.33
2	−16.67	52	11.11	110	43.33	360	182.22	620	326.67	1600	871.11
3	−16.11	53	11.67	115	46.11	365	185.00	630	332.22	1650	898.89
4	−15.56	54	12.22	120	48.89	370	187.78	640	337.78	1700	926.67
5	−15.00	55	12.78	125	51.67	375	190.55	650	343.33	1750	954.44
6	−14.44	56	13.33	130	54.44	380	193.33	660	348.89	1800	982.22
7	−13.89	57	13.89	135	57.22	385	196.11	670	354.44	1850	1010.00
8	−13.33	58	14.44	140	60.00	390	198.89	680	360.00	1900	1037.78
9	−12.78	59	15.00	145	62.78	395	201.67	690	365.56	1950	1065.56
10	−12.22	60	15.56	150	65.56	400	204.44	700	371.11	2000	1093.33
11	−11.67	61	16.11	155	68.33	405	207.22	710	376.67	2050	1121.11
12	−11.11	62	16.67	160	71.11	410	210.00	720	382.22	2100	1148.89
13	−10.56	63	17.22	165	73.89	415	212.78	730	387.78	2150	1176.67
14	−10.00	64	17.78	170	76.67	420	215.56	740	393.33	2200	1204.44
15	− 9.44	65	18.33	175	79.44	425	218.33	750	398.89	2250	1232.22
16	− 8.89	66	18.89	180	82.22	430	221.11	760	404.44	2300	1260.00
17	− 8.33	67	19.44	185	85.00	435	223.89	770	410.00	2350	1287.78
18	− 7.78	68	20.00	190	87.78	440	226.67	780	415.56	2400	1315.56
19	− 7.22	69	20.56	195	90.55	445	229.44	790	421.11	2450	1343.33
20	− 6.67	70	21.11	200	93.33	450	232.22	800	426.67	2500	1371.11
21	− 6.11	71	21.67	205	96.11	455	235.00	810	432.22	2550	1398.89
22	− 5.56	72	22.22	210	98.89	460	237.78	820	437.78	2600	1426.67
23	− 5.00	73	22.78	215	101.67	465	240.55	830	443.33	2650	1454.44
24	− 4.44	74	23.33	220	104.44	470	243.33	840	448.89	2700	1482.22
25	− 3.89	75	23.89	225	107.22	475	246.11	850	454.44	2750	1510.00
26	− 3.33	76	24.44	230	110.00	480	248.89	860	460.00	2800	1537.78
27	− 2.78	77	25.00	235	112.78	485	251.67	870	465.56	2850	1565.56
28	− 2.22	78	25.56	240	115.56	490	254.44	880	471.11	2900	1593.33
29	− 1.67	79	26.11	245	118.33	495	257.22	890	476.67	2950	1621.11

EQUIVALENTS OF TWADDELL, BAUMÉ AND SPECIFIC GRAVITY SCALES

Twaddell	Baumé	Specific Gravity	Twaddell	Baumé	Specific Gravity	Twaddell	Baumé	Specific Gravity	Twaddell	Baumé	Specific Gravity
0	0	1.000	44	26.0	1.220	88	44.1	1.440	131	57.1	1.655
1	0.7	1.005	45	26.4	1.225	89	44.4	1.445	132	57.4	1.660
2	1.4	1.010	46	26.9	1.230	90	44.8	1.450	133	57.7	1.665
3	2.1	1.015	47	27.4	1.235	91	45.1	1.455	134	57.9	1.670
4	2.7	1.020	48	27.9	1.240	92	45.4	1.460	135	58.2	1.675
5	3.4	1.025	49	28.4	1.245	93	45.8	1.465	136	58.4	1.680
6	4.1	1.030	50	28.8	1.250	94	46.1	1.470	137	58.7	1.685
7	4.7	1.035	51	29.3	1.255	95	46.4	1.475	138	58.9	1.690
8	5.4	1.040	52	29.7	1.260	96	46.8	1.480	139	59.2	1.695
9	6.0	1.045	53	30.2	1.265	97	47.1	1.485	140	59.5	1.700
10	6.7	1.050	54	30.6	1.270	98	47.4	1.490	141	59.7	1.705
11	7.4	1.055	55	31.1	1.275	99	47.8	1.495	142	60.0	1.710
12	8.0	1.060	56	31.5	1.280	100	48.1	1.500	143	60.2	1.715
13	8.7	1.065	57	32.0	1.285	101	48.4	1.505	144	60.4	1.720
14	9.4	1.070	58	32.4	1.290	102	48.7	1.510	145	60.6	1.725
15	10.0	1.075	59	32.8	1.295	103	49.0	1.515	146	60.9	1.730
16	10.6	1.080	60	33.3	1.300	104	49.4	1.520	147	61.1	1.735
17	11.2	1.085	61	33.7	1.305	105	49.7	1.525	148	61.4	1.740
18	11.9	1.090	62	34.2	1.310	106	50.0	1.530	149	61.6	1.745
19	12.4	1.095	63	34.6	1.315	107	50.3	1.535	150	61.8	1.750
20	13.0	1.100	64	35.0	1.320	108	50.6	1.540	151	62.1	1.755
21	13.6	1.105	65	35.4	1.325	109	50.9	1.545	152	62.3	1.760
22	14.2	1.110	66	35.8	1.330	110	51.2	1.550	153	62.5	1.765
23	14.9	1.115	67	36.2	1.335	111	51.5	1.555	154	62.8	1.770
24	15.4	1.120	68	36.6	1.340	112	51.8	1.560	155	63.0	1.775
25	16.0	1.125	69	37.0	1.345	113	52.1	1.565	156	63.2	1.780
26	16.5	1.130	70	37.4	1.350	114	52.4	1.570	157	63.5	1.785
27	17.1	1.135	71	37.8	1.355	115	52.7	1.575	158	63.7	1.790
28	17.7	1.140	72	38.2	1.360	116	53.0	1.580	159	64.0	1.795
29	18.3	1.145	73	38.6	1.365	117	53.3	1.585	160	64.2	1.800
30	18.8	1.150	74	39.0	1.370	118	53.6	1.590	161	64.4	1.805
31	19.3	1.155	75	39.4	1.375	119	53.9	1.595	162	64.6	1.810
32	19.8	1.160	76	39.8	1.380	120	54.1	1.600	163	64.8	1.815
33	20.3	1.165	77	40.1	1.385	121	54.4	1.605	164	65.0	1.820
34	20.9	1.170	78	40.5	1.390	122	54.7	1.610	165	65.2	1.825
35	21.4	1.175	79	40.8	1.395	123	55.0	1.615	166	65.5	1.830
36	22.0	1.180	80	41.2	1.400	124	55.2	1.620	167	65.7	1.835
37	22.5	1.185	81	41.6	1.405	125	55.5	1.625	168	65.9	1.840
38	23.0	1.190	82	42.0	1.410	126	55.8	1.630	169	66.1	1.845
39	23.5	1.195	83	42.3	1.415	127	56.0	1.635	170	66.3	1.850
40	24.0	1.200	84	42.7	1.420	128	56.3	1.640	171	66.5	1.855
41	24.5	1.205	85	43.1	1.425	129	56.6	1.645	172	66.7	1.860
42	25.0	1.210	86	43.4	1.430	130	56.9	1.650	173	67.0	1.865
43	25.5	1.215	87	43.8	1.435						

Relation of Capacity, Volume and Weight

1 pint	= 28.875 cubic inches
1 quart	= 57.75 cubic inches
1 gallon (U. S.)	= 231 cubic inches
1 gallon (English)	= 277.274 cubic inches
7.4805 gallons	= 1 cubic foot

1 gallon water at 62 Fahr. weighs 8.3356 lbs.

REFERENCES CONSULTED

Aircraft Engineering
Agr. Gas N. S. Wales
Allg. Oel v. Gettzeitung
Amer. Dyestuff Reporter
Amer. Electrop. Society
Amer. Gas Assoc. Proc.
Amer. Machinist
Amer. Paint Jol.
Amer. Paint & Varnish Mfrs. Assn.
Amer. Perfumer
Amer. Photography
Amer. Society Testing Materials
Anal. Fis. Quim.
Atelier Photography
Ault & Wiborg Varnish Wks. Handbook
Austrian Patent Office

Bakers' Review
Belgian Patent Office
Berichte Ges. Kohlentech.
Better Enameling
Bied. Zentralblatt
Brass World
Brewers' Tech. Review
Brewery Age
Brit. Indus. Finishing
Brit. Jo. of Photography
Brit. Medical Jol.
Brit. Patent Office
Brit. Plastics
Brit. Soap Mfr.
Bull. Imp. Hyg. Lab.
Bulletin of Imperial Institute
Bull. Soc. Franc. Phot.
Bureau of Standards Publications

Camera
Camera (Luzern)
Canadian Patent Office
Chemical Abstracts
Chemical Analyst
Chemical Formulary
Chemical & Metallurg. Engin.
Chemiker-Zeitung
Chemist Analyst
Chemist & Druggist
Chem. Zent.

Chimie Industrie
Combustion
Cotton
Cramer's Manual

Dansk. Tids. Farm
Der Chemisch Technische Fabricant
Der Parfumer
Deutscher Zuckerind
Devt. Part. Zeitung
Drug & Cosmetic Industry
Druggists' Circular
Drugs, Oils, & Paints
Drug Trade News
Dutch Patent Office
Dyestuffs

Eastman Kodak Co.
Electric Journal

Farbe v. Lacke
Farben Zeitung
Fein Mechanic v. Prazision
Fettchen, Umscham
Fils & Tissus
Focus
Food Products Jol.
French Patent Jol.
Fruit Products Jol.

Gas Journal
German Patent Office
Glass Industry

Hawaiian Planters' Record
Hide & Leather
Hungarian Patent Office

Idaho Agri. Experiment Station
India Rubber World
Indian Lac Research Inst.
Industrial Chemist
Industrial Finishing
Industrial Woodworking

Japanese Patent Office
Journal American Ceramic Society
Journal Amer. Dental Assn.
Jol. Amer. Medical Assn.
Jol. Appl. Chem. Russ.

Jol. Chemical Industry
Journal Council Sci. Industrial Research
Journal Dept. Agriculture Ireland
Journal Econ. Etomology
Jol. Federation Curriers
Jol. Ind. & Chemistry
Jol. Institute of Metals
Jol. of Society of Chemical Industry
Jol. Society of Chemical Industry
   (Japan)
Jol.Soc. Leather Trades
Jol. Soc. Rubber Ind., Japan

Keram., Stekle
Khimstroi
Korrosion
Kunstdinger, Undlam

Lakokras, Ind.
Lancet
Laundry Owner's Natl. Assn.

Malayan Agric. Jol.
Manufacturing Chemist
Melliand Textile Monthly
Metal Industry
Metall und Erz
Mettallurg
Metallurgist
Metals & Alloys
Minn. Agric. Experiment Station
Monats-Bull. Schwelz. Ver. Gass
   Wass.
Munic. Eng. San. Record
Museum Technique

Nat'l Butter & Cheese Jol.
New York Agric. Experiment Station
Nitrocellulose

Ober Flachen Tech.
Oil & Color Trade Jol.
Oil & Soap
Oils, Drugs & Paints

Paint Mfr.
Paint & Varnish Production Mgr.
Paper Maker
Paper Trade Jol.

Parfum Mod.
Peinture, Pigments Vernis
Perf. and Ess. Oil Record
Phar. Acta Helva
Pharmaceutical Jol.
Phot. Abstracts
Phot. Chronik
Photofreund
Phot. Ind.
Phot. Korr.
Photog. Kronik
Phot. Rev.
Photo Rundschau
Phytopathology
Plater's Guide Book
Portland Cement Assn.
Practical Druggist
Printing Industry
Prob. Edelmetalle
Proc. World Petroleum Congress
Purdue Agric. Experiment Station

Quart-Jol. Phar. Pharmacologie

Rayon & Mell. Tex. Monthly
Refiner & Nat. Gas Mfr.
Revue Appl. Mycology
Russian Patent Office

Science
Seifen Sieder Zeitung
Soap
Soap Gazette
Soap Gazette & Perfumer
Solvent News
Sovet-Sakhar
Spirits
Swedish Patent Office
Synthetic & Applied Finishes

Tex. Agric. Exp. Station
Textile Colorist
Textile Mfr.
Textile Recorder

U. S. Bureau of Mines
U. S. Dept. of Agric.
U. S. Patent Office

Welsh Agricultural Jol.

Zeit. Untersuch. Lebensm.

"TRADE NAMES" AND SUPPLIERS

In the following pages will be found a list of ''Trade Names'' and the companies who supply them. The number against the ''Trade Name'' refers to the supplier with the corresponding number in the suppliers' section.

Trade Names are names given by manufacturers and suppliers, which in themselves are not descriptive chemically of the products they represent.

The reasons for giving products trade names instead of chemical names are (1) Many chemical names are extremely long, difficult to spell and much more difficult to pronounce. (2) Many trade name products are not single chemical entities and (3) Many manufacturers feel much freer to give information about a product on which they have the protection afforded them by a trade name.

It was, therefore, felt advisable to include certain trade name products in some of the formulae in the book, so that the reader will obtain the latest information on commercial formulation which otherwise would not be available.

TRADE NAME LISTINGS

293

Pearl Essence 46
Peerless Carbon Black 9
Pentacetate 67
Petrohol 71
Plastogen 78
Proflex 30
Pyla-White 60
Pylakromes 60
Pylam Dyes 60

Q

Quakersol 57

R

Resin C 8
Resinox 23
Rezinels 30
Rezyls 3
Rhodamine BX 60
Rhodine 26
Rhodol 26
Rozolin 30
Rubber Orange 2 R 26

S

Salamac 32
Sapinone 30
Schultz Silica 18
Sheragum 30
Sicapon 30
Silex 73
Soligen Dryers 1
Stearoricinol 30
Sono-Jell 69
Super Spectra Black 9
Suspendite 30

T

Tanak 3
Teglac 3
Tetralin 26
Texavac 3
Thinnex 26
Thionex 26
Thylox 40
Ti-Tone 41
Titanox B 74
Triclene 26
Tripoli 73
Tuads 78
Turpenol 3

U

Urazine 26
Ureka C 64

V

Vandex 78
Varnolene 72
Vaselene 20
Vinylite Resins 13
Viscogum 30

W

Wetting Oil S F 3

Y

Yarmor Pine Oil 35

Z

Zerone 26
Zimate 7a
Zyklon B 3

SUPPLIERS OF "TRADE NAME" CHEMICALS

24	Darco Sales Corp.	New York City
25	Dow Chemical Co.	Midland, Mich.
26	Du Pont de Nemours, E. I. & Co.	Wilmington, Del.
27	Franco-American Chemical Works Corp.	Carlstadt, N. J.
28	General Electric Co.	Bridgeport, Conn.
29	General Plastics, Inc.	N. Tonawanda, N. Y.
30	Glyco Products Co., Inc.	Brooklyn, N. Y.
31	Goldschmidt, Th. Corp.	New York City
32	Grasselli Chemical Co.	Cleveland, Ohio
33	Hall, C. D. Co.	Akron, Ohio
34	Harshaw Chemical Co.	New York City
35	Hercules Powder Co.	Wilmington, Del.
36	Hooker Electrochemical Co.	New York City
37	Industrial Chemical Sales Co.	New York City
38	International Pulp Co.	New York City
39	Johns Manville Corp.	New York City
40	Koppers Products Co.	Pittsburgh, Pa.
41	Krebs Pigment & Chemical Co.	New York City
42	Lehn & Fink.	New York City
43	Lewis, John D.	Providence, R. I.
44	Mallinckrodt Chemical Works	New York City
45	Mathieson Alkali Co.	New York City
46	Mearl Corp.	New York City
47	Monsanto Chemical Wks.	New York City
48	Mutual Chemical Co. of America	New York City
49	National Aniline and Chemical Co.	Buffalo, N. Y.
50	Naugatuck Chemical Co.	New York City
51	Neville Chemical Co.	Pittsburgh, Pa.
52	Nulomoline Co.	New York City
53	Nuodex Products Co.	Newark, N. J.
54	Pacific Coast Borax Co.	New York City
55	Paramet Chemical Corp.	Long Island City, N. Y.
56	Penn Salt Mfg. Co.	Philadelphia, Pa.
57	Penn. Sugar Co.	New York City
58	Pfaltz & Bauer	New York City
59	Philadelphia Quartz Co.	Philadelphia, Pa.
60	Pylam Products Co.	New York City
61	R & H Chemical Co.	New York City
62	Resinous Products & Chemical Co.	Philadelphia, Pa.
63	Robeson Process Co.	New York City
64	Rubber Service Labs. Co.	Nitro, W. Va.
65	Scott-Bader & Co.	London, England
66	Seeley & Co.	New York City
67	Sharpless Solvents Corp.	Philadelphia, Pa.
68	Solvay Sales Corp.	New York City
69	L. Sonneborn Sons, Inc.	New York City
70	Spencer, Kellogg Co.	New York City
71	Stanco, Inc.	New York City
72	Standard Oil Co. of New York	New York City
73	Swann Chemical Co.	New York City
74	Titanium Pigment Co.	New York City
76	U. S. Gypsum Co.	Chicago, Ill.
77	U. S. Industrial Chem. Co.	New York City
78	Vanderbilt, R. T. Co.	New York City
79	Will & Baumer Candle Co.	New York City
80	Wolf, Jacques Co.	Passaic, N. J.

INDEX

INDEX